Modeling Random Processes for Engineers and Managers

Modeling Random Processes for Engineers and Managers

JAMES J. SOLBERG
Purdue University

John Wiley & Sons, Inc.

PUBLISHER	Don Fowley
ACQUISITIONS EDITOR	Jennifer Welter
SENIOR PRODUCTION EDITOR	Valerie A. Vargas
MARKETING MANAGER	Christopher Ruel
SENIOR DESIGNER	Jeof Vita
PRODUCTION MANAGEMENT SERVICES	Elm Street Publishing Services
EDITORIAL ASSISTANT	Mark Owens
MEDIA EDITOR	Lauren Sapira
COVER PHOTO	© Blend Images/Fotosearch, LLC

This book was set in Times Roman by Thomson Digital and printed and bound by Hamilton Printing Company. The cover was printed by Phoenix Color.

This book is printed on acid free paper. ∞

Copyright © 2009 John Wiley & Sons, Inc. All rights reserved. No part of this publication may be reproduced, stored in a retrieval system or transmitted in any form or by any means, electronic, mechanical, photocopying, recording, scanning or otherwise, except as permitted under Sections 107 or 108 of the 1976 United States Copyright Act, without either the prior written permission of the Publisher, or authorization through payment of the appropriate per-copy fee to the Copyright Clearance Center, Inc. 222 Rosewood Drive, Danvers, MA 01923, website www.copyright.com. Requests to the Publisher for permission should be addressed to the Permissions Department, John Wiley & Sons, Inc., 111 River Street, Hoboken, NJ 07030-5774, (201)748-6011, fax (201)748-6008, website http://www.wiley.com/go/permissions.

To order books or for customer service please, call 1-800-CALL WILEY (225-5945).

ISBN 978-0-470-32255-0

Printed in the United States of America

10 9 8 7 6 5 4 3 2 1

Contents

Preface ix

1
Probability Review 1

1.1 Interpreting and Using Probabilities 2
1.2 Sample Spaces and Events 3
1.3 Probability 4
1.4 Random Variables 6
1.5 Probability Distributions 6
1.6 Joint, Marginal, and Conditional Distributions 11
1.7 Expectation 14
1.8 Variance and Other Moments 16
1.9 The Law of Total Probability 18
1.10 Discrete Probability Distributions 20
1.11 Continuous Probability Distributions 23
1.12 Where Do Distributions Come From? 26
1.13 The Binomial Process 28
1.14 Recommended Reading 29

2
Formulating Markov Chain Models 32

2.1 An Example 33
2.2 Modeling the Progress of Time 34
2.3 Modeling Possibilities as States 36
2.4 Simplifying Assumptions 38

2.5	Modeling Changes of State as Transitions	40
2.6	Obtaining the Data	45
2.7	Another Example	46
2.8	A Case Study	47
2.9	Higher Order Markov Chains	50
2.10	Reducing the Number of States	52
2.11	Nonstationary Markov Chains	53
2.12	Other Example	54
2.13	Summary	67
2.14	Recommended Reading	67

3
Markov Chain Calculations — 72

3.1	Walk Probabilities	73
3.2	Transition Probabilities	74
3.3	State Probabilities	78
3.4	A Numerical Example	79
3.5	Expected Number of Visits	80
3.6	Sojourn Times	82
3.7	First Passage and Return Probabilities	83
3.8	Computational Formulas for All Markov Chains	86
3.9	Special Classes of Markov Chains	86
3.10	Steady-State Probabilities	87
3.11	The Uses of Steady-State Results	92
3.12	Mean First Passage Times	93
3.13	Computational Formulas for Ergodic Markov Chains	96
3.14	Terminating Markov Chains	96
3.15	Expected Number of Visits	98
3.16	Expected Duration of a Terminating Process	99
3.17	Absorption Probabilities	100
3.18	Hit Probabilities	102
3.19	Conditional Mean First Passage Times to Absorbing States	103
3.20	Computational Formulas for Terminating Processes	105
3.21	Call Center Calculations	105
3.22	Classification Terminology	106
3.23	Additional Complications in Infinite Chains	111
3.24	Dealing with a Reducible Process	112
3.25	Periodic Chains	113
3.26	Ergodic Chains	114
3.27	Recommended Reading	115

4
Rewards on Markov Chains — 119

4.1	Formulation	120
4.2	A Numerical Example	120
4.3	Expected Total Reward	121
4.4	Random Variable Rewards	124
4.5	Semi-Markov Processes	126

4.6	Limiting Results for Ergodic Processes	126
4.7	Total Reward for Terminating Processes	130
4.8	Case Study	132
4.9	Discounting	133
4.10	Case Study	135
4.11	Recommended Reading	137

5
Continuous Time Markov Processes — 140

5.1	An Example	141
5.2	Interpreting Transition Rates	146
5.3	The Assumptions Reconsidered	149
5.4	Aging Does Not Affect the Transition Time	150
5.5	Competing Transitions	152
5.6	Sojourn Times	153
5.7	Embedded Markov Chains	154
5.8	Deriving the Differential Equations	155
5.9	Solving the Differential Equations	157
5.10	State Probabilities	159
5.11	First Passage Probability Functions	159
5.12	State Classification	160
5.13	Steady-State Probabilities	161
5.14	Other Computable Quantities	163
5.15	Case Study	165
5.16	Birth-Death Processes	167
5.17	The Poisson Process	169
5.18	Properties of Poisson Processes	171
5.19	Khintchine's Theorem	172
5.20	Phase-Type Distributions	173
5.21	Conclusions	175
5.22	Recommended Reading	175

6
Queueing Models — 179

6.1	An Example	180
6.2	General Characteristics	182
6.3	Performance Measures	186
6.4	Relations Among Performance Measures	188
6.5	Little's Formula	190
6.6	Markovian Queueing Models	191
6.7	The M/M/1 Model	193
6.8	The Significance of Traffic Intensity	198
6.9	Unnormalized Solutions	200
6.10	Limited Queue Capacity	202
6.11	Multiple Servers	204
6.12	Is It Better to Merge or Separate Servers?	207
6.13	Which is Better: More Servers or Faster Servers	208
6.14	Case Study: A Grain Elevator	209
6.15	The M/M/c/c and M/M/∞ Models	210

6.16	Finite Sources	212
6.17	The Machine Repairmen Model	214
6.18	Numerical Calculations Using a Spreadsheet	214
6.19	Queue Discipline Variations	217
6.20	Non-Markovian Queues	218
6.21	The M/G/1 Model	219
6.22	Approximate Solutions for Other Models	220
6.23	Conclusion	221
6.24	Recommended Reading	221

7
Networks of Queues — 225

7.1	Open Networks of Markovian Queues	226
7.2	An Example Open Network	227
7.3	Extensions	228
7.4	Closed Networks	229
7.5	A Preliminary Example	229
7.6	Relative Arrival Rates	230
7.7	Unnormalized Solutions for Individual Stations	232
7.8	Assembling the Pieces of the Solution	234
7.9	Calculating the Normalization Constant	235
7.10	Performance Measures for Closed Networks	237
7.11	Creating a Closed Model	239
7.12	Case Study	242
7.13	Extensions	247
7.14	Approximate Methods	247
7.15	Recommended Reading	248

8
Using the Transition Diagram to Compute — 251

8.1	An Example	252
8.2	Definitions	254
8.3	Steady-State Probabilities	258
8.4	How to Generate All In-trees	259
8.5	Check Your Understanding	262
8.6	Generalization to Other Quantities	263
8.7	Mean First Passage Times	264
8.8	Results for Terminating Processes	265
8.9	How to Simplify the Arithmetic	266
8.10	How to Systematically Generate r-Forests	267
8.11	Summary of Results	267
8.12	How to Remember the Formulas	268
8.13	Advanced Topics	268
8.14	Recommended Reading	269

Appendix 1 — 271

Appendix 2 — 278

Index — 300

Preface

There is an underappreciated kind of mathematics that offers enormous opportunities for application for a wide range of problems in the real world. Despite its value, the subject of stochastic process models has been largely overlooked. In fact, most people have no idea that this treasure exists.

Let me give you one illustration by means of a personal anecdote. Many years ago, not long after I had learned a certain principle that is covered in this book, I had a chance to visit a factory where an automated system had been in operation for several years. The managers were disappointed that it had never achieved the level of productivity that had been expected. Manufacturing engineers had attempted various modifications with only minor success. Being aware of the principle that I had learned in school, I noticed immediately there was a possible change that might improve the productivity. Although I was hesitant to suggest it—after all, I was just an inexperienced outsider with no knowledge of the complexities of their operations— I worked out some easy calculations that seemed to confirm that the change would produce a big improvement. By then they were almost desperate to try anything and it cost almost nothing to implement, so they gave my suggestion a try. To their great surprise, the simple change in how they were using the resources that were already in place produced about 30 percent improvement in productivity. It was worth millions to them. There was nothing brilliant about my insight; I simply applied a principle that I had learned as a student. What surprised me most about the experience was that no one else had exploited this opportunity before I visited. It was "low-hanging fruit," ready to be picked, because none of the people in the plant were aware of the simple principle of pooling (which is explained in Chapter 7).

I continued to have similar experiences involving that same principle and others throughout my career, and many of my former students reported stories along the same lines. Despite clear and compelling evidence of this kind, even now relatively few people seem to recognize how useful this particular body of knowledge is in the real world. I see examples every day of flawed systems and bad decisions that could have been avoided if those people who were responsible had any understanding of the material in this book.

How could such treasure be overlooked? I think that the explanation has to do with how the material is usually presented. Although there are some fine textbooks that explain the theory, they are somewhat inaccessible to beginners. The concepts, terminology, and notation give the impression that the subject matter is quite esoteric. For example, the technical description of what you will learn is "modeling of stochastic processes." That phrase is meaningless to most people. If you replace the word "stochastic" with "random," you come a little closer to suggesting what the subject is about, but it may still seem far from the mainstream of real-world human experience. Explaining that it relies upon probability theory does not help, because few people are really comfortable with that subject. Getting past those barriers takes some work. Moreover, the typical examples in most introductory textbooks seem to reinforce the notion that the entire subject is unimportant. Illustrating the concepts with coin flips or buttons in jars (as many authors do) undermines their significance.

My mission is to provide an easier entry into the subject with enough depth to get to interesting results. By "interesting," I mean "useful but not obvious." The examples, case studies, and problems are all designed to suggest actual important applications, so that you can appreciate the value of what you are learning at the time you have to invest the effort. There are a lot of explanations of the thought process that goes into creating a model, and of the intuitive logic behind certain equations. These unusually wordy explanations are intended to take some of the mystery out of otherwise arcane derivations.

I want it to stress that this is not intended to be a mathematics textbook. If you flip through the pages, you will find a lot of symbols and equations, along with explanations of why certain things are true, but there will be nothing even close to a rigorous proof. Of course, the need to define concepts carefully requires some mathematical precision. If you want to understand the mathematics per se (which is highly recommended), there are other books that serve that purpose well. The emphasis here is on learning how to create and use a model of real world phenomena, rather than the math itself.

Intended Audience

This book was designed for a one-semester course having at least one course in probability as a prerequisite. It is suitable for undergraduates or beginning graduate students having no previous exposure to stochastic processes. Although calculus is necessary for a few results in continuous time models, for the most part linear algebra is sufficient. The presentation of material in this book has been rewritten and reorganized several times, always with the intent of making it easier for the novice to learn.

Unique Features

Aside from a few things that I developed myself, most of the methods and results presented have been around for a long time. The theory is standard, but the presentation is not. You will find that there is more explanation of concepts and less formal mathematics than in most textbooks in the same area. In particular, there is an emphasis on formulation of models - in essence, how to think about a problem in order to reach the point where the mathematics captures what is needed. Few textbooks discuss this aspect directly because it requires some creativity, and that is something that is difficult to teach.

The book also covers quite a few formulation tricks to get around what might appear to the novice to be very limiting assumptions. Once you know the tricks, you open a very wide range of applications. The many examples and case studies represent realistic applications, albeit somewhat simplified for purposes of clarity. Most of the problems at the ends of chapters involve additional applications. These are meant to expand upon the text, not just reinforce what has already been covered.

The solution algorithms, for the most part, are simple. In almost every case, the work will amount to solving a set of linear equations. There are many ways to do that, and any method that you know will be fine. For larger problems, you will want to use computer software, but again there are many packages that will suffice.

Chapter 8 presents for the first time in print, an unusual computational approach for obtaining many results for Markov chains and Markov processes. In an odd way, it unifies the discrete and continuous time cases because the approach works exactly the same for both cases, despite the fact that the usual algebraic treatments are distinctive. It works equally well for both numerical and symbolic solutions, and is particularly suitable for hand computation when the number of states is small.

A Note to Instructors

There are many styles of learning and many styles of effective teaching. Of course, you will have to find what works best for you and your students. However, I am glad to share with you my practices, for whatever value they may have.

Despite requiring a prior course in probability, I have always found it necessary to allow the first week for reviewing the material in chapter 1. However, I would caution against dragging it out for much longer. The real work starts in Chapter 2. In my experience, learning how to formulate a model is the most difficult challenge for students who have never attempted it before. Often they will want to skip that step to get on to something they feel more comfortable doing (such as memorizing formulae). That is why I took the unconventional approach of addressing the issue in a separate chapter. I spend several lectures going over the material in the first part of Chapter 2, but leave the examples for self-study. I strongly emphasize that the learning goal is not to memorize anything, but to acquire the skill to create new models.

Usually, the first test that requires them to construct a Markov chain model to fit a verbally described situation comes as a shock. Then I have to explain again that the goal is to learn a skill, not to recall a collection of facts. My grading policy has always included a provision to delete one bad test grade, which almost always turns out to be the first one. Despite many attempts in many different ways to convince the students in advance that they would have to approach the material and studying in a way that is different from what they are accustomed to, I never found an approach that avoided that initial shock. For that reason, I tried to schedule that first exam early, so that the necessary adjustment could come in time.

The many formulas that appear in the various chapters are easily confused or distorted. I never saw much value in testing whether students can remember them all correctly. Consequently, I allow them to bring one page of their own notes (a crib sheet, as they call it) to each exam. I explain that my tests are, for all practical purposes, open book exams, except that I do not want them fumbling through the text searching for the appropriate help. The single page has three advantages. First, it condenses into a manageable form the essential information that they might want to have at hand. Second, the process of constructing it helps them to organize the material in their own way. Third, it should convince them that producing the formulas is not the point. Later in life, they can always look up a formula that they find a need for, but the skill of creating the model has to be learned through practice.

I assign all of the problems, because each of them was designed to extend or clarify an important concept. However, I never had them count toward a grade because I did not want to have to guard the answers. I would rather make them available with just a little resistance so that a struggling student can get help at the appropriate time. I emphasize that it is important that they attempt the problems, but it less important that they succeed on their own. Working on the problems in groups is not discouraged, provided that each member of the group makes an intellectual effort. Getting the answer from someone else simply defeats the purpose of the practice exercise.

If you have to abbreviate the coverage, Chapters 4, 7, and 8 can be omitted without harming the overall treatment. Chapter 8, in particular, is a very unique computation technique that is published here for the first time. Although I have found that many students like it, you may not find it to your taste. In any case, it is so unconventional that it can easily be omitted without harm. On the other hand, all of the others are linked in critical ways. For example, Chapter 7 on queueing theory relies upon Chapter 6 on continuous time Markov processes, which in turn depends upon concepts that are explained in Chapters 2 and 3.

I personally cover everything in the book in one semester, but I do not allocate lecture time evenly over the material. For example, I would spend at most one lecture each on Chapters 4, 7, and 8. I want the students to be aware of the opportunities in these areas, but I do not hold them responsible for any of this material on tests. Similarly, some of the sections within chapters (such as 2.9, 2.10, 5.19, 5.20, 6.19, 6.20, 6.21, and 6.22) can be mentioned only briefly. I prefer to retain flexibility to answer questions and go over problems to ensure that the skill-building takes effect, rather than cover every detail.

Website

The publisher maintains a website with supplementary materials for downloads. These include problem solutions, spreadsheet implementations of models, and figure artwork. Any errata that are discovered will be posted there as well. The site address is www.wiley.com/college/solberg.

A Special Note to Students

Although there is significance to the sequencing of chapters and of sections within chapters, you do not have to read the book from beginning to end like a novel. It is perfectly acceptable to skim quickly through a chapter or even look ahead into later chapters to get an overview of where you are going. You will want to read individual sections carefully for deep understanding, but not necessarily all at once or even in the order they are presented. If you think you can skim through a section, feel free; you can always come back if you need to. The real test of whether you understand is whether you can do the problems. If you can do them without struggling, you are fine to proceed. If you get stuck, just go back and read a little more carefully within the chapter.

There will be a lot of notation. We will use up most of both the Roman and Greek alphabets, with subscripts, superscripts, italics, and parentheses mixed in liberally. I have made a strong effort to keep the notation easy to remember, but it will still be easy to confuse symbols. If you can avoid procrastinating, it will be easier. If you wait too long to straighten out your confusion, it will only get worse because there will be more to get confused. Notation is not the central point, of course. Notation is useful only as a shorthand way to write something you are supposed to understand. Often students want to accept the notation as given, memorize some formulas, and be tested on whether they remember those formulas. That will not be nearly enough to benefit from this book. You must understand what all of the notation means in order to know which formulas to use.

An introductory book cannot cover everything in the area, nor can it achieve the depth that we would like. There are many more advanced textbooks that build upon what is introduced here. That material becomes more difficult, obviously, but also more beneficial. We are sometimes forced in this text to use overly simple examples because we are working at the introductory level. But you should not infer from the simplicity of those examples that the theory is incapable of supporting much more realistic and complicated problems. It is just a matter of getting to that level gradually. Many fine books are available once you have overcome the initial obstacles. Each chapter has some recommended references for continuing your study. If, as I hope, you are inspired to learn more, you should first seek out a more rigorous treatment of the basic theory before going on to more advanced topics. You may have to wade through some more difficult presentations of theory, but at least you will be fortified with the realization that the pay off will far exceed the investment.

You may be accustomed to reading a problem and looking in the text for a similar one that you can adjust just a little. In most cases, that approach will not be effective here. To get to the point where this material will be useful in the real world, you have to develop the skill to create your own models. If that sounds difficult, it can be at first, but it gets easier with practice. You may feel frustration; that seems to be typical. The problems are arranged to increase in difficulty within each chapter. If you get stuck, as you probably will, you should not just give up nor continue to repeat your failures. You should look for hints in the back of the book, ask questions of your instructor, or talk things over with another student. In any case, you should work to achieve full understanding, not just to find the answers. Usually, obstacles will be nothing more serious than the wrong point of view, which can be easily corrected by a fresh perspective. At some point, most students achieve a conceptual breakthrough when the lights go on. One of my great joys as an instructor has been to witness those moments. Although it does not come as soon or easily for some as for others, in my experience it has always come to those who persist.

The most valuable thing to take away from the book is an altered way of thinking. Chances are, you will forget most of the mathematics within a few months unless you use it regularly. That is not a tragedy; you can always look it up when you realize you need it. But when you learn to think in terms of stochastic process models, you will see the world differently and more clearly. Specifically, you will have a better handle on dealing with risk, uncertainty, incomplete information, and variability. Think about it. If you properly absorb what is being presented, and acquire the skills that are offered, you will be able to predict the future more accurately and make decisions much better than most people-even most highly educated people. That is a skill worth acquiring, even if it takes a bit of hard work.

I would never pretend that the information in this book is essential to success in life; there are lots of people who have done just fine without it. However, I can promise that the effort you put into learning it will be repaid many times over. It is not just a bunch of abstract theory that some academic scholars find interesting; it is a powerful tool for dealing with reality. In short, if this book does not change your life, I will be disappointed and so should you.

Acknowledgments

Finally, I want to acknowledge the contributions of the several hundred students who used various versions of the class notes that led to this final published version, as well as the instructors and reviewers who made helpful suggestions. These include Professors Huann-Sheng Chen, Chuck Carson, Bruce Schmeiser, Mohamed Aboul-Seoud, and Michael Taaffe.

James J. Solberg
West Lafayette, Indiana

CHAPTER 1

Probability Review

CHAPTER CONTENTS

1.1. Interpreting and Using Probabilities 2
1.2. Sample Spaces and Events 3
1.3. Probability 4
1.4. Random Variables 6
1.5. Probability Distributions 6
1.6. Joint, Marginal, and Conditional Distributions 11
1.7. Expectation 14
1.8. Variance and Other Moments 16
1.9. The Law of Total Probability 18
1.10. Discrete Probability Distributions 20
1.11. Continuous Probability Distributions 23
1.12. Where Do Distributions Come From? 26
1.13. The Binomial Process 28
1.14. Recommended Reading 29

Although anyone tackling this book should have some previous exposure to probability, this chapter serves as a concise, convenient "refresher" summary of probability concepts that are important later. Students who are confident of their abilities in this area may elect to skim this chapter; those who feel weak on the subject should consult an introductory textbook, do all of the problems provided, and also find other problems to do. All of the subsequent material in the book depends heavily upon the basic concepts of probability.

It is worth mentioning early that we will be using a lot of probability, but not much statistics. Most introductory textbooks are so oriented toward statistics that they leave a slanted impression about probability. Students who have any confusion about the difference between the two subjects may have trouble with this book. Very concisely: Statistics is about interpreting data; probability is about representing uncertainty and/or variability. The two subjects converge when comparing data to what could have been expected from hypothetical assumptions. For example, hypothesis testing compares some numbers computed from data to numbers from a table of a specified probability function, such as the Normal, F, or Chi-squared distributions. The comparison tells whether the assumed hypothesis (corresponding to the tabulated number) is reasonably consistent with the measured data. But that is not the use of probability theory that will come into play here, so we will have little use for those tabulated distributions. Instead, we will need to understand the basic properties of random variables, distributions, and parameters so that we can manipulate them to produce predictions. We will have almost no use for the discrete uniform distribution (where permutations and combinations are used), the Normal distribution, or most of the other distributions that have names.

1.1 Interpreting and Using Probabilities

Some people learn all of the notation, rules of manipulation, and formulas, but never really "get" probabilities. Unless they are told directly to calculate a probability, they would never think of doing it on their own. So, although they may know *how* to calculate with them, they lack a full understanding of *why* probabilities are useful. Many introductory textbooks unintentionally contribute to the problem by using examples—cards, dice, colored balls, and such—that have no relevance to ordinary life. This text strives to provide more realistic examples that matter to engineers and managers. Obviously, we still have to begin with very simple situations and gradually work up to full realism.

The most immediate and obvious use of probabilities is to quantify uncertainty. Some people are very uncomfortable with uncertainty, preferring everything to be black or white, true or false, one way or the other, with no ambiguity. But most people understand that reality is not so simple—that sometimes people have to accept the fact that certainty cannot be achieved. Probabilities provide a way to quantify the "shades of gray" between impossibility and certainty. For what purpose? Generally speaking, the numbers in help to improve decision making.

When is uncertainty such an important factor that it demands quantitative treatment? Of course, if there is little at stake, you can make any choice without fear of making a big mistake. Or, if you have absolute certainty about the consequences of your actions, there is no need to assess probabilities. But let's face it—very few of the important decisions you make in life will be blessed by complete and accurate information. Almost always, you will be forced to choose with less information than you would like. On the other hand, if you have absolutely *no* information to work with, there is not much you can do with probabilities. So, we can conclude that probabilities are most useful when you need to make decisions about matters of importance and you have only partial information with which to work. That description still covers an enormous range of opportunities. In particular, engineers and managers deal with such issues routinely, because they design and control complex systems.

If you have only two choices, and want to favor the more likely event (such as betting on the winner in a two-team contest), there are ways—other than probabilities—to represent the comparative likelihoods. In sports competitions, it is common to use *point spreads*. For example, one team may be favored by three points, which means that (in someone's judgment) the first team is just as likely to score three more points than the second team's total as it is to score less than that. So, for purposes of betting, it is considered to be a fair bet when the weaker team is "given" an extra three points. Handicaps and headstarts are similar notions for equalizing chances.

Another way to express uncertainty is to use odds. If the success and failure of a certain outcome are equally likely, the odds would be 1:1 (spoken as "one to one"). If the odds against an outcome are given as 3:2 (spoken as "three to two"), it means that a bet of two units should win three units if the outcome is realized. That situation would reflect the fact or belief that the outcome is less likely to succeed than it is to fail, so the payoff should be greater than the bet to make the wager fair.

Although point spreads and odds are common in gambling situations, they do not serve very well elsewhere. In a business situation, for example, you cannot equalize the competition by handicapping the leader. Furthermore, you are commonly interested in more than winning or losing, so the two-outcome range of possibilities is far too limited. Although it is true that odds can always be translated to probabilities and vice versa (you may want to figure out how), probabilities are much easier to manipulate than odds when the situations grow complicated.

We usually speak of uncertainty as something describing the future—something that we are unsure about because it has not yet happened. But there are other sources of uncertainty that are also worth attention. Sometimes you need to answer questions about something that has already occurred, but you do not know the result. For example, a business competitor may have already taken some action that is hidden from you. Or perhaps you are in the process of conducting investigations and do not have complete answers yet.

There is another use of probability that does not involve uncertainty at all: We may have complete and accurate information about something, but that something is a set of values, rather than a single value. That is, probabilities are useful in describing a particular measurable property of individuals in the population. (Here, a population is any collection, not necessarily a biological one.) For example, if we determine the year of birth of every student in the class and then ask, "What year was the class born in?" we may not be able to answer with a single number even though we have all of the information. To give a full answer, we would typically have to specify a set of values—several years—and also the count of the number of students born in each year. Those counts, or frequencies, are equivalent to probabilities when they are normalized; that is, divided by the total number in the class. Hence, probabilities are useful in describing the *variation* in the population, even when that variation is fully known. In this case, the probability of any particular value corresponds to the fraction of the total population having that value. You can use ordinary probability rules to manipulate these fractions and always recover absolute numbers by multiplying the fractions by the total size of the population.

The greatest value from understanding probability comes from gaining a conceptual framework and vocabulary for dealing with uncertainty and variation. Even if you lack sufficient data to calculate anything, you can mentally weigh the factors better than people who lack that understanding. Those who have learned the concepts well use them every day.

1.2 Sample Spaces and Events

We turn now to a more formal presentation of the concepts. An *experiment* is a well-understood procedure or process whose outcome can be observed but is not known in advance with certainty. Reread that sentence; there is a lot that is said and left unsaid in those few words. For example, nothing is mentioned about being able to control anything. For this word and others that are defined here, mentally compare the formal definition and the informal use of the same word in ordinary language to be sure you understand the difference.

The set of all possible outcomes of an experiment is called the *sample space*. Whenever the sample space consists of a countable number of outcomes, it is said to be *discrete*; otherwise, it is *continuous*. An *event* is any subset of the sample space, including the empty, or null, set and the entire sample space. When the result of the experiment becomes known, we would say that a specified event *has occurred* if the observed outcome is contained in the subset which is the event. The empty set is an event that can never occur; the entire sample space is an event that is certain to occur. A set consisting of any single outcome is called an *elementary event*.

Of course, we want events to correspond to what you would ordinarily consider them to be. Often the most natural way to specify them is to describe them in words. However, the reason for defining them formally as sets is to establish a mathematical way to combine and manipulate them. The basic algebra used to manipulate events is set theory. You need to know all the rules to be able to express the real world events that we are going to model.

As a short reminder, the set theoretic union of two events produces another event. If $C = A \cup B$, we would say in words that event C had occurred if event A *or* event B (or both) occurred. Similarly, the intersection of two events corresponds to the word *and*. The complement of any event is another event; we would say that $\bar{A}$ had occurred if A had not occurred. Two events are *mutually exclusive* if their intersection is the empty set, which can be thought of as the impossible event. In other words, two events are mutually exclusive if they could not both occur. There are several other basic rules of set theory that you should know (DeMorgan's laws, the distribution rules, and so forth); look them up if you need help in recalling them.

1.3 Probability

When the "probability of an event" is spoken of in everyday language, almost everyone has a rough idea of what is meant. It is fortunate that this is so, because it would be quite difficult to introduce the concept to someone who had never considered it before. There are at least three distinct ways to approach the subject, none of which is wholly satisfying.

The first to appear, historically, was the frequency concept. If an experiment were to be repeated many times, then the number of times that the event was observed to occur, divided by the number of times that the experiment was conducted, would approach a number that was defined to be the probability of the event. The ratio of the number of chances for success out of the total number of possibilities is the concept with which most elementary treatments of probability start. This definition proved to be somewhat limiting, however, because circumstances frequently prohibit the repetition of an experiment under precisely the same conditions, even conceptually. Imagine trying to determine the probability of global annihilation from a meteor collision.

To extend the notion of probability to a wider class of applications, a second approach involving the idea of "subjective" probabilities emerged. According to this idea, the probability of an event need not relate to the frequency with which it would occur in an infinite number of trials; it is just a measure of the degree of likelihood we *believe* the event to possess. This definition covers even hypothetical events, but seems a bit too loose for engineering applications. Different people could attach different probabilities to the same event.

Most modern texts use the third concept, which relies upon a purely axiomatic definition. According to this notion, probabilities are just elements of an abstract mathematical system obeying certain axioms. This notion is at once the most powerful and the most devoid of real-world meaning. Of course, the axioms are not purely arbitrary; they were selected to be consistent with the earlier concepts of probabilities and to provide them with all of the properties everyone would agree they should have.

We will go with the formal axiomatic system, so that we can be rigorous in the mathematics. We want to be able to calculate probabilities to assist in making good decisions. At the same time, we want to bear in mind the real-world interpretation of probabilities as measures of the likelihood of events in the world. The whole point of learning the mathematics is to be able to use it in everyday life.

A *probability* is a function, $P(.)$, mapping events onto real numbers, and satisfying

1. $0 \leq P(A) \leq 1$, for any event A.
2. $P(S) = 1$, where S is the whole sample space, or the "certain" event.
3. If $A_1, A_2, A_3 \ldots$ are a set of pairwise mutually exclusive events (finite or infinite in number), then $P(A_1 \cup A_2 \cup A_3 \cup \ldots) = P(A_1) + P(A_2) + P(A_3) + \ldots$.

Although probabilities have a number of other properties well worth mentioning, these three axioms are sufficient to derive the others.

These three axioms are not enough to determine uniquely the probability of any event. For all but trivial sample spaces, there will exist an infinite number of ways to assign probabilities to events while satisfying the three axioms. At this point, we are merely establishing properties or rules required of *any* assignment of probabilities to events.

Some of the additional basic laws of probability (which can be proved from the above axioms) are

4. $P(\varphi) = 0$, where φ is the empty set, or the impossible event. In words, an event that cannot occur must be assigned a probability value of zero. Usually the converse is true also; namely, if an event has a probability of zero then it cannot occur. However, that statement is not always true. When there are an infinite number of outcomes in S, there are times when possible (though extremely unlikely) events have a probability value of zero.

5. $P(\overline{A}) = 1 - P(A)$. In words, the probability that an event does not occur is 1 minus the probability that it does occur. Another way to look at it is that the probability of any event plus the probability of its complement must sum to 1.

6. $P(A \cup B) = P(A) + P(B) - P(A \cap B)$, for any two events, A and B. This is the relation that seems to give students trouble. The tendency is to want to add probabilities without considering whether the events are mutually exclusive. When they are not—that is, when there is some possibility for both A and B to occur—then you have to subtract off the probability that they both occur.

7. $P(A|B) = P(A \cap B)/P(B)$ provided $P(B) \neq 0$. This "basic law" is, in reality, a definition of the conditional probability of an event, A, given that another event, B, has occurred. The notation for this conditional probability is $P(A|B)$, (read as "the probability of A given B").

   Conditional probabilities are very important in modeling, and we will see a great deal more of them. The notion of conditional probability conforms to the intuitive concept of altering our estimate of the likelihood of an event as we acquire additional information. That is, $P(A|B)$ is the new probability of A after we know that B has occurred.

   It is common in modeling applications to know $P(A|B)$ directly but not to know $P(A \cap B)$. For that reason, rule 7 often appears in the equivalent form shown in rule 8 below.

8. $P(A \cap B) = P(A|B)P(B)$. Conditional probabilities are useful only when the events involved, A and B, have something to do with one another. If knowledge that B has occurred has no bearing upon our estimate of the likelihood of A, we would say that the two events are independent and write rule 9, shown next.

9. $P(A|B) = P(A)$ if and only if A and B are independent. This rule can be taken as the formal definition of independence. Combining axioms 8 and 9, rule 10 immediately follows.

10. $P(A \cap B) = P(A)P(B)$ if, and only if, A and B are independent. Alternatively, rule 10 could be taken as the definition of independence and rule 9 would immediately follow. Rules 7, 8, 9, and 10 are all closely related; you should see them as variations of the same "fact" about dependent events. You should also realize that you will rarely be given numbers and asked to check the formulas to see whether dependence or independence applies. Almost always, you will have to decide for yourself whether or not the events are related, and then use the appropriate formula.

    A set of events $B_1, B_2, \ldots, B_n$ constitute a *partition* of the sample space S if they are mutually exclusive and collectively exhaustive, that is,

$$B_i \cap B_j = \varphi \text{ for every pair i and j}$$

and

$$B_1 \cup B_2 \cup B_3 \cup \ldots \cup B_n = S$$

In simple terms, a partition is just any way of grouping and listing all possible outcomes such that no outcome appears in more than one group. When the experiment is performed, one and only one of the B_i will occur. It is easy to prove that with rule 11.

11. $\sum_i P(B_i) = 1$ for any partition $B_1, B_2, \ldots, B_n$

12. $P(A) = \sum_i P(A|B_i)P(B_i)$ for any partition $B_1, B_2, \ldots, B_n$. This is one of the most useful relationships in modeling applications. It is one expression of the so-called law of total probability, which will be discussed in detail later in the chapter.

1.4 Random Variables

Although events may be directly assigned probabilities, more commonly the events are first associated with real numbers, which are then in turn associated with probabilities. For example, if the experiment involves observing the number of heads appearing when four coins are tossed, it would be natural to associate the possible outcomes with the integers 0, 1, 2, 3, and 4. These integers are not, in themselves, events, but the event corresponding to each value is obvious. The function that assigns numbers to events is called a *random variable*. You can think of it as a coding of the events, much like identification numbers that are used for convenience but do not in any way alter the events themselves.

In most cases, the rule that provides the value in the range of the random variable to go with each real-world event is so obvious that no special attention need be given to it. It is important to realize, however, that values of random variables have probabilities associated with them only because the values correspond to events that possess the probabilities directly. Using random variables gives us an indirect way to refer to events.

It is interesting to note that a random variable is, technically speaking, neither random nor a variable. It is conceptually convenient, however, to suppress all references to the real-world events and to regard a random variable as an ordinary variable whose value is randomly selected. In other words, once the random variable is well defined, we may speak of any value in the range of the random variable as if it were actually the event. It makes sense, thereby, to speak of the probability that a random variable, X, equals some particular number. (We really mean the probability of the event having that code value.)

If the values in the range of a random variable are integers (or, more precisely, a countable subset of the real numbers), the random variable is *discrete*. If the range consists of all values over an interval of the real numbers, the random variable is *continuous*. A discrete random variable could be either finite or infinite, depending on the number of values in the range. A continuous random variable would always have an (uncountably) infinite number of possible values, though the range could be bounded below, above, or both.

A word of caution is in order with respect to the use of the word "random." Sometimes, particularly in statistical applications, the word carries the connotation of equal likelihood. For example, when we say, "Take a random sample," we mean (among other things) that each member of the sampled population should have an equal chance of being selected. In general, however, the word "random" does not carry any such connotation.

1.5 Probability Distributions

Any rule that assigns probabilities to each of the possible values of a random variable is a probability distribution. The term is used somewhat generally, because there are several different ways to specify such a rule. More precise terms are used when a particular form is intended. However it is described, the rule essentially tells you how the total probability value of 1—that is, the amount of available probability for the entire range of possible values—is spread over those values.

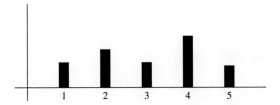

■ **FIGURE 1.1** A typical pdf

For discrete random variables, the most obvious and commonly used method of specifying the rule is to indicate the probability for each value separately. The function p(x), defined as

$$p(x) = \boldsymbol{P}(X = x)$$

is called the *probability distribution function*, or pdf for short. (Note the different use of uppercase and lowercase letters. Uppercase is used for names of random variables; lowercase is used for values.) Although it is not essential, many people find it helpful to think of distributions in graphical terms. A discrete pdf would look something like Figure 1.1.

The values along the horizontal axis of Figure 1.1 correspond to the possible values of the random variable (which could extend to infinity in either or both directions), and the heights of the bars indicate the values of the probabilities. The overall shape of the pdf is not important; it could look like almost anything. The only necessary features are that the heights of the bars are never negative and the sum of all of them add up to 1.

An alternative, equally sufficient method to specify a probability distribution is to give the *cumulative distribution function*, or cdf for short, F(x), defined as

$$F(x) = \boldsymbol{P}(X \leq x)$$

A third choice would be the *complementary cumulative distribution function*, or ccdf for short, G(x), defined as

$$G(x) = \boldsymbol{P}(X > x)$$

If any one of these—the pdf, cdf, or ccdf—is known, the others can be easily obtained in obvious ways. For example,

$$F(x) = 1 - G(x) \quad \text{and} \quad p(x) = F(x) - F(x - 1)$$

Graphs of the cdf or ccdf have characteristic forms. The cdf "steps upward" from 0 to 1, where the height of the step at x corresponds to the probability value at x. It can never step down because that would imply a negative probability value, which is not allowed. So the cdf is a monotonically nondecreasing function. The cdf for the pdf shown in Figure 1.1 would look like Figure 1.2.

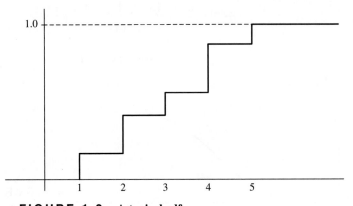

■ **FIGURE 1.2** A typical cdf

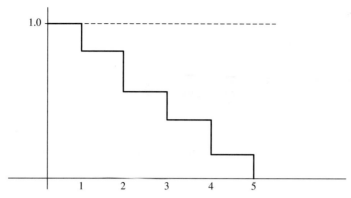

■ **FIGURE 1.3** A typical ccdf

Similarly, the ccdf in Figure 1.3 "steps down" from 1 to 0 and must be a monotonically nonincreasing function. Notice that, for any value of x, the cdf and ccdf sum to 1.

For continuous random variables, the situation is somewhat complicated by the fact that range of possible values is uncountably infinite. It is not consistent with the axioms of probability to allow each individual value to have positive probability. In fact, with the possible exception of a countable number of points, each individual value must be assigned the probability of zero! In contrast to the discrete case, a probability of zero does not necessarily imply that the corresponding event is impossible; it could merely mean that any one particular value is so unlikely, when considered next to the uncountably infinite set of alternatives, that the probability must be negligibly small. Consequently, it is fruitless to speak of the probabilities of particular values of random variables in the continuous case.

On the other hand, it makes perfect sense to speak of the probability that the value will fall within some interval. In particular, the cumulative distribution function, F(x), is well defined by

$$F(x) = \boldsymbol{P}(X \leq x)$$

Also, because the probability that X will exactly equal x may be zero, it can happen that

$$\boldsymbol{P}(X \leq x) = \boldsymbol{P}(X < x) + \boldsymbol{P}(X = x) = \boldsymbol{P}(X < x)$$

In other words, sometimes in the continuous case no distinction between strong and weak inequalities, or between open and closed intervals, need be made. Of course, the distinction must be scrupulously maintained in the discrete case or in continuous cases where some specific values have nonzero probability.

A typical continuous cdf will look something like Figure 1.4.

From the definition, it is apparent that F(x) must have the following properties:

$$0 \leq F(x) \leq 1 \qquad \text{for all x}$$

$$\lim_{x \to -\infty} F(x) = 0$$

$$\lim_{x \to \infty} F(x) = 1$$

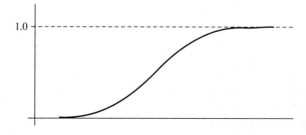

■ **FIGURE 1.4** A typical continuous cdf

and

$$F(y) \geq F(x) \quad \text{for any } y > x$$

In words, the function F(x) must be bounded between 0 and 1, must approach zero at the left extremity of its range and one at the right extremity, and must be monotonically nondecreasing. (Actually, the last three imply the first.) Conversely, any function having these properties will qualify as a cumulative distribution function for some continuous random variable. Notice that although the figure shows a continuously rising function (which is typical), there is nothing to require that property; it could take sudden steps upward at some points.

Given the cumulative distribution function, one can easily express the probability that the random variable will assume a value within any specified region. For example,

$$P(a \leq X \leq b) = P(X \leq b) - P(X \leq a) = F(b) - F(a)$$

The *complementary cumulative distribution function*, G(x), defined by

$$G(x) = P(X > x)$$

or by

$$G(x) = 1 - F(x)$$

would also serve to describe fully the distribution. A typical continuous ccdf, as shown in Figure 1.5, would look like the cdf turned over.

The *probability density function*, f(x), is a function that, when integrated between a and b, gives the probability that the random variable will assume a value between a and b. That is,

$$P(a \leq X \leq b) = \int_a^b f(x)dx$$

The relation between the density function and the distribution function is direct

$$F(x) = \int_{-\infty}^{x} f(y))dy$$

and

$$f(x) = \frac{d}{dx}F(x)$$

Although it may not seem to be the most natural way to describe a probability distribution, the density function is used more often than the cumulative or complementary cumulative distribution functions. In the case of a few distributions, only the density function

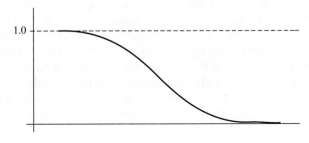

■ **FIGURE 1.5** A typical continuous ccdf

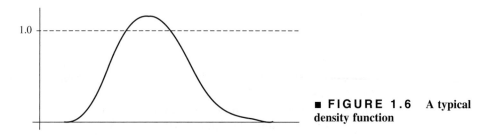

■ **FIGURE 1.6** A typical density function

can be expressed in closed form; the others must be expressed as integrals of the density function. It is important, therefore, that you learn to think in terms of density functions. One of the first things to get straight is that the value of the density function at some point is not a probability. The only way to get a probability from a density function is to integrate it.

The appearance of a density function is often something like Figure 1.6.

Any density function will have the properties

$$\int_{-\infty}^{\infty} f(x)dx = 1$$

and

$$f(x) \geq 0 \qquad \text{for all x}$$

The first property is a direct consequence of the definition, but the second requires a brief argument. If f(x) were negative at any point, then there would exist two points, a and b, such that the integral of f(x) between a and b was negative. This would imply that

$$P(a \leq X \leq b) < 0$$

which is impossible because probabilities cannot be negative. Therefore f(x) must be non-negative everywhere.

Any function f(x) having the two properties mentioned above will qualify as a probability density function for some continuous random variable. Notice, in particular, that there is no requirement that f(x) be bounded above. The second property sometimes leads students to the mistaken presumption that f(x) cannot exceed 1. In fact, f(x) can be much greater than 1 over a narrow range, provided only that the integral over any interval does not exceed 1. Notice also that there is no requirement that f(x) be continuous. Functions that are discontinuous, or abruptly "jump" from one value to another, can be integrated without difficulty, provided only that the points of discontinuity are limited in number. The method, of course, is to separate the interval that you want to integrate into a sequence of intervals over each of which the density function is continuous.

Although, as already noted, it is important to keep in mind that f(x) is not a probability, it is useful in many applications to be able to substitute something involving f(x) into expressions as if it were a probability. A generally reliable device is to think of the notation f(x)dx as representing the probability that the random variable equals x. The dx part of the expression can be regarded as an interval of infinitesimal width, so the product of f(x) and dx is (roughly speaking) an area under the curve, or a probability. The presence of dx will indicate that an integration must be performed before an exact expression can be inferred.

Although the distinction between the discrete and continuous random variable cases is important, there are occasions when it is convenient to have a unified terminology to cover both cases. The letters pdf may be used to refer to either the probability distribution function, in the discrete case, or the probability density function, in the continuous case. Similarly, the letters

1.6 Joint, Marginal, and Conditional Distributions

Whenever more than one random variable is involved in a single problem, there is a possibility that they are related. If so, it would not be sufficient to describe the probability distribution of each random variable in isolation; the relation between or among them must also be described. There are two methods in common use.

Suppose that two random variables, X and Y, are involved. The joint cumulative distribution function, or joint cdf, F(x, y), is defined by

$$F(x, y) = P(X \leq x, Y \leq y)$$

Here, the comma represents the same as the intersection of the events implied by the terms. In words, F(x, y) is the probability that X takes on a value less than or equal to x *and* that Y takes on a value less than or equal to y. The same definition will suffice whether the random variables are both discrete, both continuous, or mixed. Conceptually, the basic idea is to extend the notion of a cumulative distribution function to two dimensions. Obviously, the same basic idea can be used to extend the notion to higher dimensions.

If both X and Y are discrete, the joint probability distribution function is defined by

$$p(x, y) = P(X = x, Y = y)$$

If both are continuous, the joint probability density function is defined by

$$f(x, y) = \frac{\partial}{\partial x} \frac{\partial}{\partial y} F(x, y)$$

The latter must be integrated twice in order to obtain a probability. In particular,

$$P(r \leq X \leq s, t \leq Y \leq u) = \int_r^s \int_t^u f(x, y) \, dy \, dx$$

Each of these is just a two-dimensional extension of the appropriate function for single random variables, and can be extended to higher dimensions in the obvious way. The term "joint pdf" will describe either function.

EXAMPLE 1.1

When X and Y are discrete and there are only a small number of possible values for each, it can be convenient to express the joint pdf in a table. For example, Table 1.1 shows a case where X can assume any of three values and Y can assume any of four. You read the cell entry directly to get the joint probability for any pair of values. For example, $P(X = 1, Y = 10) = 0.2$.

■ **TABLE 1.1**
A Joint pdf

	Y = 5	Y = 10	Y = 15	Y = 20
X = 1	0.1	0.2	0	0
X = 2	0	0.25	0.25	0
X = 3	0	0	0.1	0.1

Sometimes a joint pdf for two or more random variables is given, but you want to know the pdf for just one of the random variables. That is, you might want to make a probability statement about, say, X, without regard to the value of Y. When both X and Y are discrete, the marginal probability distribution function of X is obtained from the joint pdf by

$$p(x) = \sum_y p(x, y)$$

When both are continuous, the marginal probability density function of X is given by

$$f(x) = \int_y f(x, y) dy$$

A marginal pdf is just an ordinary pdf, with all of the usual properties and interpretations. The word "marginal" merely conveys the information that it was obtained from a joint pdf.

If you have the joint pdf in the form of a table, as in Table 1.1, you get the marginal pdf by summing over rows or columns. For example, if you want the marginal pdf of X, you would sum across each of the three rows. In words, the probability that X takes on the value 1 is the sum of the probabilities that $X = 1$ and Y takes on any of its possible values. So, you just add across the first row to find $P(X=1) = 0.3$.

By symmetry, the marginal pdf of Y is obtainable from the joint pdf of X and Y by summing or integrating over all values of X. If more than two random variables are involved in a joint pdf, the marginal pdf for any one can be found by summing or integrating over all values of all random variables other than the one whose marginal pdf is sought. Although it is not often used, the marginal cdf is, if anything, even easier to obtain from a joint cdf:

$$F(x) = \lim_{y \to \infty} F(x, y)$$

Dealing with the cdf also has the advantage of permitting a single expression to cover both the discrete and continuous cases.

Independence of random variables is a property deriving from independence of the events that the random variables represent. Two random variables are independent if for all x and all y,

$$F(x, y) = F(x)F(y)$$

or, in terms of pdfs,

$$p(x, y) = p(x)p(y)$$

for discrete random variables, and

$$f(x, y) = f(x)f(y)$$

for continuous random variables. When the random variables are independent—but only then—the joint distribution can be constructed from the marginals.

Independence of random variables is an extremely important concept. Not only must you know how to manipulate the functions in the presence or absence of the property, but you also must judge whether the property can be reasonably assumed to hold in real-life situations. Because the mathematical definition may not be sufficiently revealing by itself to allow the student to grasp the concept at an intuitive level, a bit of further discussion seems warranted. When we say that the joint distribution can be obtained simply by multiplying the marginals, we are admitting that the joint distribution contains no more information than is already contained in the separate descriptions of the random variables. In other words, there is no need to account for the influence that one of the random variables might exert upon another. This would be true if, and only if, no such influence exists. Although the definition of independence of random variables is very similar in appearance to the definition of independence of events, it is actually a much stronger requirement. In order for X and Y to be independent, it is necessary that *every* event associated with X be independent of *every* event associated with Y.

The method of expressing joint pdfs or cdfs is just one of the ways to describe a relationship between two random variables. The other method is based on the idea of fixing a value for one and describing the subsequent distribution for the other. If both are discrete, the *conditional probability distribution function* of X given y (a particular value of the random variable Y) is defined by:

$$p(x|y) = \boldsymbol{P}(X = x | Y = y)$$

In p(x|y), x is the argument of the function and y can be regarded as a parameter. In other words, we may insert various values of x into the function to get the probability that the random variable equals x, but this probability will be contingent upon the value of y. Through its definition as a conditional probability, the conditional pdf is easily related to the joint pdf by the expression

$$p(x|y) = \frac{p(x, y)}{p(y)} \qquad \text{provided } p(y) \neq 0$$

An analogous function exists for continuous random variables, but cannot be defined directly in terms of a conditional probability. The *conditional probability density function* of X given Y is most simply defined in terms of the joint density function

$$f(x|y) = \frac{f(x, y)}{f(y)} \qquad \text{provided } f(y) \neq 0$$

This function must be integrated with respect to x in order to yield a probability; the y simply acts as a parameter.

The conditional pdf of X given Y reduces to the marginal pdf if, and only if, X and Y are independent. In notation,

$$p(x|y) = p(x) \qquad \text{for all } x, y$$

or

$$f(x|y) = f(x) \qquad \text{for all } x, y$$

if, and only if, X and Y are independent. These expressions are entirely consistent with our earlier discussion of independence. If knowledge of the value of Y contributes nothing to a probability statement involving X, it must be that X and Y are unrelated.

Whenever a conditional distribution and one marginal distribution is given, the other marginal can be obtained. The procedure is first to obtain the joint distribution and then use that to get the desired marginal. In the discrete case, the expressions would be

$$p(x, y) = p(x|y)p(y)$$
$$p(x) = \sum_{y} p(x, y)$$

Therefore,

$$p(x) = \sum_{y} p(x|y)p(y)$$

The analogous formula in the continuous case would be

$$f(x) = \int_y f(x|y)f(y)dy$$

Both of these expressions are very useful in modeling applications.

1.7 Expectation

To describe a random variable completely requires a probability distribution in one of its various forms. If we were to require, however, a single number that best "summarized" the information contained in the distribution, we would almost certainly want to specify the "center" of the distribution. There are several ways to define "center," but the most useful is the expectation.

The *expectation* of a random variable X, denoted $\boldsymbol{E}(X)$, is defined by:

$$\boldsymbol{E}(X) = \sum_{-\infty}^{\infty} xp(x) \qquad \text{when X is discrete}$$

and

$$\boldsymbol{E}(X) = \int_{-\infty}^{\infty} xf(x)dx \qquad \text{when X is continuous.}$$

The same quantity may be called the *expected value of* X (although this term is quite misleading), the *mean* of the distribution, or the *first moment* of the distribution of X. All of these terms refer to the same thing. However, it should *not* be confused with an arithmetic average or a sample mean. The latter are statistical entities; we would compute them from data. An expectation is calculated from, and is an attribute of, a probability distribution. It can be regarded as a weighted average of the values of X, in which each possible value is weighted by the probability of its occurrence.

Although $\boldsymbol{E}(X)$ is often called the expected value of X, one should be on guard against "expecting" $\boldsymbol{E}(X)$ to occur as the value of X. Indeed, when X is discrete, $\boldsymbol{E}(X)$ might not even be a *possible* value of X. It is true that if the experiment for which X is defined were to be repeated independently many times and the observed values of X were collected and averaged, then this average would be "close" to $\boldsymbol{E}(X)$, in a certain probabilistic sense. However, this fact is a theorem of statistics (one form of the Law of Large Numbers) and has little significance for any single trial.

Typically in decision making, when one is forced to rank the options by some preference, the expectation is the value that is compared. One can easily criticize that approach, because the center of the distribution (or any other single value, for that matter) is a poor representation of the full range of possibilities. However, it is usually the most practical approach, and it can be justified at least to the extent that the expectation weights the outcomes "fairly."

One of the reasons that the expectation is so useful as a measure of centrality is that it has a number of very convenient properties. For any random variable X and any constants a and b,

$$\boldsymbol{E}(aX) = a\boldsymbol{E}(X)$$

and

$$\boldsymbol{E}(X+b) = \boldsymbol{E}(X) + b$$

In words, both multiplicative and additive constants can be "pulled out" of the expectation. For any two random variables X and Y,

$$E(X+Y) = E(X) + E(Y)$$

In words, the expected value of a sum is the sum of the expected values. The same relation can be extended to sums of more than two random variables and will hold whether or not the random variables are independent. The fact that sums "separate" and constants "pull out" of expressions in the obvious ways without any complications imply that the expectation is a linear operator. It is *always* linear, with no additional requirements on the random variables. (The same cannot be said for variances or other moments.)

Whenever X and Y are independent, the expected value of a product of random variables will decompose; that is,

$$E(XY) = E(X)E(Y)$$

but this relation does not generally hold when the random variables are dependent. We will come to the more general case shortly.

Another convenience associated with using the expectation is the fact that the expectation of an arbitrary function of a random variable is easily expressed. Let h(X) be any function of X. Then if X is discrete,

$$E(h(X)) = \sum_x h(x)p(x)$$

and if X is continuous,

$$E(h(X)) = \int_x h(x)f(x)dx$$

In other words, h(x) merely replaces x in the definition of $E(X)$. These expressions are not a new definition or an obvious fact, but are derived by considering a random variable $Y = h(X)$ and relating the distribution of Y to the distribution of X.

A concept used repeatedly in the book is that of *conditional expectation*. Formally, the conditional expectation of a random variable X given the value of a related random variable Y is defined by

$$E(X|y) = \sum_x xp(x|y) \qquad \text{when X is discrete}$$

and

$$E(X|y) = \int_x xf(x|y)dx \qquad \text{when X is continuous.}$$

The conditional expectation of X given y can be combined with the distribution of Y to yield the unconditional expectation of X. In notation,

$$E(X) = \sum_y E(X|y)p(y) \qquad \text{when X is discrete}$$

and

$$E(X) = \int_y E(X|y)f(y)dy \qquad \text{when X is continuous.}$$

A concise way to express both of these is

$$E(X) = E[E(X|y)]$$

but this form does not suggest how useful the relation is as a technique for formulating an expression for $E(X)$. The other forms suggest that the expectation of X can be thought of as a weighted average of the conditional expectations of X given y, taken over all possible conditions y, with each possible $E(X|y)$ weighted according to the probability of occurrence. We will discuss this relation further in Section 1.9, "The Law of Total Probability," page 18.

1.8 Variance and Other Moments

After you know something about the central location of a distribution, most commonly expressed as an expectation, the next most valuable summary information would be about the spread or dispersal of the values that the random variable takes on. You could use the *range* (the interval between the highest and lowest value) if it is finite, or any of several other ways to measure the spread. But the most common measure is the *variance*, or its square root, the *standard deviation*. Computing it involves what may seem to be a nasty calculation, but the properties justify the definition.

The *n*th moment of a random variable is defined as the expectation of the *n*th power of the random variable. Since X^n is just a special case of a function of X, the *n*th moment can be expressed as

$$E(X^n) = \sum_X x^n p(x) \qquad \text{when X is discrete}$$

and

$$E(X^n) = \int_X x^n f(x) dx \qquad \text{when X is continuous.}$$

The first moment is, of course, the expectation. The *n*th central moment or the *n*th moment about the mean is defined as

$$E([X - E(X)]^n) = \sum_X [x - E(X)]^n p(x) \qquad \text{when X is discrete}$$

and

$$E([X - E(X)]^n) = \int_X [x - E(X)]^n f(x) dx \qquad \text{when X is continuous.}$$

In words, it is the expectation of the *n*th power of the random variable after it has been "shifted" by subtracting the expectation.

After the expectation, the next most important single number used to summarize distributions is the second moment about the mean, more commonly known as the *variance*. Denoting the variance of X by $V(X)$,

$$V(X) = \sum_X [x - E(X)]^2 p(x) \qquad \text{when X is discrete}$$

and

$$V(X) = \int_X [x - E(X)]^2 f(x) dx \qquad \text{when X is continuous.}$$

In both the discrete and continuous case, the variance can be shown to equal the second moment minus the expectation squared. That is,

$$V(X) = E[X^2] - E(X)^2$$

This form is often more convenient to use in algebraic manipulations. Of course, if you have the expectation, it is easy to convert between the variance and the second moment in either direction.

The variance, being defined as a weighted average of the squared deviations from the expectation, is a measure of the spread, or dispersion, of a probability distribution. One of the objections to its use for this purpose is that the units are not those of X but of X^2. The *standard deviation*, defined as the square root of the variance, overcomes this objection. We conventionally use a lowercase sigma for standard deviations, so the definition would be

$$\sigma_X = \sqrt{V(X)}$$

It is usually a little easier to work with variances than with standard deviations (just because they avoid the square root), but there are certainly times when the standard deviation is more meaningful. Either is a simple one-to-one transformation of the other, so they both convey the same information.

Sometimes we may want to express the *relative* amount of variation in a random variable, rather than an absolute measure. For example, suppose we had two random variables X and Y, which are measured on different scales (say meters and kilograms), and we wanted to say which was more variable than the other. It would make no sense to compare the variances or even the standard deviations, because the dimensional units (meters and kilograms) are not consistent.

One way to express a relative measure of variability is the *coefficient of variation*, defined as the ratio of the standard deviation to the mean,

$$C_X = \frac{\sigma_X}{E(X)}$$

This is a dimensionless value because the units in the numerator and denominator cancel out. A value close to zero would mean that the standard deviation is much less than the mean, and a value greater than one (or less than -1) would mean that the standard deviation is more than the mean. The standard deviation is always positive, but the mean could be negative, so the coefficient of variation could take on a negative value. However, the interpretation of relative variation would be same.

The properties of variances, standard deviations, and coefficients of variation are not so obvious as those of expectations. Whereas the behavior of expectations conforms to what intuition would suggest, considerable care must be exercised in dealing with the others. The rules for dealing with multiplicative and additive constants are

$$V(aX) = a^2 V(X)$$

and

$$V(X + b) = V(X)$$

In words, a multiplicative constant can be "pulled out" of a variance, but must be squared; an additive constant can be "dropped out." When considering a sum of random variables, the variance of the sum will be the sum of the variances, if the random variables are independent. For two independent random variables X and Y,

$$V(X + Y) = V(X) + V(Y)$$

On the other hand, if the random variables are dependent, this relation will not generally hold. The correct expression for the general case requires another definition.

Given two random variables X and Y, the *covariance* of X and Y is defined by

$$COV(X, Y) = E([X - E(X)][Y - E(Y)])$$

but this expression can be shown to equal

$$COV(X, Y) = E(XY) - E(X)E(Y)$$

It will be recalled that when X and Y are independent, $E(XY) = E(X)E(Y)$, so the covariance of independent random variables is zero. The converse does not always hold; that is, the mere knowledge that the covariance of random variables is zero would not be enough for one to conclude that they are independent. Indeed, examples can be constructed of dependent random variables for which the covariance equals zero. On the other hand, a nonzero covariance definitely implies a relationship between the random variables, so the covariance is used as a (somewhat imperfect) measure of the degree of dependence. Another related measure of dependence is the *correlation coefficient* between X and Y, usually denoted by ρ, (lowercase Greek letter rho), which is defined as

$$\rho = \frac{COV(X, Y)}{\sqrt{V(X)V(Y)}}$$

Returning to the variance of a sum of random variables, the general equation for two random variables is

$$V(X + Y) = V(X) + V(Y) + 2COV(X, Y)$$

1.9 The Law of Total Probability

There is an extremely useful equation relating conditional probabilities, sometimes called the law of total probability. We will also refer to the concept as "conditioning," because it is a common way to develop expressions that are helpful in computations of either probabilities or expected values. That is, when we are faced with the need to find a complicated probability or expectation, we "condition" on some other random variable to simplify the task. This is a very important idea—it is probably not exaggerating to call it the key idea in stochastic processes—so you should be sure that you understand what is going on. It will not be enough to remember a formula, because the notation will change with the circumstances. You must understand the idea and adapt the notation to whatever situation you are in when you need to use it.

You have already seen it in one form, back in relation 12 on page 6. It was

$$P(A) = \sum_i P(A|B_i)P(B_i) \text{ for any partition } B_1, B_2, \ldots, B_n$$

Another form, expressed in terms of discrete random variables, would be

$$P(X = x) = \sum_y P(X = x|Y = y)P(Y = y)$$

or, with shortened notation,

$$p(x) = \sum_y p(x|y)p(y)$$

The same relation for continuous random variables would be

$$f(x) = \int_y f(x|y)f(y)dy$$

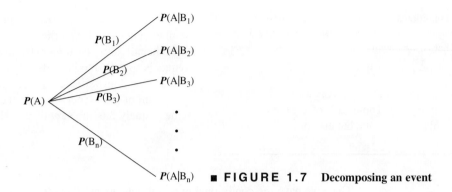

■ **FIGURE 1.7** Decomposing an event

Both of these expressions also appeared earlier in the chapter. There can be a lot of variations in the way the law appears, but they are all based on the same idea.

EXAMPLE 1.2

Here is the thinking you go through: Suppose, to take a concrete example, that we need to find the probability of some event A. Suppose further that the situation is too complicated to know $P(A)$ directly, but we could know the probability of A if we knew which of a number of possible conditions held. That is, we know the conditional probabilities $P(A|B_i)$ for a set of mutually exclusive, collectively exhaustive conditions, B_i. It would make some sense to "average" these various possible values for the probability of A. But if the conditions B_i, are not all equally likely, the various $P(A|B_i)$ should not be given equal weight in the average; each should be weighted according to the probability that the condition B_i does in fact hold, or $P(B_i)$. This logic produces relation 12.

Another way to see the relation is to imagine breaking the event down into a set of alternative possibilities, as shown in Figure 1.7. (This is something like a decision tree.) Assuming that we have some way to get the separate conditional probabilities, all we have to do is to weight the branches by the probabilities that the separate conditions hold—the $P(B_i)$. Of course, the related B_i events must be such that one and only one of them will hold, which corresponds exactly to the requirement that they form a partition.

The reason that the conditioning argument is so useful in the context of stochastic processes is that the events we condition upon—the events indicated by B_i—are events that happen sometime before A in the progress of time. For example, we can often say something about the probability of A if we know what happened just before.

A very similar idea can be used to find expected values. You have seen some expressions of the idea under conditional expectations, but let's frame the issue in more intuitive terms here. Faced with the problem of expressing $E(X)$, for some random variable X, one might try to find another random variable, Y, whose distribution is known or can be found, and which has the property that when the value of Y is specified, the expectation of X is easy to obtain. Usually, however, our use of the concept will be such that the conditional expectation may be known directly.

EXAMPLE 1.3

For example, suppose that we are interested in an inventory problem and X represents the number of units of some product sold during a specified period. If Y represents the number of customers who purchase some number of units during the period, and if the expectation of the number of units purchased is the same for each customer, say 3.6 units,

> then the conditional expectation of the number of units sold given that the number of customers is y would be 3.6y, for any y. That is, we obtain
>
> $$E(X|y) = 3.6y$$
>
> without having to use the conditional probability distribution p(X|y). The details of the logic involved are probably unnecessary, but just to verify rigorously that the result is correct, we may argue as follows: The number of units sold, X, is the sum of the amounts sold to each individual customer. If the number of customers is specified to be y, then X will consist of the sum of y random variables. The expectation of a sum is the sum of the expectations; if each of these is the same, namely 3.6, the sum of y of them is 3.6y.

Once we have the conditional expectations, we put them together in the obvious way, namely, by taking a weighted sum, where the weights are the probabilities for the respective values of y. In notation, this expression is

$$E(X) = \sum_y E(X|y) p(y)$$

which is the expression shown earlier for conditional expectations.

Again, in the context of stochastic processes, the typical use of the conditioning argument is to condition on events at some prior time. You will see the idea applied numerous times over the next few chapters.

1.10 Discrete Probability Distributions

There is an infinite variety of functions that satisfy the requirements to be probability distributions. Only a few occur so commonly that they have been given names. In statistical applications, you will almost always find yourself using named distributions (for example, Normal, Student's t, F), but in real-world modeling applications you will more frequently construct distributions that are unnamed. Here we will mention just a few of the most commonly used distributions.

The Discrete Uniform Distribution

When a random variable X has only a finite number of possible values, each of which can occur with equal likelihood, the distribution is called *discrete uniform*. Without serious loss of generality, we may assume that the range of X is x = 1, 2, ..., N, in which case the probability distribution function is

$$p(x) = \frac{1}{N} \quad \text{for } x = 1, 2, \ldots, N$$

When X has this range, the mean and variance are

$$E(X) = \frac{N+1}{2}$$

and

$$V(X) = \frac{N^2 - 1}{12}$$

Of course, a shift or scaling of the range of X will have a corresponding effect upon the pdf, mean, and variance. In any case, the pdf is just 1 divided by the total number of possible values, for each value, and the expectation falls at the midpoint of the range.

Although it has many uses, the discrete uniform distribution is not so important as it is frequently thought to be by beginners in probability. Elementary textbooks often give so much emphasis to combinatorial probability—using permutations and combinations to count the number of ways that events could occur and using these counts (together with the assumption of equal likelihood) to form probabilities—that it is easy to develop a concept of probability theory that is limited to this one special case. It is important to realize that the discrete uniform distribution is just one of many useful distributions.

The Bernoulli Distribution

If a random variable must assume one of two values (usually, but not always 0 or 1), it is said to be a Bernoulli random variable. The corresponding experiment, which has only two possible outcomes, is called a Bernoulli trial. Usually the outcome that is mapped by the random variable onto the value 1 is called a *success* and the other is called a *failure*. The distribution is given by

$$p(1) = p$$

and

$$p(0) = 1 - p \text{ or } q$$

where p is the only parameter of the distribution, often referred to as the "probability of success." (Note that the p on the left is a pdf, while the p on the right is a parameter.) The mean of a Bernoulli defined on 0 and 1 is p. The variance is pq.

The distribution may seem so trivial as to be undeserving of special attention. Although it is true that direct applications are limited, it turns out that a number of more important distributions can be derived from considering a sequence of independent Bernoulli trials. We will return to this subject after we see some more distributions.

The Binomial Distribution

Let X be a discrete random variable defined over the range x = 0, 1, 2, ..., n. If

$$p(x) = \binom{n}{x} p^x (1-p)^{n-x}$$

then we say that X has a binomial distribution with parameters n and p, where n is a positive integer and $0 \leq p \leq 1$. The notation

$$\binom{n}{x}$$

refers to the so-called binomial coefficient defined by

$$\binom{n}{x} = \frac{n!}{x!(n-x)!}$$

A binomially distributed random variable usually can be thought of as counting the number of successes in a sequence of n independent Bernoulli trials, where the probability of success on any trial is p. Tables of binomial coefficients and the binomial distribution are readily available.

The expectation, or mean number of successes, for a binomial is np. The variance is $np(1-p)$.

The Poisson Distribution

Let X be a discrete variable defined over the range x = 0, 1, 2, ..., ∞. If

$$p(x) = \frac{\lambda^x e^{-\lambda}}{x!} \quad \text{for } x = 0, 1, 2, \ldots$$

then we say that X has a Poisson distribution with parameter λ, where λ must be positive. The Poisson distribution has a number of convenient properties that contribute to its usefulness in modeling. The expectation and variance are equal to one another, and are given simply by the parameter of the distribution:

$$E(X) = V(X) = \lambda$$

The distribution is reproductive; that is, the sum of independent Poisson distributed random variables will be another Poisson distributed random variable. The parameter of the sum random variable will be just the sum of the parameters of the constituent random variables.

One of the common usages of the Poisson distribution is as an approximation to the binomial distribution when the number of trials (n) becomes large while the probability of occurrence (p) becomes small. All that is required for the approximation is to give the two distributions the same expectation. That is, let $\lambda = np$.

Another common use of the Poisson distribution is to describe the number of events occurring within some period of time. In this context, it is the usual practice to use λt as the parameter of the distribution, where t is interpreted as the length of the period and λ is now the mean "rate" at which events occur. The Poisson process and its properties will be discussed in some detail in Chapter 6.

The Geometric Distribution

There are two common versions of the geometric distribution. If X is defined over the range x = 1, 2 ..., ∞ and has the pdf

$$p(x) = p(1-p)^{x-1} \quad \text{for } x = 1, 2, 3, \ldots, \infty$$

where $0 \leq p \leq 1$, we would say that X has the geometric distribution beginning at 1. If it is defined over the range x = 0, 1 ..., ∞ (that is, starting at zero rather than 1)
and

$$p(x) = p(1-p)^x \quad \text{for } x = 0, 1, 2 \ldots, \infty$$

we would say that X has the geometric distribution beginning at zero. It is apparent that one version is just a shifted version of the other, and that other shifts could be made without altering the form of the distribution. Both of these versions appear in applications and are easily confused.

The expectations and variance for the geometric distribution beginning at 1 are, respectively,

$$E(X) = \frac{1}{p}$$

and

$$V(X) = \frac{1-p}{p^2}$$

When the distribution begins at zero, the variance is the same, but the expectation is $(1-p)/p$.

A possible interpretation of X, when it begins at 1, is as the number of trials in a sequence of independent Bernoulli trials that will occur before the first success is observed. More precisely, it is the number of the trial on which the first success occurs. If X begins at zero, it could be thought of as counting the number of failures before the first success. In either case, X counts trials, so the geometric distribution is often regarded as a waiting-time distribution. One should not confuse this interpretation of the geometric distribution with that of the binomial distribution. The latter fixes the number of trials and counts successes.

The Negative Binomial Distribution

Let X be a discrete random variable defined over the range $x = k, k+1, \ldots, \infty$. We would say that X follows a negative binomial distribution if

$$p(x) = \binom{x-1}{k-1} p^k (1-p)^{x-k} \qquad \text{for } x = k, k+1, \ldots, \infty$$

where k is an integer >1 and $0 \leq p \leq 1$. Another name for the same distribution is the Pascal distribution. When $k = 1$, the distribution reduces to the geometric. The expectation and variance are

$$E(X) = \frac{k}{p}$$

and

$$V(X) = \frac{k(1-p)}{p^2}$$

The explanation for this distribution just extends that of the geometric. X represents the number of the trial, in a sequence of independent Bernoulli trials, on which the kth success occurs. Thus the negative binomial distribution is another waiting-time distribution. Thinking of X in this way suggests that the waiting time for the kth success ought to be the sum of k waiting times for the one success. Because the trials are independent, this logic is valid. It is a fact that the sum of k independent geometrically distributed random variables will yield a random variable whose distribution is negative binomial with parameter k.

Sometimes the negative binomial distribution is used without any waiting-time interpretation, but simply because the parameters can be adjusted so as to fit a set of data. In this case, it may be desirable to have the range of X begin at zero, rather than k. If so, the appropriate pdf would be

$$(x) = \binom{k + x - 1}{x} p^k (1-p) \qquad \text{for } x = 0, 1, 2, \ldots, \infty$$

The variance would be the same, but the expectation would be $k(1-p)/p$.

1.11 Continuous Probability Distributions

The Continuous Uniform Distribution

When a continuous random variable X is restricted to a finite range $a \leq x \leq b$ and is such that "no value is any more likely than any other," then X would be appropriately described by the continuous uniform distribution. It is the obvious analog of the discrete uniform distribution, which restricted the random variable to a finite number of equally likely values. The description, "no value more likely than any other," is somewhat loose, because, of course, the probability of any one value for a continuous random variable is zero. A better, although less intuitive, description would be, "the probability that x falls within any interval in the range of X depends only on the width of the interval and not on its location."

In any case, the distribution is rigorously defined by its probability density function:

$$f(x) = \frac{1}{b-a} \qquad \text{for } a \leq x \leq b$$

The expectation is at the midpoint of the range,

$$E(X) = \frac{a+b}{2}$$

and the variance is

$$V(X) = \frac{(b-a)^2}{12}$$

The Normal Distribution

Easily the most important continuous probability distribution, the normal distribution has been useful in countless applications involving every conceivable discipline. The usefulness is due in part to the fact that the distribution has a number of properties that make it easy to deal with mathematically. More importantly, however, the distribution happens to describe quite accurately the random variables associated with a wide variety of experiments.

The range of a normally distributed random variable consists of all real numbers. The probability density function is defined by the equation

$$f(x) = \frac{1}{\sigma\sqrt{2\pi}} e^{-\frac{(x-\mu)^2}{2\sigma^2}} \qquad \text{for } -\infty \leq x \leq \infty$$

where the parameter μ is unrestricted and the parameter σ is positive.

The two parameters μ and σ used to specify the distribution happen to correspond to the mean and standard deviation, respectively, of the random variable. Any linear transformation of a normally distributed random variable is also normally distributed. That is, if X is normal with mean μ and variance σ^2, and if $Y = aX + b$, then Y is normally distributed with mean

$$E(Y) = a\mu + b$$

and with variance

$$V(Y) = a^2\sigma^2$$

The significance of these facts is that every normal distribution, whatever the values of the parameters, can be represented in terms of the *standard* normal distribution, which has a mean of zero and variance of 1. The linear transformation required to convert a normally distributed random variable X with mean μ and variance σ^2 to the standard normal random variable Z is

$$Z = \frac{X - \mu}{\sigma}$$

The density function of the standard normal random variable is just

$$f(z) = \frac{1}{\sqrt{2\pi}} e^{-\frac{z^2}{2}}$$

Unfortunately, an integral of the density function cannot be evaluated by ordinary methods of calculus, so there is no closed form expression for it, other than as an integral of the density function. However, extensive tables of the cumulative distribution function are available. Once you become familiar with the tables, virtually any desired probability can be evaluated with little trouble.

The normal distribution is reproductive; that is, the sum of two or more normally distributed random variables is itself normally distributed. The mean of the sum is, as always, the sum of the means. The variance of the sum is the sum of the variances, provided that the random variables are independent. Even if they are not, the variance of the sum can be expressed in terms of the variances and covariances of the constituents.

An even more remarkable result is established by the famous central limit theorem, which states that (under certain broad conditions) the sum of a large number of independent *arbitrarily* distributed random variables will be (approximately) normally distributed. Since quite frequently a random variable of interest may be conceptualized as being composed of a large number of independent random effects, the central limit theorem explains why the normal distribution appears so often in real-life applications. It also provides justification for *assuming* that certain random variables are normally distributed.

The Negative Exponential Distribution

Let X be a continuous random variable defined over the range 0 to ∞. If

$$f(x) = \lambda e^{-\lambda x} \qquad \text{for } x \geq 0$$

where the parameter λ is positive, we say that X has the negative exponential distribution or, sometimes, just the exponential distribution. The cumulative distribution function has, in this case, a convenient expression

$$F(x) = 1 - e^{-\lambda x}$$

The complementary cumulative distribution function is even simpler:

$$G(x) = e^{-\lambda x}$$

The expectation of a negative exponentially distributed random variable is the reciprocal of the parameter

$$E(X) = \frac{1}{\lambda}$$

and the variance is the square of the same value

$$V(X) = \frac{1}{\lambda^2}$$

The negative exponential distribution is used extensively to describe random variables corresponding to durations. In other words, it is a waiting time distribution. It has a number of useful properties, but since these are explored fully in Chapter 6, no further discussion need be included here.

The Erlang-k Distribution

A continuous random variable defined over the range $x \geq 0$ is Erlang-k distributed if its density function is of the form

$$f(x) = \frac{\lambda^k x^{k-1}}{(k-1)!} e^{-\lambda x} \qquad \text{for } x \geq 0$$

where the parameter λ is positive and k is an integer ≥ 1. When k = 1 the density function reduces to that of a negative exponential distribution, so the Erlang-k distribution can be thought of as a generalization of the negative exponential. In fact, if we had k independent negative exponential random variables, each with the parameter λ, then the sum of these random variables would be Erlang-k distributed with parameters λ and k. If each of the negative exponential random variables is a waiting time, the Erlang-K random variable can be thought of as the time until the *k*th event.

The expectation is most easily found as the sum of the expectations of the negative exponential random variables

$$E(X) = E(X_1 + X_2 + \ldots + X_k)$$

$$E(X) = E\left(\frac{1}{\lambda} + \frac{1}{\lambda} + \ldots + \frac{1}{\lambda}\right)$$

$$E(X) = \frac{k}{\lambda}$$

and the variance is found by a similar argument

$$V(X) = \frac{k}{\lambda^2}$$

In addition to its use as a waiting time for the *k*th event, the Erlang-k distribution is often considered as a candidate to fit empirical data in queueing, reliability, inventory, and replacement applications. In this case, k has no physical interpretation; it is just a parameter that may be adjusted to obtain a better fit.

There are many other distributions that are not summarized here: the hypergeometric, student's t, Chi-square, Raleigh, Pearson, Beta, and Gamma, to name a few. All of them have practical uses, but this chapter has focused on just those that will come up in later chapters of this book. You may want to make a table of them for your own reference.

1.12 Where Do Distributions Come From?

The common distributions—the ones that have names—are used often because they are relatively simple and fit certain situations. In most cases, they were derived from assumptions (rather than from statistical observations). For example, when you assume that every outcome in a finite sample space has equal likelihood, you get the uniform distribution. When the assumption of equal likelihood makes sense, you can use the uniform distribution. In other circumstances, other assumptions and therefore other distributions fit the situation better. To become a good modeler, you have to learn which assumptions go with which distributions, so that you can make logical selections. All other things being equal, you would like to pick a distribution that is easy to work with—one that has only a few parameters, that has a convenient functional form, and that has desirable properties. However, you cannot pick an easy one if the required assumptions do not fit the situation.

If you do not know very much about a particular random phenomenon, one would ordinarily attempt to acquire data representing a large number of independent samples of the random variable one has in mind. Sometimes, of course, the acquisition of adequate data may be economically infeasible or even physically impossible. In these cases, there may exist theoretical justification for believing that a certain distribution family is appropriate. For example, if the phenomenon can be thought of as the number of successes in a sequence of independent Bernoulli trials, a binomial distribution would be appropriate; if it can be thought of as consisting of the sum of a large number of independent random variables, the central limit theorem would suggest the normal distribution. On other occasions, the choice of distribution is influenced by a need for particular mathematical properties.

Preferably, however, one would like to have real-world data to provide assurance that the distribution selected really does describe the real-world phenomenon. Because it is difficult to see any pattern in a raw list of values, one would ordinarily plot a histogram as a first step in identifying an appropriate distribution. The next step, that of selecting one or more candidate distribution types, requires a familiarity with the characteristics of various distribution families. In particular, one has to know what "shapes" a pdf is capable of assuming, in order to decide whether there is any hope of adjusting the parameters to get a pdf that looks like the histogram. *A Guide to Probability Theory and Some of Its Applications*, by C.L. Derman et al. (referenced on page 29) provides especially good descriptions of all of the distribution types summarized only briefly here, as well as a number of others that have not even been mentioned. It also provides guidance on how to fit each distribution to particular data, and gives examples.

Once a distribution type is at least tentatively selected, the next problem is to set values for the parameters that fix the distribution within the family. Unless other external factors intervene, one would usually use the data to estimate, in the formal statistical sense, values for the parameters. In a few cases, the statistics to use are obvious. For example, the parameter λ in a Poisson distribution is estimated by the sample mean, and μ and σ^2 in a normal distribution are estimated by the sample mean and sample variance, respectively. In other cases, however, the appropriate statistic is not so obvious. The Derman et al. book also is useful in providing this kind of information.

After the parameters are adjusted so as to provide the best fit to the data that a selected distribution type can provide, one is still left with the question of whether the fit is good enough. In other words, you should validate your model by checking the goodness of fit. As a bare minimum, you could graph the precise pdf over the histogram (using vertical scales that permit comparison), and observe the discrepancies. A more formal procedure would be to perform any of several available statistical tests for goodness of fit. The chi-square and the Kolmogorov-Smirnov goodness-of-fit tests are probably the best known. Descriptions of these two tests can be found in almost all intermediate-level statistics textbooks.

One of the basic points to bear in mind about statistical goodness-of-fit testing is that the null hypothesis assumes that the candidate distribution is correct. Only if the discrepancies between the data and the candidate distribution are significantly large will the test cause you to reject the candidate. In other words, the test is, by its very nature, biased in favor of whatever distribution you have selected to test. The mere fact that the test does not reject the distribution should not be taken as strong evidence that the selected distribution is correct. Others might have selected different distributions and come up with just as much confirmation that their choices were correct. This is particularly likely to occur when the data base is small.

The word to describe the capability of a statistical test to detect that a null hypothesis is false is *power*. Other factors being equal, a greater amount of data will make for a more powerful test. To obtain a very powerful test, however, may require truly enormous quantities of data—orders of magnitude greater than would be required for good hypothesis tests about parameters. It is easy to see why this is so if you think about how many total observations are required to provide enough information about the "tails" of a distribution to ensure that you have obtained a proper fit.

As a final philosophical point, it is well to keep in mind that *no amount of data can confirm absolutely that you have selected the correct distribution*. Ultimately, (and this is the main point of this discussion) there is no escape from having to make assumptions. On the other hand, remember that there is no need for a model to be perfect. It only has to be adequate to be useful.

1.13 The Binomial Process

There is a very simple stochastic process that we can begin our study with. It is really too simple to do very much with, but it relates several of the discrete distributions to one another and may help to set the stage for more practical extensions.

We obtain the process by assuming we have a continuing sequence of independent Bernoulli trials. That is, on any one trial we have only two possible outcomes, called success and failure. We can get that random variable from any sample space by considering any event and its complement. For example, we could say that something happens (success) or it does not (failure). Then we imagine repeating the same experiment and keeping a running total of the number of successes. We could graph the results from any sequence on a chart like that in Figure 1.8. Each time there is a success, the count steps up by one; for each failure, the count remains level. (In Figure 1.8, we had three failures followed by two successes, and so forth.)

We call this sequence a *binomial process*, not to be confused with a binomial distribution. Of course, there is a connection. If, in advance of observing a binomial process, we specify some fixed number of trials we are going to run, n, and ask for the distribution of the total number of successes we will experience, that random variable will have a binomial distribution. (Go back and read the definition if you do not see why.) But we could also look at the binomial process in some other ways to get different random variables with different distributions. If we start the process and ask for the number of trials until the first success, we get a geometrically distributed random variable. (Again, make sure you understand why.) Or, if we want to reach a certain number of successes, say k, and ask how many trials that will take, we get a negative binomial random variable. We can even get Poisson or normal random variables if p is small and n is large. The distribution you get depends upon what question you are asking, even though the underlying process—the sequence of independent Bernoulli trials—is the same.

The binomial process is an elementary example of a stochastic process. It tracks the (uncertain) progress of a variable over time. Although it is useful for some simple things, it is limited by two constraints: the trials must be independent, and we can only increment the variable by one (or zero) unit for each time step. By the end of the next chapter, we will be able to fully escape from both of those constraints and have a much more useful class of stochastic processes.

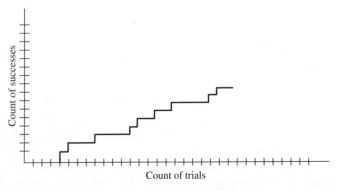

■ **FIGURE 1.8** A binomial process

1.14 Recommended Reading

If any of the topics mentioned in this chapter seems hazy, or if you would just feel more confident about proceeding if you work some problems, you should by all means devote some time to an elementary textbook on probability. There are many fine ones available. Unfortunately for the purposes of this book, the orientation of many beginning texts leans toward statistical, as opposed to modeling, applications. Also, the more recent textbooks tend to be encyclopedic in coverage, rather than concisely focused on the most important introductory topics. However, any of the older books by Clarke (1), Cramer (2), Drake (4), or Meyer (10), should serve the purpose adequately. If one does not suit your taste, feel free to select another. These older books are out of print, but can be found in the library. Feller's two volumes, (5) and (6), are classics familiar to everyone seriously interested in probability. Even beginners can find much of interest in them. The first volume deals with discrete distributions; the second, with continuous distributions. If you want to buy an inexpensive book, some of the Dover paperbacks (7), (8), and (12) are reprints of excellent older textbooks.

1. Clarke, B., and R. Disney, *Probability and Random Processes for Engineers and Scientists*. Wiley, New York, 1970.
2. Cramer, H., *The Elements of Probability Theory and Some of Its Applications*. Wiley, New York, 1955.
3. Derman, C., L. J. Gleser, and I. Olkin, *A Guide to Probability Theory and Application*. Holt, Rinehart, and Winston, New York, 1973.
4. Drake, A. W., *Fundamentals of Applied Probability Theory*. McGraw-Hill, New York, 1967.
5. Feller, W., *An Introduction to Probability Theory and Its Applications*, vol. I, 2nd ed. Wiley, New York, 1957.
6. Feller, W., *An Introduction to Probability Theory and Its Applications*, vol. II. Wiley, New York, 1966.
7. Freund, John E., *Introduction to Probability*. Dover, New York, 1973.
8. Goldberg, Samuel, *Probability: An Introduction*. Dover, New York, 1960.
9. Hsu, Hwei, *Probability, Random Variables, and Random Processes*. Schaum's Outlines, McGraw-Hill, New York, 1997.
10. Meyer, P. L., *Introductory Probability and Statistical Applications*, Addison-Wesley, Reading, MA, 1965.
11. Parzen, Emanuel, *Modern Probability Theory and Its Applications*. Wiley. New York, 1960.
12. Pfeiffer, Paul E., *Concepts of Probability Theory*. Dover, New York, 1978.
13. Ross, Sheldon M., *A First Course in Probability*. Macmillan, New York, 1976.

Chapter 1 Problems

Note: This chapter is a *review* of material you should have learned before. The following problems are designed to test your understanding of basic probability concepts and rules and to help you assess your readiness for the course. If any of them give you trouble, you should immediately begin remedial work, using some more complete introductory probability textbook.

Sets and Basic Rules of Probability

1. Imagine an experiment in which one student is selected at random from among all currently enrolled students in this university. Let A be the event that the selected student is classified as enrolled in engineering (one of the engineering schools), and let B be the event that the same selected student is

currently enrolled in this class. Express in set notation the following events,
   a. The student is not in engineering.
   b. The student is in engineering and in this class.
   c. The student is not in engineering but is in this class.
   d. The student is not in engineering and is not in this class.
   e. The student is either in engineering or is in this class.
   f. The student is either in engineering or in this class, but not both.

2. A new television show has been prepared, but has not yet been broadcast. Let A be the event that, after the first appearance, it gets good reviews by critics. Let B be the event that it is popular with the public. Let C be the event that it is liked by advertisers. Express in set notation the following events.
   a. The show is liked by critics, the public, and advertisers.
   b. Critics do not like the show, but it is popular with the public and advertisers.
   c. Critics and advertisers like the show, but the public does not care for it.
   d. None of the three target audiences likes the show.

3. Suppose there are five horses in a horse race. Describe three different sample spaces for the outcomes of the race, depending upon your interest:
   a. You bet on a single horse and care whether you win or lose.
   b. You care which of the five horses wins.
   c. You care about which horses come in first, second, and third.

4. Suppose that an experiment has five possible outcomes, which are denoted $\{1, 2, 3, 4, 5\}$. Let A be the event $\{1, 2, 3\}$ and let B be the event $\{3, 4, 5\}$. (Notice that we did not say that the five outcomes are equally likely; the probability distributions could be anything.) For each of the following relations, tell whether it could possibly hold. If it could, give a numerical example using a probability distribution of your own choice; if it could not, explain why not (what rule is violated).
   a. $P(A) = P(B)$
   b. $P(A) = 2P(B)$
   c. $P(A) = 1 - P(B)$
   d. $P(A) + P(B) > 1$
   e. $P(A) - P(B) < 0$
   f. $P(A) - P(B) > 1$

5. The sample space of a particular experiment is given by $S = \{0, 1, 2, 3, 4, 5\}$. Let three events be defined as $A = \{0, 1, 2\}$, $B = \{0, 2, 4\}$, and $C = \{1, 3, 5\}$. Assume that the probabilities of A, B, and C are given, but no further information is available. (Note, in particular, that we are not assuming equal likelihood for the elementary outcomes.) Express the probabilities of as many of the following events as you can.
   a. $A \cap B$
   b. $B \cup C$
   c. $\overline{A}$
   d. $B \cap \overline{C}$
   e. $\overline{(A \cap B) \cup C}$

6. Prove relation 4 on page 5 using only the axioms 1, 2, and 3, and the rules of set theory. (This is just an exercise in set theory, not a complicated proof.)

7. Prove relation 5 on page 5 using only the axioms 1, 2, and 3, and the rules of set theory. (This is just an exercise in set theory, not a complicated proof.)

Joint and Conditional Probabilities and Independence

8. For each of the following pairs of events, categorize them as independent or dependent and explain your choice.
   a. Rain today, rain tomorrow.
   b. Rain today, rain one month from today.
   c. Rain one year ago today, rain today.
   d. Receiving the grade of A in an introductory probability course; receiving grade of A in this course (same person).
   e. Receiving the grade of A in freshman-level physics; receiving same grade in this course.

9. If two events are known to be mutually exclusive, could they also be independent? Could they be dependent? If they are known to be independent, could they also be mutually exclusive? Could they be not mutually exclusive? If they are *not* mutually exclusive, could they be independent? Could they be dependent? If they are *not* independent, could they be mutually exclusive? Is is possible that dependent events could be not mutually exclusive? (Some of these questions are actually the same question, expressed in different words. The questions are meant to help you get the distinctions straight in your mind.)

10. If A, B, and C are events, and we know that the pair A and B are independent, and that B and C are independent, can we conclude that A and C are independent?

11. A graduating senior seeking a job has interviews with two companies. After the interviews, he estimates that his chance of getting an offer from the first company is 0.6. He thinks he has a 0.5 chance with the second company, and that the probability that at least one will reject him is 0.8. What is the probability that he gets at least one offer?

12. About 10 percent of the population is left-handed. Of those who are right-handed, about 40 percent own dogs. If you were to select a person at random, what is the joint probability that the chosen person is a right-handed and does *not* own a dog?

13. If A and B are two events, with neither being empty sets or the entire sample space, prove that if $P(A|B) > P(A)$ then $P(B|\overline{A}) < P(B)$.

Distributions

14. Here is a table giving the joint distribution of two random variables X and Y.

XY	1	2	3	4	5
2	0.10	0.05	0.15	0.10	0.10
4	0.04	0.02	0.06	0.04	0.04
6	0.04	0.02	0.06	0.06	0.02
8	0.02	0.01	0.03	0	0.04

What is the conditional probability $P\{Y = 6 | X = 2\}$?

15. Using the same joint distribution as shown above, are X and Y independent? (Give a yes or no answer, and explain why or why not.)

16. Using the same joint distribution as shown above, give the marginal distribution of Y.

17. Using the same joint distribution as shown above, what is the covariance of X and Y?

18. Suppose that the density function of a continuous random variable is f(x) = 2x, for values of x in the range [1, a] and f(x) = 0 elsewhere. What is the value of a?

19. Suppose that two random variables, X and Y, have a joint density function given by f(x, y) and by checking we find that $f(x, y) \neq f(x)f(y)$. That is, the two are not independent. Some of the calculated moments are: $\boldsymbol{E}(X) = 10$, $\boldsymbol{E}(Y) = 8$, $\boldsymbol{V}(X) = 9$, $\boldsymbol{V}(Y) = 4$, $\boldsymbol{E}(XY) = 84$. What is the expectation of the random variable W = X + Y + 4?

20. Using the same information as the previous problem, what is the variance of the random variable W?

21. Using the same information as the previous problem, what is the correlation coefficient of X and Y?

Common Distributions

22. In a sequence of 10 independent Bernoulli trials, where the probability of success is 0.4, what is the expected number of failures? (Variance?)

23. If a Bernoulli random variable X is defined so that success is given the value 10 and failure is given the value 5, and P(success) = 0.6, what is the expected value of X?

24. The geometric distribution describes, among other things, the waiting time until the first success in a sequence of independent Bernoulli trials. What distribution gives the corresponding waiting time to first success in continuous time?

25. If X is a Poisson distributed random variable with a mean of 3, and Y is another Poisson distributed random variable with a mean of 2, and the two are independent, what is the variance of the sum X + Y?

26. If X is a random variable having a binomial distribution with n = 20 and p = 0.4, and Y is a transformed version of X, where Y = 2X + 3, what is the expected value of Y?

Optional

27. See if you can figure out the rules that convert odds to probabilities and vice versa. For example, if you are given odds of 5:1, what should that mean in terms of probabilities? And if you are told that the probability of winning a bet is 0.25, what should the odds be? Assume that the odds reflect fair payout (rather than distorted values that leave some profit for the people handling the bet).

28. If you found problem 27 to be easy, you may want to try to express the odds equivalents of the rules of probabilities. For example, how do you combine the odds for two mutually exclusive events to get the odds of the union? (From doing this, you will learn why probabilities are so much nicer to deal with than odds.)

CHAPTER 2

Formulating Markov Chain Models

CHAPTER CONTENTS

2.1. An Example 33
2.2. Modeling the Progress of Time 34
2.3. Modeling Possibilities as States 36
2.4. Simplifying Assumptions 38
2.5. Modeling Changes of State as Transitions 40
2.6. Obtaining the Data 45
2.7. Another Example 46

2.8. A Case Study 47
2.9. Higher-Order Markov Chains 50
2.10. Reducing the Number of States 52
2.11. Nonstationary Markov Chains 53
2.12. Other Examples 54
2.13. Summary 67
2.14. Recommended Reading 67

The basic methodology introduced in this chapter was first developed around the beginning of the twentieth century by the Russian mathematician Andrei Andreevich Markov (1856–1922). The class of stochastic processes named for him form a subclass with enough simplifying assumptions to make them easy to handle. These assumptions limit the range of applications, but not so severely that they restrict the theory to trivial cases. With the aid of a few tricks, you can get Markov theory to work for almost any kind of application. Markov processes also provide the foundation for more advanced theory, so it is essential to begin with a good understanding of these assumptions.

This chapter and Chapter 3 deal with Markov processes in which time is measured discretely, or counted. This subclass is usually referred to as Markov chains. The concepts seem to be easier to learn when the discrete time restriction is made, perhaps because the only mathematics required is algebra. It is not very difficult to add rewards or costs to the model, so that we could evaluate consequences of random behavior as well as track the changes. However, that topic can be separated and is reserved for Chapter 4. Many of the concepts established in the discrete time case hold also for continuous time Markov processes, which are the subject of Chapter 5.

We will spend this entire chapter on how to formulate Markov chain models. This seems to be the most difficult task for beginners, because it involves new concepts and terminology, as well as a certain amount of creativity. Once we have covered the basics, if you are impatient to see what you can calculate, proceed immediately to the next chapter to see computations for a wide variety of results. However, it still will be helpful to read through the end of the chapter as it provides specific applications to illustrate tricks and variations in formulation. Those practical examples may help you understand a concept or a solution method, as well as broaden your perspective of the range of applications.

2.1 An Example

EXAMPLE 2.1

We begin with one simple example that illustrates in concrete terms some of the basic Markov chain model concepts. This is just a starting point, because you will want to develop an abstract understanding of the concepts so that you can apply them to very different situations. The central idea is to develop a model that represents something that changes over time, taking into account the random nature of the dynamics.

Suppose that you want to track the random movements of an independent trucker. He owns his own rig (the tractor, or the front end that pulls various kinds of trailers). His home base is in the Atlanta area and he hauls loads to Atlanta, Baltimore, Chicago, and Detroit (conveniently represented by the letters A, B, C, and D). (Figure 2.1) Having driven the routes many times, the driver knows the best roads and always takes the same paths (though we really do not care about the details of any route). When he arrives at a destination, he drops his load (disconnects from the trailer) and then—perhaps after a sleep break—calls a dispatcher who assigns him another load. For example, if the driver is in Baltimore, the dispatcher could assign him a trailer that could have Atlanta, Chicago, or Detroit as its destination. In Atlanta and Chicago, it is even possible that the load would be local, so that the next stop might be within the same city.

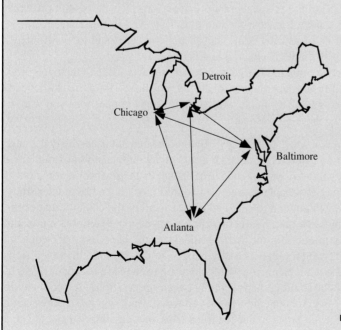

■ **FIGURE 2.1** The trucking routes

The driver does not get to choose and cannot refuse an assignment (these variations can be handled, but they complicate the model). The dispatcher has his own concerns, including satisfying priorities and balancing assignments. From the driver's perspective, it is a matter of chance where the assignments will take him. When his travels take him back to his home base in Atlanta, he takes two or three days off to spend with his family.

We are deliberately simplifying this description of trucking practices so that we will have a very small model to work with initially. We could easily expand the model (for example, by extending the number of destinations) to make it more realistic. Note that we are modeling the path of a single vehicle, not a fleet. There are more advanced methods in Chapter 7 to treat randomly circulating populations of things, at which time we will be able to model a fleet.

Some representative examples of the kinds of questions we might want to answer include:

1. If the driver is currently in Atlanta, what is the probability that he will be back in Atlanta after three trips?
2. If the driver is in Chicago, how many trips on average will occur before he next reaches his home base in Atlanta?
3. Over a long period of time, what fraction of the trips are to Detroit?
4. At an arbitrary time in the future, what is the likelihood that the driver will be on a trip to Chicago?
5. Out of the next ten trips, what is the expected number of times the driver will visit Baltimore?
6. What is the expected number of trips between visits to the driver's home in Atlanta?

These are questions that relate only to the random nature of the trips themselves. With additional data about, say, the average profit or distance traveled by trip type, we could address further questions, such as the total revenue collected or the total distance covered. Such additions certainly are feasible, but for now we confine our attention to just the trips. Notice that some of the questions are about probabilities and some are about expected values. In fact, there are many different ways to ask questions about the same stochastic process, and we eventually will show ways to compute answers for them. But there is a danger of confusion if you do not develop a clear understanding of the differences.

2.2 Modeling the Progress of Time

Normally, we think of "natural" time as continuous. That is, it flows without breaks, and any time interval can be subdivided infinitely. Conceptually, there is no limit to the resolution of real time, even if our ability to measure it is coarser than that. However, in this chapter, we are treating time discretely; that is, measuring it in "steps" that we can count off with integer values. (Time as a continuous variable is covered in Chapter 5.)

There are three common approaches to treat time discretely. We will call them the *equal interval*, *random interval*, and *event count* methods of discretizing time. The one to use depends partly on the kind of data that may be available, partly on the kinds of questions you want to answer, and partly on the modeler's preference.

Equal Interval Method: Perhaps the most obvious, though not necessarily the best, way to think of measuring time discretely is to slice it into equal intervals. For example, we could count off time in, say, one-day increments. The choice of the interval is a matter of discretion, but should reflect a reasonable degree of resolution for the process under study. A wise choice would be frequent enough to catch all or nearly all of the changes that occur, but not so frequent that many steps pass before anything changes. For example, you would not want to choose seconds or minutes for the truck problem because those intervals would be so brief relative to the rate at which things actually happen in the example that most steps would show no change. On the other hand, you would not want to choose a month or a year as the interval, because too much would be happening between observations. A good way to think of choosing an interval is to imagine how often you would want to take an observation if you were collecting data for the process. For the truck trips, one-day intervals would be a good choice.

Random Interval Method: A less common but occasionally useful way to count off time is to select random moments. This convention might match a data set of observations that are taken at random times—a fairly common technique for gathering statistical data over time. Time would then be indexed by a count of the sequence, with no implication for the duration of the intervals.

Event Count Method: A third and often preferred way to measure the passing of time is to count the trips, as if we were observing the process only at the instants when trips either begin or end. This convention will avoid any difficulties having to do with the differences in the durations of separate trips. It also has the advantage of ensuring that one and only one event will occur at each time increment. (The other two methods could have multiple events or no events occurring between observation times.) Of course, the event count method prevents us from dealing with any issues related to actual clock time. For example, we could answer questions about the mean number of trips until something happens, but not the mean number of days.

> **Definition** The discrete points of time at which the system is observed are called *epochs*.

When we want to graph something changing over time, the normal convention is to express the time axis as horizontal and running from left to right as time increases. It would not have to be done that way, but it virtually always is. So, using that convention, we can illustrate in Figure 2.2 the three ways of discretizing time, approximating what we normally think of as continuous "real" time. Take particular note of how the epochs in the event count method align with the real time events. Of course, we lose any information about the lengths of intervals in each of the discretizing methods.

For purposes of modeling the truck problem, we will elect the event count method. To be specific, we shall regard the *n*th point of time as the instant at which trip number n ends. Those instants are the only ones that our model will capture. We will not, for example, be able to say anything about how long a driver waits for an assignment, where the truck might be in the roadway system at any intermediate time, or how long the driver spends driving from one place to another. Take note of the decision to look at the *ending* points of trips; we could have chosen starting points instead. Either is permissible, but you need to be clear about what you decide so that there is no ambiguity in interpretation.

We need to assume something about how long the process goes on. If we want to imagine that it terminates at an uncertain time based upon something that eventually happens, we would need to use some methods that will be introduced later. For now, we will make the easier assumption that it just continues forever. We will be able to compute probabilities for events at any specific time and/or in the limit as time goes to infinity.

We also need to decide where time starts; that is, the point in time that corresponds to the zero epoch. In some cases, that time is fixed by circumstances beyond our control, but more commonly, we can choose it. Sometimes we may want the starting point to correspond to the actual beginning of the process; other times we may want to model a process that has been operating for a

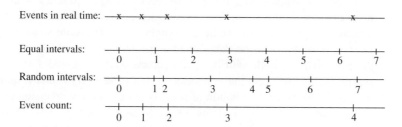

■ **FIGURE 2.2** Ways of counting off time

long time and set the zero epoch as the beginning of our observations. Often, we will choose the zero epoch to represent now—the present—so that negative values would signify the past and positive values would signify the future. For the truck problem, it makes sense to do it that way.

2.3 Modeling Possibilities as States

After deciding how time is to be treated, we turn attention to what can happen in the process and decide how to formalize that. There are only four possible destinations in our example. It will be convenient to both name and number these for quick reference. There is no particular ordering of the possibilities in this case (although a natural ordering does appear in many applications), so let us arbitrarily make the following assignment, which is based on alphabetical order.

$$\text{Atlanta} = 1, \text{Baltimore} = 2, \text{Chicago} = 3, \text{Detroit} = 4$$

At each time step of our constructed model, a trip terminates at one of these locations. Obviously, we could have more destinations, at the expense of making the problem larger. We are modeling the process *only* at the termination times, so all of the details of what occurs between these epochs is ignored. This may seem unnatural or improper until you get used to the idea that modeling is an art of selective simplification.

> **Definition** The numerical values that represent all of the possible conditions of the process over all time are called *states*.

It is important to realize that the numbers we use to designate states are just convenient indicators. They do not necessarily represent any measurable quantities. Therefore, it would not make sense to perform arithmetic with them, such as adding or multiplying the state values. They will show up as subscripts or arguments in our notation.

The set of states must be mutually exclusive and collectively exhaustive. That is, the states must be distinct, so that the process could not be in more than one at any particular epoch, and the set must be complete, so that all possibilities are covered. One way to make sure that the set of states is complete is to define one to represent "everything else." The *set* of all states will be designated by $\mathcal{S}$ (notice the script form), and the *number* of states by S. The state space $\mathcal{S}$ could be finite or infinite; we will see examples of both kinds. For the methods we are explaining in this chapter, they must be discrete, that is, countable with integers. There are such things as continuous state stochastic processes, but we will not be dealing with them in this book. You may use zero, or even negative numbers, if you like. Sometimes it is useful to specify states with multidimensional labels, such as $(0, 0, 1)$, to represent in a natural way the condition of several things at once. As long as the states can be matched to integers, as opposed to real numbers, the states will be discrete.

In the truck example, it is easy to identify the four obvious states, but sometimes you may have to think carefully to choose an appropriate resolution for the state space. That is, you will have to address the trade-off between having more states, which gives a more precise representation of the possibilities, versus having fewer states, which gives a coarser representation but is easier to handle in computations. The state space resolution and the time interval resolution can be determined separately, but should be reasonably compatible in the sense that they both contribute to the precision of the model. For example, it would not be sensible to choose a very fine resolution of states and a very coarse resolution of time, because inaccuracy in the representation of time would defeat any precision that you were trying to obtain from the choice of states. Other factors affecting the definition of states are the availability of data and the requirements to satisfy certain assumptions, which we will be coming to shortly.

2.3 Modeling Possibilities as States

By the way, the assignment of numbers to the real-world events would constitute the definition of a random variable if we were only interested in a single occurrence. For example, if we said that we wanted to sample at some particular moment in time to discover the destination of the truck at that moment, just one random variable would suffice. However, we want to model the movements of the truck over time, so a single random variable will not be enough. What we need is an infinite set of random variables—one for each time step—and, furthermore, these random variables must be linked to one another by probabilistic dependency. That is, if we are to represent how events up to some point in time affect events beyond that point in time, we must account for the relationship. This is starting to sound like a nasty mess; it certainly is in the most general case. But we will introduce simplifications that make calculations very practical. First, however, we want to establish the most general definitions.

For a real-world process, we usually know what has happened in the past. So, for example, the ordered sequence of observed states { 1, 2, 1, 3, 4, 2, 3, 1, 4, 3, 2, 3, 1, 2, 1, 4, . . . } would be a particular *realization*—just one of many, many possibilities. It would correspond to the particular sequence of trips starting in Atlanta, going next to Baltimore, then back to Atlanta, then to Chicago, and so forth.

> **Definition** A record of observed states over time is called a *realization* of the process. A realization is also called a *sample path* in some books. We consistently use *realization*, in order to avoid confusion with the term "path" in the graph theoretic sense.

A realization can be expressed as a list of state values, or it can also be portrayed in a graph as in Figure 2.3. The epochs appear on the horizontal axis, and the states appear on the vertical. Reading the graph from left to right, we can trace the sequence of trips as they occurred in this particular realization. Of course, we can only draw such a graph if we have data from the past, so a realization does not predict the future. Sometimes, we may connect the dots to emphasize the changes, but the connecting lines would have no other significance. Also, the equal spacing of epochs does not necessarily imply the equal interval treatment of time; any of the three treatments could be shown this way.

If we want to predict what a stochastic process will do in the future (which is uncertain), we would have to answer in terms of probabilities. A random variable capable of assuming any of the state values can be defined for each future epoch. In other words, the entire uncertain future can be described as a sequence of random variables $(X_1, X_2, X_3, \ldots)$. It is customary to include another (degenerate) random variable, X_0 in this sequence, to represent the information about the last known state. Epoch 0 is usually interpreted as "the present," but could be any time that either the process itself or our observation of the process is considered to start. It is a degenerate random variable in the sense that we usually know its exact value and would not really need a distribution to describe it. The entire sequence $(X_0, X_1, X_2, \ldots)$, consisting of an

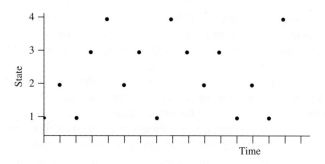

■ **FIGURE 2.3** A realization

ordered sequence of arbitrary random variables indexed by time and extending out to infinity is the stochastic process.

> **Definition** A discrete time, discrete state *stochastic process* is an ordered sequence of random variables indexed by a time counter.

One way to think of a stochastic process is to imagine it as the set of *all* possible realizations that *could* occur, along with all of the information necessary to describe those possibilities. In the most general case, each random variable in the sequence could be dependent on all previous ones.

2.4 Simplifying Assumptions

To this point we have no data or probability information. Because the process is described by a sequence of random variables, $(X_0, X_1, X_2, X_3, \ldots)$, we must *at least* specify the probability distribution of each X_i. However, this will not be enough. In general, the random variables, X_i, are related to one another. (If they were all independent, the stochastic process could be decomposed into a set of unrelated random variables, which could be analyzed separately.) Whenever one has two or more dependent random variables, either a joint distribution function or a conditional distribution function is necessary to express the relationship completely. At worst, then, the complete specification of a stochastic process would require joint distribution functions, taken jointly over all of the random variables. But this would be absurdly impractical to attempt for any real-world process.

Markov Assumption

At this point we need Markov's simplifying assumption. A Markov chain is a discrete time stochastic process in which each random variable, X_i, depends only on the previous one, X_{i-1}, and directly affects only the subsequent one, X_{i+1}. The term "chain" suggests the linking of the random variables to their immediately adjacent neighbors in the sequence.

> **Definition** A discrete time stochastic process is a *Markov chain* (or satisfies the *Markov property*) if, and only if,
>
> $$P\{X_n = j_n | X_{n-1} = j_{n-1}, X_{n-2} = j_{n-2}, X_{n-3} = j_{n-3}, \ldots, X_0 = j_0\}$$
> $$= P\{X_n = j_n | X_{n-1} = j_{n-1}\}$$
>
> for all epochs $n, n-1, n-2, \ldots, 0$ and all state sequences $j_n, j_{n-1}, j_{n-2}, \ldots, j_0$.

In words, the conditional probability of any state occurring at any fixed time, given the entire prior realization of the process up until that time is the same as the conditional probability for the same event given just the identity of the state at the previous epoch. In other words, the information about everything prior to the most recent observation makes no difference; it can be dropped from consideration without any loss. The Markov assumption is basically an independence assumption. It says that knowledge of the state at any time is sufficient to predict the future of the process from that point on; information about how that state was reached is superfluous. Simpler ways to say the same thing are, "given the present, the future is independent of the past," and "the process is forgetful."

There is a way to "extend the memory" of the process to include the states at two or three (or, in principle, any finite number) of previous epochs. Such Markov chains are called second- or third- or higher order Markov chains. We can do that using a trick involving the redefinition of states, so that all of the basic theory remains intact. That is, higher-order Markov chains are really just the same as ordinary Markov chains insofar as the theory and calculation methods are concerned. What this means in practice is that the Markov assumption is not really as limiting as it might first appear. The trick is in formulating the model correctly. An example will be shown later in this chapter.

You should understand that the Markov assumption is not always satisfied, but when it is, a lot of complications vanish. In our truck example, the Markov assumption is perfectly justified, because the destination for each successive trip is chosen by the dispatcher, who does not pay any attention to the drivers' previous stops. The driver's memory of prior trips or possible preference for one destination over another have no influence. However, the process would *not* be Markovian if the destination of the truck were somehow influenced by where it had been previously. Suppose, for example, that the driver had not been home for the last several trips and now wanted to return to his home base. In such a situation, the driver might refuse any load except that destined for Atlanta. This situation would clearly violate the Markov assumption. Our example was described, however, to provide a case where the assumption would be satisfied.

Sometimes, as in our example, you can make a firm determination of whether or not the Markov assumption is satisfied. More commonly, it is satisfied (if at all) in the weaker sense that one is unable to *detect* any relationship between future and past states. You may sometimes suspect that a relationship could exist but at the same time be satisfied that the relationship is so weak that you are content to make use of the Markov assumption. In other words, it may be a reasonable approximation of reality, which is really all you can ask of a model. Still, you have to be careful about making the assumption when it is clearly false; the model would give meaningless information in such cases.

Stationarity Assumption

A second simplifying assumption will help to keep the data requirements to a minimum. Intuitively, we want to be sure that the data we use to describe the process stays the same as time progresses, so that we are free to use the same descriptive data throughout.

> **Definition** A Markov chain is *stationary* or has the *stationarity property* if, and only if,
> $$P\{X_n = j | X_{n-1} = i\} = P\{X_m = j | X_{m-1} = i\}$$
> for all epochs n and m, and all state pairs j and i.

In words, the definition says that the probability of going from any state i to any state j in one step is the same everywhere in time. The stationarity property is one of "constancy" over time, although of course it does not imply that the process remains in a fixed state. It is the *probability values* that are assumed fixed over time. Again, it is easy to imagine cases for which the assumption does not hold. If the time scale of the model is very long, the process may undergo such fundamental changes—growth, evolution, aging, policy change, and the like—that one would hesitate to call it the same process. Even over the short run, many real-life processes have peak periods or slow times during which they exhibit behavior that is different from the norm. Such changes can occur gradually or all at once, but in either case would violate the stationarity assumption. In the case of the example, one might be conscious of an increase in truck traffic to Chicago at the end of month, or to Atlanta in the winter. If so, the data that accurately described the trip probabilities at one time might not be accurate at a later time.

Compared with stochastic processes in general, stationary Markov chains are very simple. Without completely eliminating the dependence between random variables (which would severely curtail our ability to represent real-life processes realistically), we have avoided the necessity of having to express joint distributions of everything all at once. Instead, it will be sufficient to express joint or conditional distributions of just two neighboring random variables at a time.

Note: Although it is easy to compute many results for Markov chains, it can be tempting to forget about *why* the methods work. In particular, one can easily forget that the Markov and stationarity assumptions are crucial. Although it is true that many real-life processes satisfy these two assumptions (or, at least well enough for the models to be considered useful), it is also true that many do not. There are no automatic safeguards built into the methods to prevent their misuse. Even if a process does *not* satisfy the Markov and stationarity assumptions, it would still be possible to formulate a Markov chain model and use it to compute results; however, the results would simply be meaningless. Thus, a certain amount of vigilance is required to avoid applying these methods when the circumstances are inappropriate.

There exist certain statistical tests to verify the reasonableness of making the Markov and stationarity assumptions. [See U. N. Bhat, *Elements of Applied Stochastic Processes*, pp. 96–102 (the full citation is on page 68).] However, the best protection is provided by *understanding* the assumptions and what they imply and by *thinking* about whether they seem appropriate to the situation. The Markov and stationarity assumptions are unrelated to one another; either could hold in the absence of the other.

2.5 Modeling Changes of State as Transitions

Once you can assume or verify the Markov and stationarity properties, you can characterize the entire stochastic process for all time by a relatively few probability values. The key is to focus on one-step transitions. Whenever we use the word transition without a modifier, we mean the single-step case. If you want to talk about longer intervals, you have to specify the number of steps (such as "five-step transition") or use the phase "n-step transition" if you want to leave it general.

> **Definition** An *n-step transition* (in a Markov chain) is the change of state that occurs over a specified time interval consisting of n steps. A *(one-step) transition* is the change that occurs over one time step, or between two consecutive epochs.

When we use the equal interval or random interval methods for treating time, it is possible that the state does not change from one epoch to the next. Even when we use the event count method, there are times when we want an event to leave the state unchanged. For example, the truck problem with a local assignment would leave the city at the end of the trip the same as at the end of the previous trip. There is no difficulty in allowing transitions of this kind (that is, ones that leave the state unchanged), but we do have adjectives—*real* and *virtual*—to distinguish the two kinds of transitions when necessary.

> **Definition** A *real transition* is a transition in which the ending state is different from the initial state. A *virtual transition* is one in which the ending state is identical to the initial state.

The truck example has two virtual transitions, corresponding to the possibilities of local trips in Atlanta and Chicago. Of course, that is only because we defined the problem that way; we could just as easily have permitted local deliveries in every city, which would correspond to virtual transitions for every state. There is no reason to prefer either the presence or absence of virtual transitions; they do not make any significant difference to the computational methods.

Transition Probabilities

The data that we need to describe a stationary Markov chain are contained in transition probabilities. Informally, these are the probabilities of "going" from one state to another in one time step. For example, in the truck example, if we know the probability that the dispatcher in Atlanta assigns the next load to Baltimore or Chicago or Detroit, and similar information for the other dispatchers in the other cities, we have all we need to know to predict everything about future states. Formally, the transition probabilities are defined as conditional probabilities.

> **Definition** The *transition probability*, p_{ij}, is the conditional probability
> $$P\{X_1 = j | X_0 = i\}$$

In words, p_{ij} is the probability that the process is next in state j, given that it is now in state i. By the stationarity property, p_{ij} is also $P\{X_{n+1} = j | X_n = i\}$ for any n. In other words, the stationarity property allows us to use the same transition probabilities anywhere in time. If we did not have the stationarity property, we would have to use different values for the transition probabilities at different points in time and therefore would have to index the values in some way. But since we do assume stationarity and can use the same values anywhere in time, there is no need to include any time parameter in the notation.

How many transition probability values do we need? There is a transition probability defined for every ordered state pair (i,j), including the virtual transition pairs (i,i). Some of these may be zero because the corresponding transition is impossible. Hence, if there are S states in the state space, there will be at most S^2 transition probabilities describing the Markov chain. That could be a large number, but those are all of the data you need for the entire process for all time. Virtually everything you could want to calculate to answer questions about the process can be derived from the values given by the set of transition probabilities. Also, in typical cases, many of the transition probabilities are zero, so the actual number of data values required may be much less than S^2.

Transition Diagrams

We can display visually all of the possible transitions of a Markov chain in pictorial form. An arrow from point 1 to point 2, for example, would mean that if the process is in state 1 at some epoch n, then it is possible for it to be in state 2 at epoch n + 1.

> **Definition** A *transition diagram* is a directed graph in which points or nodes represent states and the directed lines or arcs represent possible transitions. An arc from point i to point j means that a transition from state i to state j is possible; that is, its probability is nonzero. The *weight* or value associated with an arc from i to j is the nonzero value p_{ij}.

In our truck example, there are four states with no restrictions on possible transitions, so the transition diagram would be as shown in Figure 2.4. The arrow from 1 to 2, for example, corresponds to the possible trip from Atlanta to Baltimore. Notice that the transition diagram for the example looks like a somewhat more abstract version of the map shown in Figure 2.1. The loops at points 1 and 3 correspond to the local trips in Atlanta and Chicago (which were possible, according to the original description).

Transition Diagram Terms: Any virtual transition is represented in the transition diagram by a loop—that is, an arc from a point to itself. One special case of a virtual transition has particular significance. If the *only* transition out of a state is a virtual transition (a loop in the

42 Chapter 2 ■ Formulating Markov Chain Models

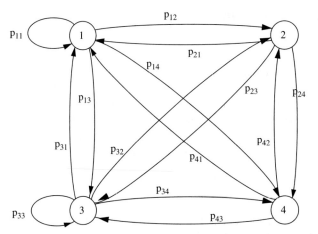

■ **FIGURE 2.4** A transition diagram

diagram), then if the process ever reaches that state, it will remain trapped in that state forever. Such a state is called an *absorbing state*, for obvious reasons.

> **Definition** An *absorbing state* is any state which has no possible transitions to any state but itself.

For reasons having to do with consistent handling of the math, we want every state to have outward transitions, so an absorbing state will always have the loop. This example has no absorbing states, but we will see one that does shortly. Absorbing states are useful in representing processes that terminate, and we will have a good deal more to say about them later.

A term for the set of all arcs leaving a point is known as a *bundle*. In a transition diagram, the bundle B_i represents the set of possible transitions that can occur when the process is in state i. Especially when the diagram is complex, it is helpful to "see" the diagram one bundle at a time.

> **Definition** The *bundle* of a point i, denoted B_i, is the set of arcs emanating from the point i.

It will also be helpful to establish formal terminology to describe a particular sequence of states in the transition diagram. We start with a general definition of a *walk*. A walk is what you get if you trace with a pencil over the arcs of the directed graph, always going in the direction of the arrows and never lifting the pencil. When the directed graph is a transition diagram, a walk corresponds to a feasible sequence of states in the Markov chain. It could be an observed sequence that is known to have occurred—a realization—or a future hypothetical sequence. We can specify a walk by naming the points/states in the sequence.

> **Definition** A *walk* in a directed graph is an alternating sequence of points and arcs, beginning and ending with points; the intermediate arcs are oriented from the prior to the following points.

Constructing Transition Diagrams: In case it is not obvious how to construct the transition diagram, here is some advice that may prove helpful in more difficult problems. Begin with a single state and think about all of the possible transitions out of that state. Those will give you the bundle of the corresponding point. After you have drawn those, proceed to the next state and repeat. By systematically covering the bundle of each point one at a time, you can be sure that you have represented every possible transition. If you are not systematic, you might miss some transitions in larger and more complicated problems.

Instead of thinking of a random process as a sequence of random variables, you can visualize it as the random walk of a particle over the transition diagram. Occupying a state is equivalent to the particle being at a particular point; a transition corresponds to a "jump" or an instantaneous movement of the particle along one of the arcs in the bundle of that point.

The transition diagram has the advantage of displaying the structure of the process visually. It is often useful in the early stages of modeling, when a picture can help translate words into mathematics. (The human mind is good at seeing and understanding visual patterns.) This is a good place to mention that Markov chains come in different varieties and that some of the variations can cause problems. In Chapter 3, we develop criteria to distinguish the well-behaved categories from those that present problems. Those criteria are related to the structure of the process, which is most clearly shown by the transition diagram. In many cases, you can tell just by looking at the diagram whether there are any problems to be concerned about when you perform calculations. There are even computational formulas that relate all of the quantities you might want to calculate directly to the transition diagram. On the other hand, transition diagrams have the disadvantage of growing increasingly messy as the number of states and transitions increase, unless there is some repetitive structure that can be made clear.

The Transition Matrix

Another way to express the description of a stationary Markov chain is by a transition matrix. The transition matrix for our truck example—or for any Markov process having four states—would be of the following form:

$$\mathbf{P} = \begin{bmatrix} p_{11} & p_{12} & p_{13} & p_{14} \\ p_{21} & p_{22} & p_{23} & p_{24} \\ p_{31} & p_{32} & p_{33} & p_{34} \\ p_{41} & p_{42} & p_{43} & p_{44} \end{bmatrix}$$

> **Definition** The *transition matrix* of a stationary Markov chain is the matrix whose elements are the transition rates. That is, it is a square matrix with as many rows and columns as there are states, and the $(i,j)^{th}$ element is the transition probability p_{ij}.

The first row of this matrix contains the probabilities for trips starting from state 1 (Atlanta). Generally, the first subscript of p_{ij}, which identifies the row, corresponds to the state at epoch 0. The second subscript, indicating the column, corresponds to the state at epoch 1. The diagonal elements correspond to the virtual transitions, or loops in the transition diagram. The nonzero elements in row i correspond to the values associated with the arcs in bundle B_i. It should be obvious (with just a little thinking) that the transition diagram and transition matrix contain exactly the same information, and that you can easily construct one from the other.

The transition matrix has some comparative advantages over the transition diagram. It can scale up to large size without adding complexity; adding more states just increases the size of the matrix. It conforms nicely to the notation and methods of linear algebra, so that all of those tools can be brought to bear. It also fits the conventions of spreadsheets, which provide convenient access to computational power. On the other hand, the matrix does not expose the structure of the process as clearly to the human eye.

Both the transition matrix and transition diagram have a few certain properties that derive from the meaning of the elements. First, the transition matrix must always be a square matrix, because there are exactly as many columns as rows (both equalling the number of states). The elements of the matrix and the arc weights will be nonnegative because each represents a probability. None could be larger than 1, again because they are all probabilities. The set of probability values associated with any row in the matrix or bundle in the diagram is a probability distribution (because every possible transition out of that state is included), which implies that the sum of the elements of each row or weights of each bundle must equal 1. A nonnegative matrix in which every row sums to 1 is called a *stochastic matrix*. Any stochastic matrix could be the transition matrix for a Markov chain.

Result For every row i of a transition matrix,

$$\sum_j p_{ij} = 1$$

The same property can be expressed more compactly in matrix form if we define a column vector of 1s.

Result Any transition matrix **P** must satisfy

$$\mathbf{P1} = \mathbf{1}$$

If this form is new to you, it may seem a bit strange. Multiplying any matrix on the right by a column of 1s has the effect of adding across the rows of the matrix. You can see this more clearly if you look at the form

$$\begin{bmatrix} p_{11} & p_{12} & p_{13} & \cdots \\ p_{21} & p_{22} & p_{23} & \cdots \\ p_{31} & p_{32} & p_{33} & \cdots \\ \cdots & \cdots & \cdots & \cdots \end{bmatrix} \begin{bmatrix} 1 \\ 1 \\ 1 \\ \cdots \end{bmatrix} = \begin{bmatrix} 1 \\ 1 \\ 1 \\ \cdots \end{bmatrix}$$

Note: We will have more productive use for this trick in later chapters, so learn it now. Be careful that you do not confuse this column vector of 1s with the identity matrix.

Initial States: The transition matrix or, equivalently, the transition diagram, provides all or nearly all of the basic data needed to describe a Markov chain. The only other data that is sometimes needed is information about the initial state. For most applications, we will know the initial state exactly. On those rare occasions when we do not, we may use a probability distribution for the initial state. The notation we will use for this distribution is

$$\mathbf{p}^{(0)} = (p_1^{(0)}, p_2^{(0)}, p_3^{(0)}, \ldots)$$

That is, we will use a row vector with as many terms as there are states, with the probability that the initial state is i represented by $p_i^{(0)}$. Later we will show how to compute state probabilities, which give the probability distribution for the state at an arbitrary time n, which will be represented by the natural extension:

$$\mathbf{p}^{(n)} = (p_1^{(n)}, p_2^{(n)}, p_3^{(n)}, \ldots).$$

Note that lowercase letters and single subscripts distinguish these state probabilities from the transition matrix and its elements. Also the superscripts identify the time epoch; we use parentheses in the superscript to distinguish the time index from an exponent or power.

2.6 Obtaining the Data

The transition matrix and, in some cases, the initial state distribution, constitute all of the data needed to compute anything at all for the entire Markov chain. Before going into those computations, we should discuss where that data might come from.

In some applications, the data will come from assumptions. For example, we could assume that all of the transitions out of a state are equally likely, so each would have a value equal to 1 divided by the number of successor states. Often, certain transitions are impossible, so they will have zero probability. Various other "structural" possibilities may imply certain values. Another way to fix the values is to place them in the model as parameters, solve (parametrically) for some measurable quantity such as the expected state at some point in time, and then adjust the parameters to get a best-fit model. Most commonly, however, the transition probabilities will have to be estimated from statistical data. That is, we will have to record data from observations of the real process we are trying to model, and then transform that data somehow into the appropriate transition probabilities.

Returning to the realization of the truck problem shown in Figure 2.1, we have the sequence of states {1, 2, 1, 3, 4, 2, 3, 1, 4, 3, 2, 3, 1, 2, 1, 4, . . . }. Although that is not enough data points to estimate anything with statistical accuracy, let us use the sequence anyway to convey the idea. To focus on transitions, you have to look at adjacent pairs. For example, in this short sequence, there are two instances of a transition from state 1 to state 2. Also, there is one instance of a transition from state 1 to state 3 and two instances of a transition from state 1 to state 4. So out of the five times the truck began trips in Atlanta (state 1), it went to Baltimore (state 2) twice, Chicago (state 3) once, and Detroit (state 4) twice. So, as a very rough estimate, using the observed frequencies as estimates of the true probabilities, we could say p_{11} is 0, p_{12} is 2/5, p_{13} is 1/5, and p_{14} is 2/5. This provides the elements of the first row of **P**. By counting the transitions that start at state 2 and state 3, we can estimate the elements of the other two rows.

Generalizing now, suppose that we have a long realization of a stochastic process with an arbitrary number of states, S, and apply the same idea of counting the frequency of specific transitions. Let c_{ij} be the number of times (a simple count) that the transition from i to j appears in the realization sequence. If we complete this count for every pair, we can get the count matrix **C**.

$$\mathbf{C} = \begin{bmatrix} c_{11} & c_{12} & \ldots & c_{1S} \\ c_{21} & c_{22} & \ldots & c_{2S} \\ \ldots & \ldots & \ldots & \ldots \\ c_{S1} & c_{S2} & \ldots & c_{SS} \end{bmatrix}$$

To convert these raw transition counts to transition frequencies, we have to divide each element by the total number of times the process had the opportunity to make a transition out of the first state of the pair. We could get those denominators by going back into the realization sequence and counting, but a more direct way is to just add up the counts in each row. That is, the sum of the elements in the first row is the total number of times the process made transitions out of state 1, the sum across the second row is the number of times it made transitions out of state 2, and so forth. Let C_i represent the sum across the *i*th row of the matrix C, that is,

$$C_i = \sum_{j=1}^{S} c_{ij}$$

Then we get a statistical estimate of p_{ij} from

$$p_{ij} \approx \frac{c_{ij}}{C_i}$$

Another way to think of the same estimation process is to construct the transition diagram, then count the number of times each arrow appears in the realization to get the c_{ij} counts. Then adjust these values to fractional values by dividing each c_{ij} by the sum of all counts *out* of the same point (or the total of the counts over the whole bundle).

Obviously, the quality of this estimate depends upon the number of observations—the more the better. If there are a large number of states S, it may take a lot of data to get reasonably good estimates of the up to S^2 possible transition probabilities. Notice that the brief realization we used to illustrate the method for the truck problem did not contain any instances of local trips. Thus, our statistical estimate of the virtual transition probabilities p_{11} and p_{33} must be zero, even though the original problem stated that these were possible transitions. The lesson is that you need to be careful in distinguishing between statistical estimates (which are approximations by their very nature) and the true probabilities. It can take a huge amount of data to get good estimates, especially when some of the transitions have low probability.

2.7 Another Example

EXAMPLE 2.2

We turn now to a different example to illustrate a different kind of Markov chain. The truck problem is a small but otherwise typical example of an "endless" process. That is, the model was set up to continue as long as needed. Even as time goes to infinity, the truck would continue to make transitions from city to city in accordance with the transition probabilities. But sometimes you want to model a process that will operate for some period of time and then stop. "Terminating processes," as they are called, are an important subclass of Markov chains that you have not seen yet.

We choose a close-to-home example to which any student can relate. The progression of students through grade levels is usually thought of as occurring at a fixed rate—one grade level per year. In reality, however, a number of different things can happen to interrupt or delay normal progress. Those chance events, in turn, affect enrollment levels. Markov chains can be used to track populations of students through grades, or otherwise defined stages of education. Obviously, this is a process that should terminate with graduation.

For our time units, we will use the equal interval method, with an interval of one year. For states, we will use the four classes of college: Freshman, Sophomore, Junior, Senior. (With more states we could model progress with finer resolution, but we want to keep the example simple.) We will also want two other states: one for graduation and one for leaving college before graduation. Numbering the states in the order they were mentioned, we would have a diagram something like Figure 2.5.

Here, p_j represents the probability of having to repeat grade j, q_j is the probability of advancing to the next grade, and r_j is the probability of dropping out from grade j. The transition matrix would look like this:

$$\mathbf{P} = \begin{bmatrix} p_1 & q_1 & 0 & 0 & 0 & r_1 \\ 0 & p_2 & q_2 & 0 & 0 & r_2 \\ 0 & 0 & p_3 & q_3 & 0 & r_3 \\ 0 & 0 & 0 & p_4 & q_4 & r_4 \\ 0 & 0 & 0 & 0 & 1 & 0 \\ 0 & 0 & 0 & 0 & 0 & 1 \end{bmatrix}$$

There are several things to call to your attention in this model. First, it is a terminating process with two absorbing states. The process eventually ends, either by graduation or withdrawal. Notice that the absorbing states are indicated by loops with weight 1 in Figure 2.5, and by a 1 on the diagonal of the transition matrix. A terminating process will have one or more absorbing states, representing the possible ending condition(s). The real-world process may actually stop when one of those states is entered, but for reasons of consistent treatment, we attach the loops. It is as if the

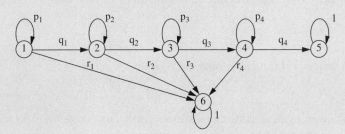

■ FIGURE 2.5
Transition diagram for college progression

process will continue to make looping transitions even though nothing happens. That may seem artificial, but it helps the mathematics to know that the rows of the transition matrix and the sum of the weights in a bundle *always* sum to 1.

Second, in this particular model, each time step (year) results in one of three transitions: staying in the same grade, advancing one grade, or exiting. There are no backward transitions to lower numbered states. That property (which is common to growth or progress models) results in a transition matrix which has all zeros below the principal diagonal. It is a special case that has some computational implications that we will not go into. In any case, we could, if we wanted or needed to, include backward transitions. We could also allow "jump" transitions, corresponding to skipping grades.

A model like this could be used to answer questions such as the following:

1. Of an entering freshman class, what fraction can be expected to eventually graduate?

2. What are the probability distribution, mean, and variance of the number of years spent in college?

3. If the four classes are of a certain size, what are the projected class sizes for the next several years to come?

4. For those students who drop out, what is the expected number of years of college experienced?

5. How many students should be admitted in order to achieve an expected graduating class of a certain size?

This is a good example to use in learning about terminating processes. You can imagine such questions for yourself and then figure out how to answer them using the calculations in Chapter 3. Similar models could be developed for selection and transfers among majors, progress through graduate programs, faculty hiring and promotion, and so forth.

But hold on a moment. Were we too hasty in building this model? Are the Markov and stationarity assumptions justified in this case? Notice that we did not have to check; we could build the model and go on to calculate answers without being forced to consider the assumptions. Therein lies the danger.

In order to accept the Markov assumption, we have to believe that the likelihood of advancing (or dropping out or repeating the grade) depends only on the current grade, and not on prior states. For example, would a student who repeated the freshman year have the same probability of dropping out as one who did not? Would a senior who has taken eight years to reach that level have the same probability of graduating as one who took only four? The Markov property seems doubtful, to say the least. However, there is more to be said on the subject. If we wanted to use the Markov chain model to predict what the *population* of students will do (as opposed to any individual student), we may (or may not) find that the Markov assumption fits. We obtain the data to construct the matrix from many students, not just one. If we are to use this data for prediction, we have to assume we are predicting for an average student—one who behaves according to the averages for the whole population that has been sampled. That will be acceptable, provided that we remember not to expect any particular student to behave that way. (In fact, this is a fairly common trick in modeling. We cannot realistically get information about the behavior of any one individual, so we take it from a population, then use the population data to model a "typical" individual. We have to remember that the typical member of a population is not the same as any particular member.)

We should also consider whether the stationarity assumption is reasonable. Can we use the data collected over, say, the last ten years to predict for the next ten? Such factors as the job market, the rising cost of tuition, and other fundamental influences on progression or retention could change the transition probabilities. They probably do not change very fast, so we might be comfortable using them for the next ten years, but perhaps not longer than that.

The important point is not whether this particular model is good, but that you have to stop and think about whether the Markov and stationarity assumptions fit before you trust anything that the model tells you. You have to do this on a case-by-case basis. There will be times when you struggle to formulate a problem as a Markov chain and manage to get it set up. You may be eager at the point to proceed, because you know what to do next. But that is just when you should start to worry about whether you are justified in making the two key assumptions.

2.8 A Case Study

EXAMPLE 2.3

Before moving on to more advanced formulation techniques, let us apply what has been presented so far to a particular case. The situations you will encounter in real life will be so varied and distinctive that you will have to address each one individually; we cannot present a general template to fit every situation. So the purpose of this example is to walk you through the *thinking process*. There will be many more formulation examples later in the chapter, and there will be many other variations explored in the problems. The skill that you should try to develop is the ability to *create* your own *original* models, using the principles that are described here.

The case involves the design of a customer service call center to support a company that is marketing a new software product. The call center will receive telephone calls from both existing and potential customers; some of the inquiries will be requesting help in using the software, while others will be about purchasing or billing. Different kinds of expertise are needed to answer these questions, so the company intends to install an automated answering system to route calls to human operators. A recorded message will say something such as, "If you are interested in purchasing our product, press 1; if you have a question about billing, press 2; if you need technical support, press 3." Routine questions will be handled in the obvious way, and the (presumably satisfied) customer will hang up. More complicated issues will require transferring the call to a supervisor who has more knowledge and authority to deal with problems.

Before anything else, we should understand why a model is wanted and what it is to be used for. If the system already existed, there would be little incentive to get the performance information through modeling; we could instead get statistical measurements from the real system. However, that is not possible in this case because the system does not yet exist. The purpose of the model is to support design decisions by predicting the quality of service that can be expected. Every call is unique, but there are enough regular patterns to suggest that a model might be helpful in predicting performance. The data will be very rough, so we should not expect precise answers. However, even some "quick and dirty" modeling will be better than guessing.

We will assume that the management of this proposed call center is primarily concerned about the quality of service provided (as opposed to cost or some other factor). For the most part, the quality of service will depend upon the people answering the telephones, but to some degree it is affected by the design of the system. The customers are not likely to feel that they are well served if there are too many transfers or excessive delays.

Once we have an idea in broad terms of what we want, we should think strategically about how to achieve our goals. A common beginner's mistake is to include too much in a model, in an attempt to copy reality as faithfully as possible. As you gain experience, you will learn to omit all but the essential elements. In general, we want the smallest, simplest model that will enable us to answer the questions of interest. In the real world, there will be many calls occurring simultaneously, competing for service and bouncing from one category to another. They will include wrong numbers, calls from companies wanting to sell something, requests for charity or political donations, and so forth. For the purposes of the model, we ignore all calls but those that concern providing services to customers. Just pretend that they do not occur. Also, we will not attempt (in this chapter) to model the competition among calls for resources and the resulting delays, but focus instead on just tracking the phases of a typical call. Hence, we should limit our thinking to the possibilities for just one call. To get to the whole collection of calls, we will imagine replaying the same process many times. This approach is similar to what was done for the trucking example at the beginning of the chapter.

To summarize what we have to do to construct a Markov chain model, we must identify the following basic elements:

1. The time units. These will be discrete and counted off as integers, but we must clearly understand what these counts correspond to in the real world.

2. The states. These categorize the distinct possibilities that can occur over time.

3. The transitions. The transitions account for all of the possible changes that can occur.

4. The probabilities. We need data to quantify the probability of occurrence of each possible change of state.

In addition, we need to be satisfied that the Markov and stationarity assumptions hold, as least to an acceptable degree. You should take these steps one at a time, in the order suggested, in order to avoid confusion and wasted effort.

Time Units: The first thing to think about is appropriate time units. In this case, we want to track what happens to calls when they enter the system. In particular, we want to estimate how many people the customer has to talk to in order to get his or her questions answered. Hence, we will want to think of a time step occurring each time a call is received at a new station or when it ends. Time step 0 will be when the call is answered by the automated system. Time step 1 will be when it is transferred to one of the three service stations. If the call is transferred to other stations, each transfer will count as another step, until the final step when the call ends. Notice that we will be modeling the transitions of individual calls, not the entire collection of calls simultaneously. We will not be able, in this model, to represent how one call might interfere with another. We will be able to do that later, when we have learned more, at which time this same situation will be treated in a more advanced way. For now, we will be content with modeling the transfers.

The States: The second thing to think about is the states. We need to distinguish among the departments, so there will clearly be a state to represent sales, one to represent billing, and one to represent support. There will also be at least one state to represent the supervisor. We might actually want three distinct supervisor states in order to keep track of where the transfers came from so that the transfer back will go to the right department. That is, we could have a supervisor state for transfers to and from sales, another for transfers to and from billing, and a third for the transfers to and from support. If we fail to distinguish in that way, a call that comes to the supervisor from sales might be returned to billing, for example. The Markov assumption requires that a call "forget" where it came from. If we were really serious about creating the best model we could, it would be important to have those three supervisor states. But since we are just illustrating the approach and we want to keep the model small, we will proceed with just one supervisor state. We might want to include a fifth state to represent the call's first appearance. An alternative approach would be to have a distribution for the initial state. It is really just a matter of preference whether you would like to have a fixed initial state and send the call to any of the three primary departments according to some distribution, or to just start the process in one of those three states according to the same distribution. Of course, it would make a difference of exactly 1 in the count of the number of transfers. We will assume that we want to count that initial transfer, so we will define a state for the automated answering machine. At this point, we have

The Call Center States

1. Initial
2. Sales
3. Billing
4. Support
5. Supervisor
6. Satisfied
7. Unresolved

The Transitions: We next need to enumerate every possible transition. It is usually best to just create the structure first—that, is to identify the transitions without worrying about the probabilities. A good way is to draw a transition diagram. Another way is to construct an empty transition matrix of the right dimension and mark the cells that are nonzero. Whichever way you do it, you need to consider every possible transition out of each state. For the transition diagram, go point by point and think about all of the outward transitions; for the matrix, go row by row and think about each column.

Here is a starting transition matrix for this model, presented as a table so the rows and columns can be labeled with the descriptive terms.

	Initial	Sales	Billing	Support	Supervisor	Satisfied	Unresolved
Initial	0	x	x	x	x	0	0
Sales	0	0	x	x	x	x	x
Billing	0	x	0	x	x	x	x
Support	0	x	x	0	x	x	x
Supervisor	0	x	x	x	0	x	x
Satisfied	0	0	0	0	0	1	0
Unresolved	0	0	0	0	0	0	1

five states, but have no way to indicate how a call ends. We want to distinguish between calls that resolve all questions in a way that satisfies the customer and those that do not. Hence, we will want to define two terminal states. Call one of them "satisfied" and the other "unresolved."

It is wise to pause long enough to consider whether we have a complete state space before thinking about transitions or probabilities. Ask yourself, "Is there anything that can happen in the process we want to model that is not covered by the states I have defined?" For example, what if a call "dropped" before the customer got an answer? Again, we want to keep the model small and simple, so we will proceed with just these seven states. To summarize, they are

The last two rows are known because states 6 and 7 are absorbing states. The first column is known because the initial state will never be revisited. The diagonal terms are known because a time step is defined as a call transfer, and an operator would not transfer to another operator within the same category. All of the cells that contain xs are possible transitions, unless we decide otherwise or the data tells us otherwise.

The Probabilities: Now we need some data to fill in the cells that contain positive values. Although we cannot obtain that data directly from the real system (because it does not exist), the management team can make some educated guesses, perhaps bolstered by experience with similar systems or reports from industrial surveys. Let us say that about

60 percent of the incoming calls will be directed to sales, and the others are equally likely to go to billing or to support. Of course, none of the incoming calls will go directly to the supervisor or to either of the terminal states. That information provides the first row of the matrix, namely, (0, 0.6, 0.2, 0.2, 0, 0, 0).

Moving to the second row, we imagine a call that is currently being handled by sales. Most of the time, say 80 percent, the call will be routine and will end with no attention from anyone else. We have no way to know if the customer is really satisfied, but we will count all such routine calls as ending in the satisfactory state because we have no indication otherwise. (You could also choose to estimate that some fraction of them are satisfactory and the others are not.) Of the 20 percent of the calls to sales that have to be transferred, most will go to billing and almost all of the others will go to the supervisor (for issues such as education discounts, volume purchases, or other unusual requests). Only rarely will a call that starts in sales go to support. Let us say that 75 percent to go to billing, 20 percent to the supervisor, and only 5 percent to support. Notice that we are making statements about conditional probabilities. For example, *given* that a call is currently in sales, and *given* that it has to be transferred, the probability that it will go to billing is 0.2. In order to express the probability that a call in sales goes next to billing, we multiply 0.8 by 0.75 (using $P(A|B) = P(A|B \cap C)P(C|B)$, where A is the event that it goes to billing, B is the event that it is in sales, and C is the event that it has to be transferred. You should be able to think this way intuitively, without resorting to the formal rules of probability. One way is to think of dividing the 0.8 probability among the three transfer possibilities in accordance with the appropriate percentages (three-quarters of it to billing and so forth). Of course, we can translate freely between percentages, fractions, and probabilities. According to this description, the second row of the transition matrix will be (0, 0, 0.15, 0.04, 0.01, 0.8, 0). Make sure you understand where all of these values came from and why they go where they do.

Next, we think about calls that go initially to billing to construct the third row. Let us say that 40 percent of the billing calls have to be transferred to the supervisor, 10 percent go to sales, and the others all end routinely (in the satisfied state). Those assumptions give us the row (0, 0.1, 0, 0, 0.4, 0.5, 0).

Calls to support are sometimes inquiries about technical capabilities or other details prior to purchase, in which case the call would be transferred to sales after the questions are answered. Let's say that 30 percent of the calls are of that nature. Of the remainder, 40 percent are answered to the satisfaction of the customer, 20 percent end without resolution, and 10 percent get transferred to the supervisor. Those assumptions give us the fourth row: (0, 0.3, 0, 0, 0.1, 0.4, 0.2).

The calls that make it to the supervisor are the difficult ones. We will assume that three-quarters of those calls can be dealt with in a way that is satisfactory to the customer, and the others cannot. The supervisor is the final authority, so would not transfer calls back to any of the other departments. The final transition matrix looks like this:

$$\mathbf{P} = \begin{bmatrix} 0 & 0.6 & 0.2 & 0.2 & 0 & 0 & 0 \\ 0 & 0 & 0.15 & 0.04 & 0.01 & 0.8 & 0 \\ 0 & 0.1 & 0 & 0 & 0.4 & 0.5 & 0 \\ 0 & 0.3 & 0 & 0 & 0.1 & 0.4 & 0.2 \\ 0 & 0 & 0 & 0 & 0 & 0.75 & 0.25 \\ 0 & 0 & 0 & 0 & 0 & 1 & 0 \\ 0 & 0 & 0 & 0 & 0 & 0 & 1 \end{bmatrix}$$

It is a good idea to double check the row sums to make sure that each one totals 1, as a check on both your thinking and your arithmetic. Once you have the matrix, the model formulation is complete. You are free to use the matrix to calculate various quantities, as the next chapter will explain. Of course, that will not be entirely mechanical, because we still have to think about the context of the application in order to know what to compute to answer the questions we have. Also, we must remember that the calculations will make use of the Markov and stationarity assumptions, so any results that we get will be no better than the quality of those two assumptions (as well as any that we used to create the data in the matrix).

We will explore some variations of this same case study in subsequent chapters. Aside from demonstrating a couple of calculations in the next chapter, we will show how to add costs and some other features in Chapter 4. Eventually, in Chapter 7, you will learn how to model the congestion that results from having a limited number of operators and how to calculate delays.

2.9   Higher-Order Markov Chains

This is a good place to discuss what you can you do if you think the Markov assumption is unwarranted. Of course, there are more advanced categories of stochastic processes which might apply, but we are not dealing with them here. Instead, we want to employ a device that redefines states in such a way that the Markov assumption will be suitable for the transformed model, after which we can use all of the simple Markov methods.

The trick is to define a new set of states consisting of fixed-length sequences of states. If the sequence is of length 2, we call the Markov chain a second-order Markov chain; if it is of length 3, we call it third-order, and so forth. The transitions in this modified process must follow the rule that the end sequence of one must match the beginning sequence of the next.

It will be easier to understand the idea with an example, so let's go back to our truck problem. Suppose that all of the dispatchers have a policy that they will not send a truck back to the city from which it just came. For example, a truck that is now in Chicago and just came in from Baltimore will not be sent back to Baltimore. The immediate implication is that the Markov property will no longer apply to the model we built, because the next transition will depend not only on where the truck currently is but also on where it was just before. We need to extend the dependence back one more epoch into the past in order to include enough information to accurately predict the future. So, using the formulation trick, we define states as city pairs: {AA, AB, AC, AD, BA, BB, BC, BD, CA, CB, CC, CD, DA, DB, DC, DD}, with A for Atlanta, B for Baltimore, C for Chicago, and D for Detroit. Of these sixteen states, we can eliminate BB and DD because those sequences will never appear (they would require virtual transitions in Baltimore and Detroit, which do not occur in our problem). To begin the process, we must have an initial state, which in this case means an initial city pair—a first trip. Let's say it is AB, which means that the first trip was from Atlanta to Baltimore. Now from there, the truck cannot return to Atlanta or stay in Baltimore, but it could be assigned a load to Chicago or Detroit. Hence, the possible transitions from state AB are to states BC and BD. Note that the end sequence (B) of the first state has to match the start sequence of the next state, simply because Baltimore is the connecting city. All other transitions from AB have probability zero because they do not correspond to possible trips. If you fill out the transition matrix using this logic, you get the matrix shown in Table 2.1.

Here the + indicates a positive value, corresponding to a possible transition. All of the other entries are zeros, because the corresponding transitions are impossible. We have quite a few more states to deal with than we had before, but now the Markov assumption is satisfied. Given the current state (a pair, which tells us where we are now and where we were just before), the next state can be predicted without any knowledge of prior states.

■ **TABLE 2.1**
Truck Transition Matrix

	AA	AB	AC	AD	BA	BC	BD	CA	CB	CC	CD	DA	DB	DC
AA	0	+	+	+	0	0	0	0	0	0	0	0	0	0
AB	0	0	0	0	0	+	+	0	0	0	0	0	0	0
AC	0	0	0	0	0	0	0	0	+	+	+	0	0	0
AD	0	0	0	0	0	0	0	0	0	0	0	0	+	+
BA	+	0	+	+	0	0	0	0	0	0	0	0	0	0
BC	0	0	0	0	0	0	0	+	0	+	+	0	0	0
BD	0	0	0	0	0	0	0	0	0	0	0	+	0	+
CA	+	+	0	0	0	0	0	0	0	0	0	0	0	0
CB	0	0	0	0	+	0	+	0	0	0	0	0	0	0
CC	0	0	0	0	0	0	0	+	+	0	+	0	0	0
CD	0	0	0	0	0	0	0	0	0	0	0	+	+	0
DA	+	+	+	0	0	0	0	0	0	0	0	0	0	0
DB	0	0	0	0	+	+	0	0	0	0	0	0	0	0
DC	0	0	0	0	0	0	0	+	+	+	0	0	0	0

You will have to estimate the transition probabilities from the original realization data; there is no way to go from the four-state version to the fourteen-state version of the model. The larger model captures more and therefore needs more information than the smaller model contains. In counting the instances of transitions, you read overlapping sequences in the realization. For example, if you see the triplet {1, 2, 3}, you would count that as an instance of a transition from state 12 to state 23 (or what we labeled AB to BC in the Table 2.1 matrix).

The second-order Markov chain has the effect of making the next state dependent on the prior two states. An nth-order chain would make it dependent on the prior n states. The higher the order, the more extended the dependency and therefore, the more information about the past is captured by the Markov chain model. In principle, you can go to as high an order as necessary to capture the realism you desire, but in practice there is a price to be paid. A second-order chain has many more states than a first-order chain, because there are many more pairs of elements than there are elements (in the worst case, n^2 versus n). A third-order chain would require up to n^3 states, and so on.

The real significance of higher-order Markov chains is to establish that the Markov assumption is not really as restrictive as it first appears. You are not limited to a dependence on just the one prior time epoch, but can (in principle) make the dependency extend to any finite number of prior epochs. Of course, the practical price you have to pay, in terms of enlarging the number of states, may discourage use of the trick, but it is nice to know that it is there if you need it.

2.10 Reducing the Number of States

Sometimes you can reduce the size of a Markov chain by collapsing a number of states into one. This transformation is called *lumping*. It only works when the probability values match up appropriately, which does not happen very often. But there are times when it is very helpful, so the trick is worth learning.

Let us use the truck example to illustrate the method. Suppose that we have the four-state version of the model with these specific transition probabilities:

$$\mathbf{P} = \begin{bmatrix} 0.2 & 0.6 & 0.1 & 0.1 \\ 0.5 & 0 & 0.3 & 0.2 \\ 0.2 & 0.4 & 0.1 & 0.3 \\ 0.2 & 0.4 & 0.4 & 0 \end{bmatrix}$$

Notice that the third and fourth rows have identical values in the first and second columns. That means that both Chicago and Detroit have the same transition probability values into Atlanta and Baltimore. In other words, you do not have to know whether you are in Chicago or Detroit to determine those values; they are the same regardless. That in turn means that we could merge Chicago and Detroit into one state. Let's call it 3. It may help to think of the two cities as one region, though geography is not really important here; even distant cities could be lumped if they satisfy the transition criteria. The transition probabilities from the lumped state to the others are simple: They are the common values. That is, $p_{3'1} = 0.2$ and $p_{3'2} = 0.4$. But now the former transitions from Chicago to Detroit and Detroit to Chicago have to be merged into the virtual transition from the lumped state to itself. Adding up those terms gives us $p_{3'3'} = 0.4$. (We get the same sum whichever row we use.) The transitions from

Atlanta and Baltimore into the lumped state similarly must be added. So the final 3 state transition matrix is

$$\mathbf{P} = \begin{bmatrix} 0.2 & 0.6 & 0.2 \\ 0.5 & 0 & 0.5 \\ 0.2 & 0.4 & 0.4 \end{bmatrix}$$

It has to be emphasized that only the fact that the probabilities worked out right allowed us to merge states 3 and 4. No other pair of states allows merging.

> **Definition** A set of states is *lumpable* if the transition probability to any state outside of the set is the same for every state in the set; moreover, a similar equality must hold for every state outside the set.

The definition of lumpable is a bit complicated, so here is how it applies to the example. States 3 and 4 in the original problem are lumpable because $p_{31} = p_{41}$ (both equal 0.2) and also $p_{32} = p_{42}$ (both equal 0.4). Another way to say it is that the parts of the rows of the lumpable states that go to states outside the set must be identical. In this case, rows 3 and 4 must be identical in columns 1 and 2.

The result of lumping is a smaller model that represents the same stochastic behavior as the original, except that the resolution is coarser. All of the trips are counted, so we can use the smaller model to answer questions about how long something takes. However, we lose the ability to distinguish between the lumped states, Chicago and Detroit in our example. That means that we could not use the smaller model to answer questions about either one individually. It might seem that the payoff is hardly worth the effort, and that is probably true in most cases. However, problem 17 (p. 76) illustrates a case where there really is an advantage to lumping.

2.11 Nonstationary Markov Chains

The stationarity assumption is more for convenience than necessity. It enables you to use the same transition matrix for each step of the process. But you could, at least in principle, use a different transition matrix at each step, or use the same matrix for some time and then switch to another. There are certainly occasions when we are reluctant to use the same transition probabilities for every time step. In the truck example, we might be aware of seasonal variations (for example, the transition probabilities are different in the summer and winter) or trending in some direction (such as increasing trips from Atlanta to Baltimore). Any such variation in transition probabilities over time would conflict with the stationarity assumption.

The problem is not so much one of how to handle the theory, but the practical difficulty of modeling the changing probabilities. For example, if we want to run the truck example with one transition matrix for some period of time and then switch to a different matrix, that is easily done. However, we have to know when to switch and we need to have enough data to estimate the values in two (or more) different matrices. Inasmuch as it requires a lot of data to get good estimates for even one matrix, more matrices just makes the job a lot bigger.

Nevertheless, there are occasions when a nonstationary model is both practical and appropriate. One example would be a process that changes over the hours of a day, the days of a week, or the months of a year. In other words, there is some repeating pattern in time so that you can collect enough data to get good estimates for each of a limited number of transition matrices. If you were modeling sales of a product and knew that Mondays were different from Tuesdays and other days, you could estimate a transition matrix for each day of the week. Or, you might just want to distinguish weekdays from weekends, and could model the process with two transition matrices. There will always be a trade-off between the loss of accuracy from not having enough different matrices and from not having enough data for good estimates. You will have to use good judgment to balance those issues.

Another way that you could achieve a nonstationary model is to have another model embedded within the transition probabilities that systematically changes them over time. For example, you could have a gradually increasing probability of failure to capture aging effects in a reliability model. An example of this possibility is explored in "Aging Infrastructure," on page 64.

2.12 Other Examples

The remainder of this chapter briefly presents a selected set of cases where Markov chains have been used for serious purposes. Each of these could be expanded into elaborate detail, but we want to demonstrate a wealth of variety of applications, not to explore the depth of possibilities for any one. You will be able to see different ways in which time or state are defined, different sources of data, and different structures in the transitions. It is hoped that you will gain appreciation for the richness of possibilities for using Markov chains in practice, and will also learn some tricks for formulating problems as Markov chains. You may want to read through quickly the first time to get an overview, but be sure to come back for a more leisurely stroll, because some of the general tricks are buried within specific examples.

You will get just a taste of a vast array of applications to whet your appetite. Markov methods are tools, so the applications are scattered throughout the literature. Although there are many that require advanced knowledge in some particular field, the examples selected for this chapter are simple enough for everyone. If you want to go further, check out the references or search the Internet using "Markov chain" as one of your keywords. You will want to use some other keywords, because you will find too many sites if you do not limit the search. If you have any doubt about the applicability of the theory, a few hours of internet searching should allay your concerns.

EXAMPLE 2.4

2.12.1 Marketing

One of the classical applications of Markov chains in marketing is the so-called brand-switching model of consumer behavior. The product could be any consumer item, such as laundry detergents, toothpaste, automobiles, refrigerators, or personal computers. The idea is to track the successive purchases of an "average" consumer, where an average consumer is a fictional individual who buys in a manner that reflects the market as a whole.

To illustrate the idea, let's say that you want to track movements of purchases among four brands of breakfast cereal. A bit later, we will discuss how to construct the transition matrix from data, but it might end up looking something like this:

	Brand A	Brand B	Brand C	Brand D	Other
Brand A	0.427	0.107	0.100	0.213	0.153
Brand B	0.012	0.588	0.082	0.118	0.200
Brand C	0.085	0.117	0.617	0.085	0.096
Brand D	0.107	0.074	0.123	0.467	0.230
Other	0.177	0.135	0.227	0.149	0.312

Notice that the "Other" state is included to ensure that the sample space is complete. Even though our interest may be limited to the four cereals, we need to recognize that there are other brands available to consumers. Notice also that time is counted off in successive purchases.

With no further computation, the matrix can be interpreted to provide useful information. For example, by comparing diagonal values, we can see that Brand C is the one that commands the greatest customer loyalty or retention—that is, buyers of Brand C are more likely to stay with the same brand for the next purchase than any other brand. We could also recognize that, among all of the competitors of Brand A, the one that is most likely to capture the next purchase is Brand D (because the highest nondiagonal probability value in Brand A's row is in the Brand D column). Other direct interpretations of the absolute or relative values in this matrix may occur to you. From calculated values, you can get additional interpretations. For example, steady-state probabilities give long-run "market share"—the fraction of the total demand that each brand captures. Or, the mean sojourn time tells you how many successive purchases of the same brand can be expected.

Despite the apparent usefulness and widespread acceptance of this application, there are a couple of reservations to be noted. First, the Markov assumption makes a strong (and questionable) assertion about buyer behavior. It says that purchases are influenced only by the most recent experience; prior purchases are, in effect, forgotten. That sounds unrealistic for individuals—we surely remember both good and

The other assumption, which may also be questioned, is the stationarity assumption. Do you really believe that the transition probabilities for buying behavior remain the same over time? What about the effects of advertising, discount sales, and/or changing prices? If you think these might influence buyers, you are denying the stationarity assumption. Of course, it is also possible that all of these competitive maneuvers simply neutralize one another, in which case the probabilities might remain fairly constant. Another possibility is that the probabilities do change over time, but so gradually that the current values can be used to make predictions over, say, the next few weeks (but not years). You really need statistical data, based upon the particular product and markets you are studying, to know if the stationarity assumption is justified.

For most kinds of purchasing behavior, you would have to collect data in reverse time. That is, since you cannot expect a buyer to know at the time of a purchase what brand he or she is going to buy next, you have to ask what was bought previously. The cereal example we started with is quite typical. To collect the data to determine transition probabilities, you might have to survey customers in several grocery stores as they purchase cereal. Since you could not expect them to know which cereal they will buy next time, you would have to collect the backward data, that is, ask them if they remember the cereal that they bought the previous time. By counting the responses and organizing them in the obvious table form, you would get something like this:

	Brand of Previous Purchase				
Brand of Current Purchase	A	B	C	D	Other
A	64	1	8	13	25
B	16	50	11	9	19
C	15	7	58	15	32
D	32	10	8	57	21
Other	23	17	9	28	44

bad experiences with products. However, the assumption might (or might not) work for the buying population as a whole, even if you know you would not want to use it to predict what any single individual would do. It takes a statistical test to determine whether the Markov assumption is appropriate for brand-switching behavior for particular products. For example, careful testing showed that some buyers of automobiles are fiercely loyal to certain brands, but if those buyers are removed from the population, the others behave as the Markov assumption would imply. Breakfast cereals may be similar; perfumes are probably not. If the Markov assumption is not appropriate, there is a trick that is explained in "Text Analysis," on page 61 that might help.

The data in this form reflects time backwards. That is, the row is the purchase at time n, the column is the purchase at previous purchase, time $n - 1$. For a normal Markov chain, we need the rows to reflect the state at time n and the columns at time $n + 1$. The easiest way to convert is just to transpose the data matrix. You want to do that *before* normalizing the data, because the transition matrix must have *rows* that sum to 1. If you transpose and normalize the above data table, you will get the transition matrix shown at the beginning of this section.

There are many other opportunities for using stochastic process models in business. Marketing is just one common use. Check the following references if you want to follow up.

Day, Ralph L., and Leonard J. Parsons, *Marketing Models: Quantitative Applications*. International Textbook Company, Scranton, PA, 1971.

Harary, F., and B. Lipstein, "The Dynamics of Brand Loyalty: A Markovian Approach," *Operations Research* 10 (1962), 19–40.

Styan, G. P. H., and H. Smith, Jr., "Markov Chains Applied to Marketing," *Journal of Marketing Research*, 1 (1964), 50–55.

Whitaker, D., "The Derivation of a Measure of Brand Loyalty Using a Markov Brand Switching Model," *Journal of Operational Research*, 29 (1978), 959–970.

EXAMPLE 2.5

2.12.2 Maze Searching and Web Browsing

Imagine a labyrinth or maze with many passages and intersections. The maze wanderer (a human, robot, rat, or other creature) will wander randomly until reaching the exit (or cheese or treasure or whatever). At each intersection, there is a choice to be made among the doors or passages available to the subject. Suppose that choice is made at random, with equal likelihood for each option. It is important for the Markov property that the intersections be indistinguishable, so that the subject cannot remember having been there before and alter the choice based on that recollection. If the person had a piece of chalk or some way to mark the route taken, the Markov assumption would be violated. There is a way around this problem by expanding the number of states, but for this example we will stick with the simplest version.

The states of the Markov chain model will correspond to the intersections; the transitions are movements along the passages from one intersection to the next (even if there are turns along the way). The exit or final goal is represented by an absorbing state.

For a very small example (which could be enlarged indefinitely), consider the diagram in Figure 2.6.

The numbers in the figure at the intersections represent the states, so the transition diagram would look like Figure 2.7. Notice that dead ends translate to loops because they return you to the same state.

The maze provides a visual image of moving through space, but the same kind of model can represent the more abstract task of browsing the Internet. Imagine each Web page as an intersection where multiple choices

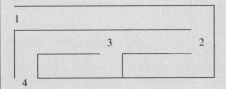

■ **FIGURE 2.6** A small maze

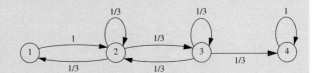

■ **FIGURE 2.7** Transition diagram maze searching

for the next page are given by hyperlinks. Obviously, if you tried to model the entire World Wide Web, you would have an enormous number of states, but you do not have to go that far. You can model a particular Web site, or perhaps a group of related Web sites, with states for each page in the set. Any departures from the set would be represented as transitions to absorbing states. That is, you would have a terminating process that lasts only as long as the viewer stayed within the set. That kind of model would be useful if you wanted to evaluate the design of a commercial Web site and needed to evaluate whether customers efficiently can find the products or services they seek.

There are other examples of navigating through abstract spaces that can be thought of as maze searching. In the early days of computer gaming, there was a primitive form of text adventure that evolved into modern graphic adventure games. These were essentially maze searching games, though the maze was disguised by colorful descriptions and the moves included options like "pick up rock" or "put key in lock." Each move altered the state of the game. Another example of navigating through abstract space is tracking chains of references through the technical literature to locate original source documents.

All of these examples can be modeling as Markov chains, provided that the choice of next move is not influenced by history. As mentioned, there are tricks to get around that limitation, but they involve enlarging the state space.

EXAMPLE 2.6

2.12.3 Social Mobility

Markov chains have been used to model the movement among social or economic classes of whole societies over generations. Thus, it can be a tool to distinguish rigid societies from those that offer more opportunity for advancement. Here is a very small, but quite genuine example. (It is taken from J. G. Kemeny and L. Snell, *Finite Markov Chains*, cited on this text's page 68. The authors obtained the example from a study of social mobility in England and Wales using 1949 data.) There are just three states: Upper, Middle, and Lower class. A person is identified with a particular class based upon occupation. The reference is not clear on how these are precisely defined, but we will take them as given. The transition matrix derived from the data is as follows:

$$\mathbf{P} = \begin{bmatrix} 0.448 & 0.484 & 0.068 \\ 0.054 & 0.699 & 0.247 \\ 0.011 & 0.503 & 0.486 \end{bmatrix}$$

This means, for example, that 44.8 percent of the children of upper-class parents managed to preserve their parents' position in society, 48.4 percent of them fell to middle-class status, and 6.8 percent plummeted to the lower classes. You can see that it was difficult, but not impossible, for people born into the middle or lower classes to rise to the upper class.

One should be extremely hesitant to use this model to predict what will happen to individuals, because there are many more important factors than parentage and luck that determine one's place in society. However, it can be a good way to measure collective tendencies. In ancient times, and in some parts of the world today, people were/are born into a station in life that is unlikely to change. In modern democratic societies, we like to think that every child is entitled to an equal opportunity to achieve whatever status his or her abilities can secure. Of course, the reality may be different from the ideal. The Markov chain model quantifies just how much opportunity is available (based on average movement in the population) and how fairly the opportunity is distributed. If parentage had no influence, the rows would be equal.

One needs to be very careful not to insert one's own prejudices into the model. For example, classifying a particular occupation as lower class may reflect a snobbish bias that is not shared by those who enjoy that occupation. For that reason, it would be preferable, it seems, to use terms such as professionals, factory workers, farm workers, self-employed, and so forth, rather than value-laden terms such as upper and lower class. If one wanted to use income categories, one would have to adjust for inflation and also remember that income is not a reliable indicator of social status.

Some of the interesting applications of Markov chain modeling of social mobility could be:

1. Comparing mobility in socialist societies to mobility in capitalist societies.

2. Comparing promotion mobility of women in American corporations to that of men.

3. Comparing employment mobility for different races before and after the passage of equal opportunity and affirmative action laws.

4. Predicting, on the assumption that equality of opportunity has been achieved (as indicated by equality of transition matrices for different groups), how many generations will be required to achieve near equality of distribution.

5. Detecting the influence of education on mobility.

6. Detecting whether urbanization promotes or hinders equality of opportunity.

7. Determining whether history is moving in the direction of more nearly equal opportunity or not.

8. Determining whether recent immigrants have the same mobility as families that have lived here for many generations.

You can probably think of some more issues in this category on your own.

EXAMPLE 2.7

2.12.4 Education

We used the progression of students through college as an example of a terminating process. There are many other ways to use Markov chains in modeling education. The admissions process, enrollment, retention, and the hiring and promotion of faculty have all received this treatment. These are all appropriate uses of the theory, because they represent population behavior. However, you should be cautious about using Markov methods to represent education of an individual. Even if you could clearly identify states for the growth of knowledge, you would be left with the conflict between the Markov assumption and the fact that humans do remember their pasts. Although you can find Markov models of individual learning in the literature, you should be highly suspicious of these. They were probably created by people who did not really understand more than the mechanics of Markov chains.

On the other hand, there are some excellent examples of Markov models in planning educational policies for universities, including those cited below. The book by Hopkins and Massey contains a number of Markov models (as well as other operations research techniques) applied to higher education. Similar models could be developed for job promotion, employment mobility, and other human resource issues in education more broadly and in the general economy.

Bessent, E. W., and A. M. Bessent, "Student Flow in a University Department: Results of a Markov Analysis." *Interfaces*, 10 (1980), 52–59.

Hopkins, David S. P. and William F. Massey, *Planning Models for Colleges and Universities*. Stanford University Press, Stanford, CA, 1981.

EXAMPLE 2.8

2.12.5 Product Flow

A very similar class of "flow" models can be used for industrial applications. When products flow through factories, jobs through stages of processing, or designs through stages of review, a structure very much like that of Figure 2.5 (page 46) often results. The branching is typically the result of some inspection or review step that determines where units go next. The final absorbing state represents completion of the product, and the other absorbing state represents scrap or failures.

There is one important difference to watch out for, however. In the education example, we were not concerned with resource constraints. That is, we can imagine many students moving through college, each operating independently and without interference from others. In industrial applications, it is common for the stages to require one-at-a-time processing from a limited number of resources (machines, people, or whatever). That contention for resources is fundamental to a different class of stochastic models, called queueing networks, which we will come to later. The simple Markov chain models cannot capture the delays that result from interference of many units moving through the stages at the same time.

On the other hand, they can still be useful for some aspects of the same applications. If, for example, we are not concerned with how much time is spent in the system, but only where things end up or how many times a stage is visited, we can use the Markov chain model. For example, we might have an electronic chip fabrication line in which several stages produce significant scrap. We could calculate the effective yield rate for the entire line by calculating the fraction that reach the completion stage (that is, an absorption probability). We could not, from the Markov chain alone, say how long it would take to manufacture a chip. (However, this can be done with "rewards" attached to the Markov chain—something that is easy to do once you know how.)

Markov chains appear in some advanced work in networks of queues. Part of the problem there is to represent the routing of customers (or jobs or whatever) from one node to another in a network. The network may have inputs and outputs in various places, branching, merging, cycling, and other structural variations. A very general way to handle all of the routing possibilities is to say that the movements occur according to a Markov chain. That is, a customer will move from station i to station j with probability p_{ij}. If one is troubled by the independence implied by the Markov assumption, one can use higher-order chains. If you can treat the queueing network analysis for arbitrary Markovian routing, you will be including nearly every conceivable possibility as special cases.

As a personal note, I have found production systems to be especially rich in useful applications of stochastic processes. In numerous cases, I have found "easy" opportunities to make substantial improvements in production systems, using results obtained from rather simple models. Most of these were not at all obvious to the management or workers who lived with the system every day. Many students have reported the same experience.

EXAMPLE 2.9

2.12.6 Growth

The past few examples have represented "flows" of various kinds. Another category of models represent "growth."

Here is a simplified version of a timber growth model that has been used in forest management. The states are size categories, as measured by tree diameter. Every three years, all the trees in a stand were measured, and from that data the following transition matrix was determined.

	0–1	1–3	3–8	>8
0–1 inch	0.92	0.08	0	0
1–3 inch	0	0.97	0.03	0
3–8 inch	0	0	0.98	0.01
>8 inch	0	0	0	1

You can see that the structure is something like the college progression model, in that the states stay the same or advance. In the matrix, the upper triangle has entries, but the lower does not. This structure is typical of growth models. Of course, there is no mathematical reason that limits us in this way, so we can just as easily handle shrinkage or other backward transitions.

The states could be almost any measurable characteristic. Models like this have been used for such things as the spread of epidemics, the acceptance of new products, and the expansion of entire markets, as well as physical growth. A popular recent application has focused on convergence of the economies of countries in the European Union. You can find many articles on the Web on this topic.

EXAMPLE 2.10

2.12.7 Population Dynamics

Markov models are used extensively by mathematical biologists (who are concerned about ecosystem and other big picture issues, as opposed to cells and other microbiology issues). Plant and animal populations exhibit variability and states can capture the variation (even if there is no uncertainty). Then, when we add dynamics, the changes in the population can be tracked over time. The Markov assumption usually works for demographic modeling, because the current state contains sufficient information to predict the state at the next time epoch. For example, if the state represents age, then, regardless of what happened up until now, everyone in the population will be one year older a year from now.

For a small example, the following matrix contains data reflecting the life cycle of dolphins. The time interval is one year and the states are 1 = yearlings, 2 = juveniles, 3 = mature, 4 = postreproductive, and 5 = dead.

$$\mathbf{P} = \begin{bmatrix} 0 & 0.98 & 0 & 0 & 0.02 \\ 0 & 0.90 & 0.07 & 0 & 0.03 \\ 0 & 0 & 0.95 & 0.04 & 0.01 \\ 0 & 0 & 0 & 0.98 & 0.02 \\ 0 & 0 & 0 & 0 & 1 \end{bmatrix}$$

This is a terminating process with just the one death state as absorbing. It can be used to calculate life expectancy, age-specific survival, the distribution of ages, and several other useful measures. You can read quite a bit more about models like this and several others in Caswell's classic book on population modeling, cited below. The same book covers Markov population models for spatial dispersion (geographical movement or migration), community composition, and the environment.

Caswell, Hall, *Matrix Population Models*, 2nd ed. Sinauer Associates, Sunderland, MA, 2001.

EXAMPLE 2.11

2.12.8 Gambling and Games of Chance

Some of the most common examples in elementary probability are games of chance—dice, cards, and so forth. Those problems make convenient examples because they are usually simple enough for beginners to understand. However, they can also mislead a student into believing that probability is primarily "for" such applications (and therefore not very useful), when in fact it is much more broadly applicable. Once you have acquired an advanced understanding, you can appreciate that those simple game examples contain the essence of the probability issues, without clouding the discussion.

Most of the simple examples involve a single play. For example, you may be asked to find the probability of drawing a certain hand of five cards out of a deck, which you would approach by finding how many ways there are to obtain that hand out of the total number of possible hands—a problem in combinatorics. However, many games of chance are sequential; that is, they involve multiple plays or rounds, continuing until some final winning or losing condition is met or until the players tire of the game. Markov chains provide a good way to model many of these sequential games. One of the problems in Chapter 3 leads you through an analysis of the game of craps. Baccarat is another casino game that can be analyzed in a similar way. Monopoly and other board games can be analyzed in principle, but they may have an enormous number of transitions (the basic transition matrix for Monopoly is 40 × 40). Tennis and baseball, which are usually regarded as skill sports rather than games of chance, have been studied with Markov chain models. If you are interested in these applications, you can check the references below or conduct a search on the Web. Before you look them up, you might try to identify the seventeen states of a game of tennis and the twenty-four states of a half-inning of baseball.

There is a well-known application called the Gambler's Ruin that represents the changing fortunes of a bettor who plays repeatedly with a constant wager. The state is the (remaining) size of the gambler's wealth, measured in integers. Each bet is for the same amount (one integer unit) and each play is independent, so the gambler's "stake" (the amount he or she has) increases or decreases by one unit with each play, according to whether he or she wins or loses that play. If we assume that the probability of winning on any one play is p and q = 1 − p is the probability of losing on any one play, the transition diagram looks like Figure 2.8.

The absorbing state on the left of Figure 2.8 represents the situation where the player has lost all of his or her money and cannot continue; hence, the "gambler's ruin." There could be a similar absorbing state on the right, corresponding to winning all of the opponent's money, but usually the model assumes that the opponent has unlimited resources. The state space consists of all of the nonnegative integers, and is therefore infinite. This is our first example of a Markov chain that needs an infinite state space.

As you might expect, it can be proven that if p < q (that is, the player is at a disadvantage), it is certain that state 0 will eventually be reached. However, we can compute the expected number of plays of the game until all of the money is lost as a function of p and the starting state, and from that function, we can determine the optimal way to bet to make the game last as long as possible. (Answer: Make the smallest bet you are allowed to make.) A lot of similar analysis of betting strategies can be carried out using this or related Markov chains.

Obviously, most casino games operate at a disadvantage to the gambler. However (not so obviously), there is one common game that you can play in a public casino that, if played perfectly, will give a slight advantage to the player. That game is Blackjack, sometimes called 21. The key to winning is to realize that the composition of the remaining deck changes as cards are played. If you can keep track of the dealt cards as they are revealed and adjust your play appropriately, you can—in principle—achieve a slightly better than 0.5 chance of winning over the long run. That is, you could if the casinos allowed you to play that way. These days, they are well aware of card counters and have effective countermeasures to protect their edge. Still, it is an interesting story. Four MIT students did the original probability calculations, and the theories were popularized by a mathematics professor named Ed Thorp in a book called *Beat the Dealer*. Dozens of variations on the system have been devised since. You can find a selection of books on this subject in any good bookstore. Just be wary of claims that you can really overcome the house advantage. Many books that are marketed to gamblers contain false claims and faulty analysis. (More fortunes have been made in selling so-called secrets to gullible readers than have been won at the gambling tables.)

A mathematically sound reference to all sorts of gambling games and systems is Richard A. Epstein's book, cited below. Among other things, you can find a proof that you can never overcome (through *any* system of varying bets) a sequential game in which the odds on each play are against you. There are many gambling systems that claim to do that.

Ash, Robert B., and Richard L. Bishop, "Monopoly as a Markov Process," *Mathematics Magazine* 45 (January), 1972, 26–29.

Epstein, Richard A., *The Theory of Gambling and Statistical Logic*, 2nd ed. Academic Press, San Diego, 1977.

Mezrich, Ben, *Bringing Down the House*. The Free Press, New York, 2002.

Trueman, R. E., "Analysis of Baseball as a Markov Process," in *Optimal Strategies in Sports*, eds. S. P. Ladeny and R. E. Machol, Elsevier-North Holland, New York, 1977.

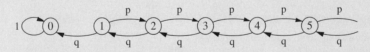

■ **FIGURE 2.8** The Gambler's Ruin

EXAMPLE 2.12

2.12.9 Financial Investments

One of the more obvious opportunities for stochastic modeling is in forecasting movements of stock prices or other investment instruments. We will start with a really simple model, then discuss some extensions.

Suppose that a stock or some other investment has values expressed in dollar increments. (That is, we will ignore cents.) Then, starting from some initial value i (the price we paid), it will move up or down in value $1 at a time. We will choose to embed our epochs at those moments when the value changes by a dollar, so as to capture every change. (We could alternatively choose to consider the end-of-day or end-of-week price, but that would require also allowing changes of more than $1 at each transition.) With those assumptions, we get a transition diagram that looks like Figure 2.9.

Notice that, structurally, the transition diagram looks just like a Gambler's Ruin model. We generalized just a little by allowing the win or loss probabilities to be state-dependent. In fact, there is a lot of similarity between investing in stocks or bonds and betting on games of chance. Of course, the financial industry does not like the comparison and goes to great lengths to convince investors that Wall Street has nothing in common with Las Vegas. A great debate was started about thirty years ago when Burton Malkiel, a professor at Princeton, published a book called *A Random Walk Down Wall Street*. He said, in effect, that stock market prices behave as a Markov chain. In his words,

> *A random walk is one in which future steps cannot be predicted on the basis of past actions. When the term is applied to the stock market, it means that short-run changes in stock prices cannot be predicted. Investment advisory services, earnings predictions, and complicated chart patterns are useless.... Taken to a logical extreme, it means that a blindfolded monkey throwing darts at a newspaper's financial pages could select a portfolio that would do just as well as one carefully selected by the experts.*

As you might imagine, many professionals took exception to his work, despite the fact that he had a great deal of statistical evidence on his side. That book still is in print and still is controversial. In fact, in recent years a great deal of effort has gone into deriving very sophisticated stochastic financial models for computerized trading. Many of these models deal with options, hedge funds, and various financial derivatives, rather than stock prices directly. It has been a matter of some controversy whether any real advantage can be gained. If the markets are truly efficient, as some argue, then the short-term up-and-down movements are no more than purely random disturbances. On the other hand, if one can find even a slight edge by detecting some signal within the random noise, you could make a lot of money. If you want to pursue this, be advised that it gets very deeply technical; you will have to learn advanced mathematics and statistics. There are no easy-to-understand models that can be proven to work. There are many books, newsletters, and consultants offering simple versions of technical analysis, but those are known to be unreliable. Remember the old adage, "If it seems too good to be true, it probably is." The following are reliable reading.

Lo, Andrew W., and Archie C. Mackinlay, *A Non-Random Walk Down Wall Street*. Princeton University Press, Princeton, NJ, 1999.

Makiel, Burton G., *A Random Walk Down Wall Street*. Norton, New York, 1999 (first published 1973, updated and republished numerous times).

Paulos, John Allen, *A Mathematician Plays the Stock Market*. Basic Books, New York, 2003.

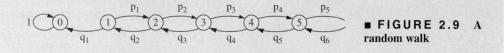

■ **FIGURE 2.9** A random walk

EXAMPLE 2.13

2.12.10 Text Analysis

Markov chains have been used to analyze language for purposes related to cryptography, speech recognition, and to help identify the authors of manuscripts. Claude Shannon, the inventor of information theory, used Markov chains in his earliest work on the information content in sentences.

Consider the sentence, "The dog chased the cat across the street." Let each distinct word and each punctuation mark be a state, and count the frequency of the transitions from one word or punctuation mark to another. From the given sentence we can build a table of frequency counts as follows (the words were first sorted into alphabetical order).

	across	cat	chased	dog	street	the	(period)
across	0	0	0	0	0	1	0
cat	1	0	0	0	0	0	0
chased	0	0	0	0	0	1	0
dog	0	0	1	0	0	0	0
street	0	0	0	0	0	0	1
the	0	1	0	1	1	0	0
(period)	0	0	0	0	0	0	0

If you then convert the frequencies to probabilities, you get the transition matrix that follows:

$$\begin{bmatrix} 0 & 0 & 0 & 0 & 0 & 1 & 0 \\ 1 & 0 & 0 & 0 & 0 & 0 & 0 \\ 0 & 0 & 0 & 0 & 0 & 1 & 0 \\ 0 & 0 & 1 & 0 & 0 & 0 & 0 \\ 0 & 0 & 0 & 0 & 0 & 0 & 1 \\ 0 & \frac{1}{3} & 0 & \frac{1}{3} & \frac{1}{3} & 0 & 0 \\ 0 & 0 & 0 & 0 & 0 & 0 & 1 \end{bmatrix}$$

Of course, this was a very short sentence with a limited number of states and transitions. Imagine doing the same thing with an entire book. (The task would be too tedious to contemplate doing by hand, but can be implemented easily with a computer program.) The result would be a large transition matrix with transition probabilities that reflect the likelihood that, in the chosen book, one particular word follows another. If you construct the matrix with a large enough sample of writing (say, a large book or several smaller ones), the matrix is a reliable "signature" or "fingerprint" of the author, because it captures the unique way that one individual puts words together. A different person would have distinctly different transition probabilities. We know from empirical tests that one author's transition matrix will remain fairly constant from one work to another, but will differ significantly from a different author's matrix. It seems to work in every language, not just English. This technique was used (along with others) to establish who actually wrote each of the Federalist papers, a series of articles published under the common pen name of Publius by Alexander Hamilton, James Madison, and John Jay, in the period when the U. S. Constitution was being formulated.

Now, does language really conform to the Markov assumption? Of course not. In order for sentences to have real meaning, the words must relate over longer spans than just adjacent pairs. For example, if we were to simulate the Markov chain we constructed from the sentence above to generate random sentences, we can produce odd sentences such as, "The cat across the dog chased the street." Each word-to-word transition is possible, according to the matrix, but the whole sentence makes no sense. With a much larger sample text, with many more words and better statistical estimates of the probabilities, you can use the transition matrix to generate interesting sentences. There is no serious purpose in doing that; it is just a whimsical generation of nonsense text that somehow almost makes sense. It can be poetic.

If we wanted a more realistic model, we could resort to higher-order Markov chains. The next word could depend on the prior two or three words. Of course, the number of states is already enormous, so extending the dependence will be costly.

The same idea of modeling the transitions from language can, and has, been used for music. Each note is a state; with some probability, it is followed by another note. The transition matrix for Bach will be significantly different from that of Beethoven or of Mozart. You can use those differences to identify composers or to generate music in the style of a particular composer. Again, the accuracy is improved with higher-order chains.

Markov models are being employed in current work to improve computerized speech recognition, handwriting recognition, and automated response systems. Again, a Web search will reveal numerous examples. The following source also is of interest.

Shannon, C. E., "Prediction and Entropy of Printed English." *Bell Systems Technical Journal* 30: 50–64, 1951.

EXAMPLE 2.14

2.12.11 Biology and Medicine

Markov chains are common in the life sciences. They have been used at the microbiology level in, for example, the analysis of DNA sequences (much like the text or music analysis of the previous section). They have been used at the human and animal scales for such things as tracking the progress of deadly diseases. And they have been used at the level of entire populations to model, for example, fluctuations in the size of the population over time or with the spread of epidemics.

Most of those applications require specific knowledge of biology, but here is a good not-too-technical example that illustrates several new points. We will model the inheritance of a genetic trait, for example, a propensity for some disease or eye color. Let's say it is eye color. The way genetics works (roughly) is that every individual has pairs of chromosomes, and an offspring gets one chromosome from each parent. A particular genetic trait may be dominant (meaning that it will be apparent in some detectable trait if it occurs in either chromosome) or recessive (meaning that it will be apparent only if it occurs in both chromosomes). We will say that a represents a brown-eye gene, and b represents a blue-eye gene, and that brown is dominant. An individual with either the pair aa or the pair ab will have brown eyes; only if the pair is bb will the individual have blue eyes. The state of an individual will be the gene pair, which could be aa, ab, or bb. (The order does not matter; we are just identifying the pair. Also, we focus only on the eye-color gene, ignoring all other genes.)

The discrete time units will be generations. Hence, the transitions will correspond to the passage of the gene state from a parent to a child. But there are two parents, and there could be many children. We will make no attempt to model the whole population, but simply trace from one parent (we'll choose the mother) to one (we'll choose the first) female offspring, continuing through numerous generations.

To set up the transition matrix, we need to determine the probability that the daughter has a particular gene pair given the knowledge of the gene pair of the mother. For example, if the mother is known to be of the type aa, the daughter could be of type aa or ab, with appropriate probabilities, but could not be of type bb (because one of the genes comes from the mother, who does not carry the blue-eye gene). Hence, one of the transition probabilities in the aa row has to be zero. For the other two, we use the law of total probability, conditioning on the gene pair of the father:

$$P\{X_1 = aa | X_0 = aa\} = P\{X_1 = aa | X_0 = aa, \text{father} = aa\}$$
$$\times P\{\text{father} = aa\} + P\{X_1 = aa | X_0 = aa, \text{father} = ab\} P\{\text{father} = ab\} + P\{X_1 = aa | X_0 = aa, \text{father} = bb\} P\{\text{father} = bb\}$$

All three terms are shown to match the requirements of the law of total probability (that is, we have to include all the possibilities for the father), but the third term will now drop out because we know the conditional probability has to be zero. That is, $P\{X_1 = aa | X_0 = aa, \text{father} = bb\} = 0$. It is also easy to simplify the first term, because if both the mother and father are aa, the daughter must also be. That is, $P\{X_1 = aa | X_0 = aa, \text{father} = aa\} = 1$. Now, for the second term, we have to consider the possibilities. Since the daughter will inherit one gene from each parent with each of the pair having equal likelihood, we have $P\{X_1 = aa | X_0 = aa, \text{father} = ab\} = \frac{1}{2}$. Figure 2.10 illustrates how the combinations produce equal likelihood for a pure aa or mixed ab daughter.

Now all that remains is to determine $P\{\text{father} = aa\}$ and $P\{\text{father} = ab\}$. If we know that, throughout the population, the probability that a randomly chosen eye-color gene is a is given by p (and therefore $P\{b\} = 1 - p$), and we assume that mate selection is independent of eye color, then the probability that the father is aa would be (from the binomial distribution), p^2. Similarly, the probability that the father is ab would be $2p(1-p)$. That gives us all we need to construct the transition probability, namely,

$$P\{X_1 = aa | X_0 = aa\} = (1)(p^2) + (1/2)(2p(1-p))$$
$$= p^2 + p(1-p) = p$$

Using similar arguments for the next transition probability (or simply recognizing that the row has to sum to 1, we find

$$P\{X_1 = ab | X_0 = aa\} = 1 - p$$

That completes the first row of the transition matrix.

Using the same logic on the other rows produces the final one-generation transition matrix (you might want to work this out for yourself).

$$\mathbf{P} = \begin{bmatrix} p & (1-p) & 0 \\ \dfrac{p}{2} & \dfrac{1}{2} & \dfrac{1-p}{2} \\ 0 & p & 1-p \end{bmatrix}$$

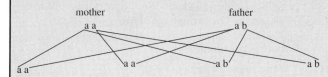

■ FIGURE 2.10 All gene pair combinations

This is a very simple version of a Markov chain model of genetic inheritance. You can find more elaborate versions in the sources cited below. Markov chain theory makes an excellent foundation for genetics because the Markov assumption is entirely justified. If you know the state of the parents, there is no additional value in knowing the state of prior ancestors; that is, whatever is inherited comes solely from parents. The stationarity assumption might be questioned over very long time periods (because of mutations, natural selection, and such), but would seem to be a reasonable assumption for at least several generations.

Biology and medicine have made extensive use of Markov models. The spread of diseases in the population, as well as the progression of particular diseases or medical conditions in individuals, have been modeled. Look for current references on the Web.

Feller, William, *An Introduction to Probability Theory and Its Applications*, vol. I, 3rd ed. John Wiley, New York, 1968.

Kemeny, J. G., and J. L. Snell, *Finite Markov Chains*. Van Nostrand, Princeton, NJ, 1960 (reprinted by Springer-Verlag, New York, 1983).

EXAMPLE 2.15

2.12.12 Binomial Process

The binomial process—the stochastic process that counts the number of successes in a sequence of independent Bernoulli trials—can be represented by a Markov chain. Of course, you do not really need a Markov chain to explain the binomial process, but representing it this way suggests some immediate generalizations that can be easily handled.

Since the binomial process counts the number of successes, the states are $0, 1, 2, \ldots$, that is, the nonnegative integers. The time count corresponds to trials, so on each transition, the state either stays the same (failure) or increases by one (success). Hence, the transition diagram looks like Figure 2.11.

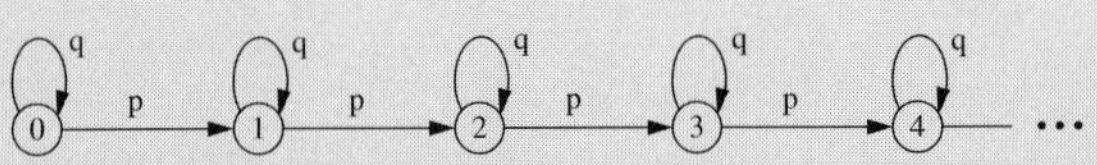

■ **FIGURE 2.11** Transition diagram for binomial process

The transition matrix looks like this:

$$\begin{bmatrix} q & p & 0 & 0 & 0 & 0 & \ldots \\ 0 & q & p & 0 & 0 & 0 & \ldots \\ 0 & 0 & q & p & 0 & 0 & \ldots \\ 0 & 0 & 0 & q & p & 0 & \ldots \\ 0 & 0 & 0 & 0 & q & p & \ldots \\ 0 & 0 & 0 & 0 & 0 & q & \ldots \\ \ldots & \ldots & \ldots & \ldots & \ldots & \ldots & \ldots \end{bmatrix}$$

As you can see, this is a very special case of the general Markov chain, so there are a lot of things we can do to generalize the binomial process. We could make the probabilities state-dependent, to represent a changing probability of success based on the number of successes so far. Or we could define a trinomial process in which each trial produces 0, 1, or 2 successes (which would add one more nonzero diagonal in the matrix). Or, we could go to a quadranomial, and so forth.

There is really not much reason to cling to the assumptions of the binomial process. With the tools of Markov chain analysis, we can handle much more general assumptions without difficulty. You can think of the binomial process as a beginner's stochastic process, which we are content to leave behind when we have more powerful tools.

EXAMPLE 2.16

2.12.13 Aging Infrastructure

This is a study of a challenging modeling situation, using several clever tricks to overcome obstacles. The problem is one of developing a reasonable model for the aging of infrastructure, such as bridges, roads, and sewers. These last a long time—typically decades—but they do deteriorate gradually and eventually fail if not maintained. We will assume that there are well-established standards for inspecting for damage, and for placing the structure into one of five states:

Infrastructure States

5 = Like-new condition.

4 = Acceptable condition.
3 = Nonurgent repairs needed.
2 = In need of immediate repair.
1 = Failure.

Those states suggest a Markov model of an ordinary terminating structure. The Markov assumption seems justified because the current state is really all we have to know; however, there are two significant problems. First, the stationarity assumption does not seem correct. As the years pass, the accumulated wear and tear ought to be reflected in the transition probabilities. Even if it has held up well, a very old structure should have a higher probability of advancing to the next stage of deterioration than a very new structure. That in itself is not an insurmountable obstacle. We can handle nonstationary Markov chains by changing the transition matrix, but that brings us to the second problem. We do not have many data points, surely not enough to statistically estimate a large number of parameter values accurately.

Let's assume we are talking about bridges of a certain design within a certain environment, subject to similar forces. We will measure time in increments of five years. There are only a dozen bridges of varying ages to obtain data from. Inspection records give us a scatter plot like Figure 2.12, in which each X represents a single data point.

What we would like is a nonstationary Markov model that in some way tracks this data, but could be used for other similar bridges to predict deterioration. With only twelve data points, we cannot hope to estimate very many parameters. The model, if we can find a way to construct it, will tell us the probability distribution for every state at every time. The initial state is 5 at time zero. The distribution of the state at time n would be $p_{5j}^{(n)}$ with j running over all five states. Hence, the expected state at time n would be

$$E(X_n) = \sum_{j=1}^{5} p_{5j}^{(n)}$$

We could plot this function of n to see how the expected state changes over time.

From looking at the data and just thinking about the nature of physical deterioration, we should expect a curve that has a "reverse S" shape. That is, it should start out fairly level, accelerate downward in the middle, and then flatten out again on the right. If we think more about how deterioration occurs, we might postulate that many small factors accumulate in a gradual way, building to a tipping point where the state makes a sudden transition. The build-up can be thought of as compounding like interest. So let's model that behavior within the transition matrix. Assume that the transition probability from any state j to state j − 1 (the next level of deterioration) increases according to the function $f(n) = (1 + r)^n - 1$, where r is some yet unspecified parameter. This is basically an interest calculation, where the principal is 1, the interest rate is r, and you want the accumulated total interest by time n. Then, putting that model into the nonstationary Markov transition matrix (and using reverse numerical order for the states), we have for the transition matrix at age n

$$\mathbf{P}_n = \begin{bmatrix} 1-f(n) & f(n) & 0 & 0 & 0 \\ 0 & 1-f(n) & f(n) & 0 & 0 \\ 0 & 0 & 1-f(n) & f(n) & 0 \\ 0 & 0 & 0 & 1-f(n) & f(n) \\ 0 & 0 & 0 & 0 & 1 \end{bmatrix}$$

That makes all of the transition matrices, for all values of n, dependent entirely on the single parameter r. For example, if r is 0.03, the first two matrices would be

$$\mathbf{P}_1 = \begin{bmatrix} 0.97 & 0.03 & 0 & 0 & 0 \\ 0 & 0.97 & 0.03 & 0 & 0 \\ 0 & 0 & 0.97 & 0.03 & 0 \\ 0 & 0 & 0 & 0.97 & 0.03 \\ 0 & 0 & 0 & 0 & 1 \end{bmatrix}$$

$$\mathbf{P}_2 = \begin{bmatrix} 0.9391 & 0.0609 & 0 & 0 & 0 \\ 0 & 0.9391 & 0.0609 & 0 & 0 \\ 0 & 0 & 0.9391 & 0.0609 & 0 \\ 0 & 0 & 0 & 0.9391 & 0.0609 \\ 0 & 0 & 0 & 0 & 1 \end{bmatrix}$$

To get the two-step transition matrix (which would cover ten years of aging, since our units are five-year increments) we multiply these two.

$$\mathbf{P}^{(2)} = \mathbf{P}_1 \mathbf{P}_2$$

$$= \begin{bmatrix} 0.9109 & 0.0872 & 0.0018 & 0 & 0 \\ 0 & 0.9109 & 0.0872 & 0.0018 & 0 \\ 0 & 0 & 0.9109 & 0.0872 & 0.0018 \\ 0 & 0 & 0 & 0.9109 & 0.0891 \\ 0 & 0 & 0 & 0 & 1 \end{bmatrix}$$

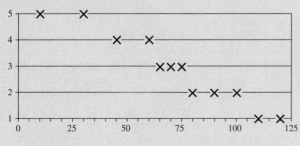

■ **FIGURE 2.12** Inspection data

Reading the first row from this matrix, we find the state distribution after ten years, from which we calculate the expected value to get $E(X_2) = 4.93$. We can continue in the same manner, multiplying the next transition matrix on the right to generate the next state distribution, and next expected value. All of those operations are easy to set up in a spreadsheet. Of course, all of the specific numerical values depend on r, which can be placed in a single cell for convenient exploration of changes.

Now if we plot the results we get with r = 0.03, we get Figure 2.13. Of course, that was just an arbitrary choice for r, so we should be able to do better. But at least the curve behaves in the general way that we want.

Now we have to adjust r to get a best fit to the data. You could do it by trial and error in the spreadsheet, but there is a more elegant and quicker way. Microsoft Excel© has a built-in "solver" that can optimize a function. We have to set up a criterion to optimize. In this case, we might choose to minimize the sum of absolute differences between the curve values and the data values, or perhaps the sum of squares of differences. Either is easily implemented, and will produce essentially the same result in this case. After we run the solver, with the cell

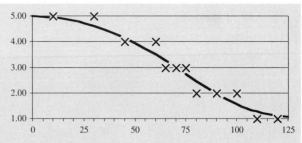

■ **FIGURE 2.14** The best-fit-curve (single parameter)

containing r as the variable, we find that the best fit is obtained for r = 0.18. Figure 2.14 shows how the corresponding curve fits the data.

Visually, we seem to have a pretty good fit. For this situation, where the number of data points is so limited, it is probably not worth more effort to improve the fit. If we had a lot more data, there are other things we could do. For example, we could use a two-parameter model for the increasing transition probability function, such as $f(n) = (1+r)^{kt}$ (letting both r and k vary). Or we could allow different function or different function parameters in different rows of the transition matrices. Those are just a few suggestions. The general point is that building a model to fit a situation calls for good judgment and a bit of creativity, as well as mastery of the tools.

Once you have a model you are satisfied with, there is a great deal that can be done with it. Since you have the complete probability distribution at every point in time, you can calculate standard deviations, the probability that the structure is unsafe, the expected remaining life, and many other metrics.

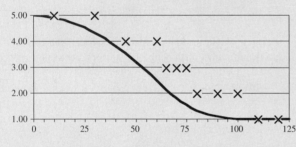

■ **FIGURE 2.13** Expected state for r = 0.03

EXAMPLE 2.17

2.12.14 Numerical Integration—Markov Chain Monte Carlo

There is a somewhat peculiar application of Markov chains that recently has become very popular among certain statisticians. In fact, if you do a search for "Markov Chain" on the Internet, you may find that most of the recent references are to "Markov Chain Monte Carlo" (MCMC). This approach could give a false impression that the primary purpose of Markov chains is for this particular use.

MCMC is really just a sampling technique that can be helpful in evaluating integrals that are otherwise difficult to obtain. Such integrals routinely occur in advanced statistics (especially Bayesian statistics, which is a very active research area, which in turn is why MCMC is so topical now). Here is how it works.

First, you should understand the ordinary Monte Carlo method of numerical integration. Suppose you had a really strange function, expressed graphically as some kind of curve (so that the integral would correspond to the area under the curve between two limits). One way to estimate the area would be to throw darts (randomly) at a picture of the curve, drawn within some rectangle of known area. Then the fraction of darts that hit within the area under the curve, relative to the total number of darts hitting within the rectangle would estimate the ratio of the two areas. Since we know the area of the rectangle, we could then multiply to estimate the area under the curve. For example, suppose the graph looked like Figure 2.15.

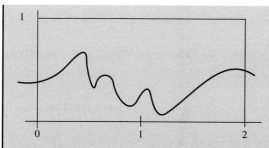

■ **FIGURE 2.15** A function to integrate

Assume we wanted the integral of this function between 0 and 2. We can get a simple enclosing rectangle of height 1, because the function does not rise above that value. So the area of the outer rectangle will be two square units. Pin the drawing to the wall, stand back, and throw darts blindly. If 500 darts hit inside the rectangle, and of these, 387 fall in the area under the curve, our estimate of the integral would be $(387/500) \times 2 = 1.584$. This, of course, is a statistical estimate, so if you want more accuracy (that is, reduce the confidence interval on the estimate), throw more darts.

Instead of throwing darts, you can accomplish the same effect with a computer by generating random number pairs that fall (uniformly) within the rectangle. That is, generate a random number between 0 and 2 for the first coordinate and another between 0 and 1 for the second. The whole process is called the Monte Carlo method after the famous casinos in Monaco, because it uses uniformly distributed random numbers (something like rolling dice) to evaluate a mathematical function that may not have anything to do with probability. The method is quite general—you only need to be able to determine whether a point falls under the curve, or whether the function evaluated at a point is greater or less than the random height. For higher dimensional integrals, an obvious extension works (though you may not be able to visualize it). The only drawback is that the pure Monte Carlo method can be quite inefficient. You have to generate a lot of points to get good estimates, and many of them may be misses.

Markov Chain Monte Carlo is an extension of the raw Monte Carlo method that uses a constructed Markov chain to generate the random points somewhat more intelligently. It is fundamentally an importance sampling technique that biases the random points to improve efficiency (that is, get good estimates with fewer points). The technique is too detailed to explain fully in this short section, but you can easily find explanations of the technique on the Internet. It is mentioned here because you will surely encounter the phrase Markov Chain Monte Carlo if you are looking for material on Markov chains, and may wonder what that subject is about.

2.13 Summary

We have only touched on a small sample of the areas in which Markov chains have been found useful. Once you catch on, you see them everywhere. They are the foundation of many more advanced stochastic processes. In fact, Markov is regarded as the father of that entire branch of mathematics. What he created was an extremely versatile and useful abstract formulation that balances realism with computational convenience.

2.14 Recommended Reading

There are many fine textbooks on the theory of stochastic processes, but few that address the issues of formulation. Most just assume that the models are given. Recently published introductory textbooks include those by Berger (3), Kao (11), Kulkarni (13), and Minh (14). You may find it easier to learn from older books that were written when the subject was less developed. Howard (10) and Kemeny and Snell (12) are suitable for beginners; however, their treatments are so unique that there is a danger of obtaining a misleading image of the subject as a whole. Used in conjunction with one of the more standard textbooks, they have a great deal to offer. Although not introductory in nature, Bailey (1) and Bartholomew (2) are useful in suggesting applications. Feller's book (7) is the classic work in the field.

1. Bailey, N. T. J., *The Elements of Stochastic Processes with Applications to the Natural Sciences*. Wiley, New York, 1964.
2. Bartholomew, D. J., *Stochastic Models for Social Processes*. Wiley, New York, 1967.
3. Berger, Marc A., *An Introduction to Probability and Stochastic Processes*. Springer-Verlag, New York, 1993.
4. Bharucha-Reid, A. T., *Elements of the Theory of Markov Processes and Their Applications*. McGraw-Hill, New York, 1960.
5. Bhat, U. N., *Elements of Applied Stochastic Processes*. Wiley, New York, 1972.
6. Cox, D. R., and H. D. Miller, *Theory of Stochastic Processes*. Wiley, New York, 1965.
7. Feller, W., *An Introduction to Probability Theory and Its Applications*, vol. I, 2nd ed. Wiley, New York, 1957.
8. Heyman, D. P., and M. J. Sobel, *Stochastic Models in Operations Research*, vol. I. McGraw-Hill, New York, 1982.
9. Heyman, D. P., and M. J. Sobel, *Stochastic Models*. North-Holland, New York, 1990.
10. Howard, R., *Dynamic Probabilistic Systems*, vols. I and II. Wiley, New York, 1971.
11. Kao, Edward, *An Introduction to Stochastic Processes*. Brooks/Cole, New York, 1996.
12. Kemeny, J. G., and L. Snell, *Finite Markov Chains*. Van Nostrand, New York, 1960 (reprinted by Springer-Verlag, New York, 1983).
13. Kulkarni, V. G., *Modeling, Analysis, Design, and Control of Stochastic Systems*. Springer-Verlag, New York, 1999.
14. Minh, D. L., *Applied Probability Models*. Duxbury, New York, 2001.

Chapter 2 Problems

For each problem in this set, all you have to do is set up the transition matrix or transition diagram. The first few are straightforward, with obvious states and treatment of time. As they become progressively more difficult, you will have to think carefully about state and time definitions. In the next chapter, you will be asked to calculate some things using those matrices, so you might want to start each problem on a new page and leave room for further work. Also, some of them will be further extended in Chapter 4. Answers for odd-numbered problems are given in the back of the book, but you cannot benefit from looking before you have made a serious attempt on your own.

1. Commercial fishermen in Alaska go into the Bering Sea to catch all they can of a particular species (salmon, herring, etc.) during a restricted season of a few weeks. The schools of fish move about in a way that is very difficult to predict, so the fishing in a particular spot might be excellent one day and terrible the next. The day-to-day records of catch size were used to discover that the probability of a good fishing day being followed by another good day is 0.5, by a medium day 0.3, and by a poor day 0.2. A medium day is most likely to be followed by another medium day, with a probability of 0.6, and equally like to be followed by a good or bad day. A bad day has a 0.3 probability of being followed by a good day, 0.2 of being followed by a medium day, and 0.5 probability of being followed by another bad day. Construct a Markov chain model to describe the way the fishing days run.

2. A self-service elevator in an old four-story building operates solely according to the buttons pushed inside the elevator. That is, a person on the outside cannot call the elevator to the floor he is on. Consequently, the only way to get on the elevator is for someone else to get off at your floor. Of the passengers entering the building at the ground floor and wanting to use the elevator, half go to the second floor, and the other half divide equally between the third and the fourth floors. Passengers above the ground floor want to go to the ground floor in 80 percent of the cases. Otherwise, they are equally likely to want to go to the other floors. Formulate the Markov chain for elevator movements.

3. The Air Quality Index (AQI) is an indicator of the air quality in a particular location, computed from a combination of standardized measurements. The numerical values fall into ranges that are coded by colors for easy comprehension. The colors are green, yellow, orange, and red, corresponding to good to very poor air quality, in order. In areas where the atmosphere stagnates, whether by temperature inversion or geographic features such as mountains, the air quality can deteriorate rapidly. Only random weather events, such as precipitation or strong winds can refresh the air in such places. A record of several years of data describing

successive days of the air quality index produced the following table.

Air Quality on First Day of Pair	Air Quality on Second Day			
	Green	Yellow	Orange	Red
Green	248	62	0	0
Yellow	29	232	29	0
Orange	20	0	140	40
Red	29	0	0	261

The numbers in the cells are counts of the number of times the corresponding pair occurred. Formulate a Markov chain model to describe the stochastic process.

4. The quality of wine obtained from a vineyard varies from year to year depending on a combination of factors, some of which are predictable and some of which are not. From historical data extending over the past 100 years, the year-to-year succession pattern was as follows:

Previous Year	Graded Year			
	Poor	Fair	Good	Excellent
Poor	2	6	8	4
Fair	6	12	20	2
Good	9	20	5	0
Excellent	3	2	1	0

Formulate the Markov chain for wine quality.

5. The Department of Agriculture wants a model to predict crop yields in Indiana. The main factor affecting how much corn (for example) will be produced is how many acres are planted with corn. Other things being equal, individual farmers try to plant those crops for which they think the market price at harvest time will be best, but other considerations are also important. They also have to take into account soil conditions, crop rotation, opportunity for irrigation, and so forth. Because of the variability in the decisions made by individual farmers, the Department of Agriculture believes that a random process model would be appropriate. There are two major crops: corn and soybeans. All other marketable crops can be lumped into a single class called "other"; land not used for marketable crops is called "fallow." Information is available on the acreage devoted to each of the four categories during the year of the survey and what was planted on the same acreage during the *previous* year. For example, of the 6 million acres planted with corn in the year of the survey, 3 million acres were planted with corn in the previous year, 1 million were planted with soybeans, 1 million were planted with some other crop, and 1 million were left fallow. Formulate a Markov chain for crop usage.

Current Year Land Usage	Previous Year Usage (in millions of acres)			
	Corn	Soybeans	Other	Fallow
Corn	3	1	1	1
Soybeans	1	1.2	1	0.8
Other	0.5	0.3	0.5	0.2
Fallow	0.2	0.2	0.1	0

6. A market survey has been conducted to determine the movements of people between types of residences. Two hundred apartment dwellers were asked if their previous residence was an apartment, a condominium, their own home, or a rented home. Similarly, 200 condominium dwellers were asked about their previous residences, and so on. The results of the survey are tabulated below.

Current Residence	Previous Residence			
	Apartment	Condominium	Own House	Rented House
Apartment	100	20	40	40
Condominium	150	40	0	10
Own House	50	20	120	10
Rented House	100	20	20	60

The data are believed to be representative of the behavior of the population at large. Formulate the Markov chain for housing movements.(Hint: Notice that the survey looks backward in time.)

7. A simple industrial robot has four axes of motion, controlled by four separate motors. We will refer to the motions as Turn (right or left), Lift (raise or lower), Extend (in or out), and Grip (contract or release). When the robot is programmed to perform a task, a sequence of these motions will take place. For example, to pick up an object, the arm would turn to align with the object, extend to reach it, grip it, and lift. We are interested in predicting how the actions will occur, in order to accelerate the controlling software. If, for example, a turn is most likely to be followed by an extension, the code that controls the extension can be readied in advance. From a sample of robot programs, representing a set of typical tasks, the following count data was derived.

Current Motion	Next Motion			
	Turn	Lift	Extend	Grip
Turn	5	40	90	15
Lift	50	2	20	80
Extend	30	40	20	60
Grip	15	90	30	15

Formulate the Markov chain for robot motions.

8. U. S. automobile manufacturers are concerned about the competition of foreign cars in the domestic market. Chrysler, in particular, seems to view its survival as being dependent upon recapturing the share of the market it lost to foreign sales over the past decade, and has begun a product design and advertising campaign directly aimed at that objective. To a somewhat lesser extent, the other U. S. manufacturers have also altered their strategies. To determine the present market share (which was determined by past buying practices) and to find out what effect the changes may have had, a (hypothetical) survey of 1,000 car owners was conducted. Of the 400 respondents who presently own a General Motors car, 200 said they would probably buy another GM car for their next purchase, 80 would buy a Ford, 80 would buy a Chrysler, and 40 favored a foreign make. There were 300 respondents who presently own foreign cars, of which 180 would buy another, 60 would buy a Chrysler, and the rest would split evenly between GM and Ford. Of the 200 Ford owners, 80 would buy another Ford and the rest would split evenly among the other possibilities. Finally, there were 100 Chrysler owners, of which 70 were loyal to Chrysler; the rest would split evenly among the other possibilities. Formulate the brand-switching Markov chain.

9. A naturalist is observing the behavior of a frog in a small lily pond. There are four lily pads in the pond, and the frog jumps from one to another. The pads are numbered arbitrarily; the state of the system is the number of the pad that the frog is on. Transitions occur only when he jumps. Although he can jump from any pad to any other, the probability of jumping to any given pad from the one he is on is *inversely* proportional to the distance (that is, he is more likely to jump to a near pad than a far one). The distances are

Between lily pads 1 and 2	6/5 feet
Between lily pads 1 and 3	2 feet
Between lily pads 1 and 4	3/2 feet
Between lily pads 2 and 3	6/7 feet
Between lily pads 2 and 4	1/2 feet
Between lily pads 3 and 4	3/4 feet

Hints: The distance between two points is the same in both directions. Remember that each row of a transition matrix must sum to 1. Formulate the Markov chain to describe the jumping frog.

10. A manufacturer of paint mixes the colors in a large vat. For our purposes, assume that he produces only three colors: red, white and, blue. If a batch of one color is followed by another of the same color or a dark one (red or blue), the vat does not have to be cleaned very thoroughly. However, changing from red or blue to white requires that the vat be well cleaned to avoid color contamination. The sequence of batches is determined by demand, availability of ingredients, and a number of other random factors. It has been determined from historical data that a batch of red has been followed by red 40 percent of the time, by white 10 percent, and by blue 50 percent. White was followed by white 20 percent, by red 40 percent, and by blue 40 percent. Blue was followed by blue 30 percent, by red 40 percent, and by white 30 percent. Formulate the Markov chain for the sequence of paint colors.

11. The accounting department of a large firm is interested in modeling the dynamics of its accounts receivable (that is, the money that is owed to it by its customers). When a charge sale occurs, a bill is sent out at the end of the month. Payment is due within thirty days, but may not occur in that time. If no payment is received within six months of the billing date, the amount is classified as a bad debt. Thus, an individual account is described by its age in months, or by "paid," or by "bad debt." For numerical values, assume that the probability that payment is made in the first month is 0.8 and declines by 0.1 each additional month. That is, the probability it is paid in the second month is 0.7, in the third month is 0.6, and so on. Formulate the terminating Markov chain that describes the debt payment.

12. A section of highway is inspected annually for damage. Based upon specific measurements of frequency and depth of cracks and other such factors, the highway section can be classified as belonging to one of six "states of deterioration." The first state, A, refers to a new condition; the last state, F, refers to a condition requiring complete reconstruction. Damage is progressive, and eventual complete deterioration is certain. However, the causes of deterioration relate to weather, traffic, and other random phenomena. Therefore, the lifetime of the highway is uncertain. Past experience suggests that the chances of deteriorating one grade between inspections is one in five, and the chances of deteriorating two grades is one in twenty. Provide a Markov chain model of the highway deterioration process, under the assumption that no repairs are made.

13. A hospital is interested in tracking the movements of heart patients through different units. The four units of interest are emergency admissions, surgery, intensive care, and recovery. Emergency patients (that is, people who have suffered heart attacks and were brought into the hospital through emergency admissions) are always assigned to the intensive care unit for a period of observation following the emergency treatment. Once a doctor determines that it is safe to do so, the patient is moved to a recovery unit. Of course, a relapse requires a return to intensive care. The surgical unit receives patients of two kinds: those who are scheduled for heart operations through non-emergency procedures (planned operations), and those who come from intensive care (semi-emergency operations). Only one patient in a hundred dies in surgery; the rest are always moved to intensive care after surgery. Of the patients in intensive care, whether or not they have previously had surgery, about 10 percent will undergo surgery; the others are moved into recovery after some period of intensive care. Once into recovery, most patients (95 percent) eventually either recover sufficiently to be discharged, but about 2 percent have to be returned to intensive care. The others die in recovery. Draw a transition diagram, and set up an appropriate transition matrix.

14. A small town purchases salt by railroad-car loads to be used for melting ice and snow on the roads during the winter. One railroad car holds twelve tons of salt. The amount of salt used in any one storm depends upon the severity and duration of the storm. Past experience shows that, of the storms which are serious enough to call for any salt, some will require only one pass of the salt trucks, some will require two passes, and a few will require three passes. Each pass (a complete coverage of all of the streets of the town) consumes five tons of salt. Also from past experience, it has been estimated that 50 percent of the storms are one-pass storms, 40 percent are two-pass, and 10 percent are three-pass. The initial supply of salt at the beginning of winter is five railroad cars, or sixty tons. Construct a Markov chain mode to show the consumption of salt over time, where time is measured discretely in the number of storms since the beginning of winter.

15. A solar heating device can recharge (store heat) in a single sunny day enough to provide for the next two days. In other words, the only times that the solar heat must be augmented by some other heat source is when there are three or more consecutive cloudy days. During the heating season, a period of 120 days, the weather patterns are described by the following Markov chain:

	Sunny	Cloudy
Sunny	1/2	1/2
Cloudy	1/3	2/3

Using this information, construct another Markov chain which is capable of indicating when the solar heat must be augmented. *Hint:* Think carefully about your state definition.

16. A print shop has three critical machines that essentially determine the productivity of the shop. They are each subject to failure every day. Assume for simplicity that if a machine fails, it is at the moment it is turned on in the morning. Let the probability that an operable machine fails on any given day be 0.1, independently of the other machines. A mechanic is called in whenever there is a failure and can always repair one machine that same day (so it is available for the next), but cannot do more than one a day. Machine A feeds both machines B and C, as shown in Figure 2.16.

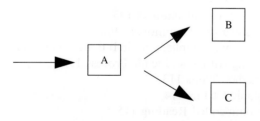

■ **FIGURE 2.16** The machine network

Thus, if machine A fails, there is no production that day. Consequently, the policy is always to repair A if it is down. If both B and C are down, and A is up, the policy is to repair B. Of course, units that are down stay down until repaired. Units that are operable today may be down tomorrow, whether or not the shop is able to produce anything. But a unit that is repaired today is certain to operate tomorrow. Develop a Markov chain model to assess the productivity of the shop. *Hint*: There are eight possible states corresponding to the combinations of up versus down conditions of the three machines. Number these in accordance with the binary numbers. That is, state 0 corresponds to (000), which means all machines are working; state 1 corresponds to (001), which means A and B are working and C is down; state 2 corresponds to (010), which means A and C are working and B is down; and so on.

17. Consider the game of craps as played in Las Vegas casinos. The basic rules of the game are as follows. One player (the shooter) rolls a pair of dice. If the outcome of that roll is a 2, 3, or 12, the player immediately loses; if it is a 7 or 11, the player wins. In all other cases (4, 5, 6, 8, 9, or 10), the number rolled on the first toss becomes the point, which the shooter must try to duplicate on subsequent rolls. If the player manages to roll the point before a 7 appears, it's a win; otherwise it is a loss. It may take several rolls to determine whether the player wins or loses. After the first roll, only a 7 or the point have any significance until the win or loss is decided. Of course, there are many ways to place bets on various possibilities, but we are concerned here with just the basic rules of the game, played only once. Model the process as a Markov chain with absorbing states for win and lose. Assume the dice are fair in evaluating the probabilities.

18. In baseball, basketball, tennis, and various other sports, the winner of a tournament is determined by a best of five or best of seven rule; that is, whichever team can win three or, respectively, four games first is the winner of the series. Such tournaments are presumed to be more fair than a single elimination game in determining the better team, since the outcome of any single game has a large element of chance. Suppose that two teams, A and B, are involved in a best of five tournament and that team A is slightly favored. In particular, the probability that team A will win in any single game against the other is 0.52. Before any games are played, one would "expect" A to win the series. However, during the series if B should happen to be ahead of A in wins, it might have the advantage. This problem is about the relative advantage of being the better team as opposed to being ahead in games. Formulate a terminating Markov chain to describe the tournament.

CHAPTER 3

Markov Chain Calculations

CHAPTER CONTENTS

- 3.1. Walk Probabilities 73
- 3.2. Transition Probabilities 74
- 3.3. State Probabilities 78
- 3.4. A Numerical Example 79
- 3.5. Expected Number of Visits 80
- 3.6. Sojourn Times 82
- 3.7. First Passage and Return Probabilities 83
- 3.8. Computational Formulas for all Markov Chains 86
- 3.9. Special Classes of Markov Chains 86
- 3.10. Steady-State Probabilities 87
- 3.11. The Uses of Steady-State Results 92
- 3.12. Mean First Passage Times 93
- 3.13. Computational Formulas for Ergodic Markov Chains 96
- 3.14. Terminating Markov Chains 96
- 3.15. Expected Number of Visits 98
- 3.16. Expected Duration of a Terminating Process 99
- 3.17. Absorption Probabilities 100
- 3.18. Hit Probabilities 102
- 3.19. Conditional Mean First Passage Times to Absorbing States 103
- 3.20. Computational Formulas for Terminating Processes 105
- 3.21. Call Center Calculations 105
- 3.22. Classification Terminology 106
- 3.23. Additional Complications in Infinite Chains 111
- 3.24. Dealing with a Reducible Process 112
- 3.25. Periodic Chains 113
- 3.26. Ergodic Chains 114
- 3.27. Recommended Reading 115

Before starting serious study of this chapter, be sure you can formulate real-world situations as Markov chains, using Chapter 2 problems as cases. That is, go through all of the problems of the previous chapter, before trying to compute anything. Most students have little trouble with the computing once the problem is correctly set up, but many have trouble with the formulation. It is generally a good idea to work lots of examples and see the variations that can occur before you risk additional confusion over all of the different things you can compute. Seeing many examples is not enough; you have to practice doing it to acquire the ability.

There are many quantities that can be computed to answer many kinds of questions. We will start with some of the most direct ones, which have no complications. The presentation will explain why the calculations work with "plausible logic," rather than formal proofs. It is important that they make sense to you each step of the way. Later, we will introduce new terminology to classify "difficult" situations that might impede or even prevent computations. We will continue to use the truck example to clarify abstractions, but you should try to see the general logic so that you can apply it in other cases.

3.1 Walk Probabilities

The simplest quantity to calculate, and the one on which all other calculations depend, is a *walk probability*. Recall that a walk in the transition diagram is just a particular sequence of states, so a walk probability is just the probability that a particular sequence of states occurs. Formally, it is a conditional probability because the beginning state is assumed known; all of the other states in the sequence are uncertain.

> **Definition** A *walk probability*, denoted $p_{j_0 j_1 j_2, \ldots, j_n}$ (where the subscripts indicate an ordered sequence of states that correspond to a feasible realization), is the conditional probability
> $$P\{X_n = j_n, X_{n-1} = j_{n-1}, X_{n-2} = j_{n-2}, \ldots, X_1 = j_1 | X_0 = j_0\}$$

Notice that the walk notation ($p_{j_0 j_1 j_2, \ldots, j_n}$) expresses the order of the sequence so that time reads from left to right (as opposed to the right-to-left order in the event description). The reason that the event notation has to run in the less-natural backward order is that the condition (which is the start of the sequence) has to be after the condition bar, which is all the way to the right. We use the event notation when we have to prove something, going all the way back to the basic rules of probability. Otherwise, it is simpler and cleaner to use the walk notation. In words, $p_{j_0 j_1 j_2, \ldots, j_n}$ is the probability that, starting from state j_0, the process goes next to state j_1, then to state j_2, etc., ending in state j_n.

To calculate a walk probability, the obvious thing is to just multiply together the transition probabilities for the steps along the walk. But remember that you need to be careful with probabilities, because the obvious approach is often wrong. In this case, the obvious approach is correct, but we need to see why.

Consider the specific walk probability for the sequence that starts in Atlanta, goes to Chicago, then to Detroit. This would be p_{134}. The initial state, Atlanta, is known, but the next two states are uncertain. The state after the first trip *could* be Chicago, but might not be, so we need a probability for that event. Even if the first trip is to Chicago, the next trip might or might not be to Detroit. Formally, we want the probability for the event that $X_1 = 3$ (the first trip is to Chicago) *and* $X_2 = 4$ (the next stop is Detroit) given $X_0 = 1$. Using the formal event notation, we want
$$P\{X_2 = 4, X_1 = 3 | X_0 = 1\}$$
This is both a joint probability (like $P\{A \cap B\}$) and a conditional probability (like $P\{A|C\}$). To expand it, we use the general probability multiplication rule $P\{A \cap B\} = P\{A|B\}P\{B\}$, while retaining the condition C. That is,
$$P\{A \cap B | C\} = P\{A|B, C\}P\{B|C\}$$
You may not have seen this form before, but it is just the ordinary multiplication rule for probabilities with the extra C condition carried along. Application of this rule to our desired walk probability gives
$$P\{X_2 = 4, X_1 = 3 | X_0 = 1\} = P\{X_2 = 4 | X_1 = 3, X_0 = 1\} P\{X_1 = 3 | X_0 = 1\}$$

Working on the right side of this equation, we can simplify the first part by applying the Markov property. It tells us we can drop the past history $X_0 = 1$ from that term ($P\{X_2 = 4 | X_1 = 3, X_0 = 1\}$). So now we have

$$P\{X_2 = 4, X_1 = 3 | X_0 = 1\} = P\{X_2 = 4 | X_1 = 3\} P\{X_1 = 3 | X_0 = 1\}$$

Then we apply the stationarity property to shift the time index of the same first part, leaving

$$P\{X_2 = 4, X_1 = 3 | X_0 = 1\} = P\{X_1 = 4 | X_0 = 3\} P\{X_1 = 3 | X_0 = 1\}$$

What is now on the right is just the product of two one-step transition probabilities.

$$P\{X_2 = 4, X_1 = 3 | X_0 = 1\} = p_{34} p_{13}$$

These are in reverse time order, but if we switch them around, we get the result we wanted, which is (in words) that the walk probability is just the product of the transition probabilities along the walk.

$$p_{134} = p_{13} p_{34}$$

Aside from the usual rules of algebraic manipulation and the basic laws of probability, the only assumptions we used were the Markov and stationarity assumptions. Both were required, but no other assumptions sneaked in. Be sure that you understand the math here, because it is the key to everything that follows.

The same logic applied to longer walks gives the general result below.

Result A walk probability is computed by

$$p_{j_0 j_1 j_2, \ldots, j_n} = p_{j_0 j_1} p_{j_1 j_2} p_{j_2 j_3}, \ldots, p_{j_{n-1} j_n}$$

In words, the walk probability is the product of the one-step transition probabilities for each step along the walk. If you visualize a walk in the transition diagram starting at the initial state, you would just multiply the weights of the arcs that you pass over as you move from state to state.

It may seem that we went to a lot of trouble to come up with an obvious answer. It is true that the final outcome is just as simple as we could possibly want; however, the details were emphasized in order to make the point that it only works because of the Markov and stationarity assumptions. Once you understand that, you may use the "obvious" logic freely; namely, to find the probability of any particular sequence of events, just multiply together the probabilities of the steps taken.

3.2 Transition Probabilities

Although walk probabilities are easy to calculate, they rarely answer a question that we care about. More commonly, we want to know the probability that the process is in a certain state at a certain time without saying anything about how it got there. In Chapter 2, we introduced the concept of an n-step transition. We used the one-step transition probabilities p_{ij} to formulate any Markov chain. We now extend that notation.

Definition An *n-step transition probability*, denoted $p_{ij}^{(n)}$, is the conditional probability $P\{X_n = j | X_0 = i\}$.

Of course, the one-step version is a special case, so $p_{ij}^{(1)} = p_{ij}$ and we can get that directly from the transition matrix. Let's turn to the two-step case.

For example, we might want to know the probability that the truck is in Detroit after two trips regardless of where it went first. If we know that it started in Atlanta, we are talking about the two-step transition probability

$$p_{14}^{(2)} = P\{X_2 = 4 | X_0 = 1\}$$

We can enumerate all of the possible ways that that event could occur. Starting from Atlanta, the truck could have made a local trip first, then gone to Detroit. Or it could have gone first to Baltimore, then to Detroit. Or it could have gone first to Chicago, then to Detroit. Or (hypothetically), it could have gone to Detroit and then made a local trip to stay in Detroit. Actually, that last possibility does not exist for our example because Detroit does not allow local trips, but we should list the theoretical possibility if we want to cover all of the cases. You can observe that there are four possible ways to end up in Detroit, corresponding to the four two-step walks that pass through each of the four cities at epoch 1. We know how to calculate walk probabilities, so it is tempting to just add up the walk probabilities for those four possibilities. But again, we should be careful to ensure that we are correctly using the rules of probability.

There are four distinct ways that the event we are interested in could come about, corresponding to the four alternative possibilities for the state at epoch 1. There is no other way that the event could occur, so we know we are not missing any possibilities. The only remaining question is whether the four walks represent mutually exclusive events. (Recall that the simple addition rule for probabilities requires that the events be mutually exclusive; if not, you have to use the version that corrects for intersections.) The four walks start and end in the same place, so you might (seeing that they have something in common) conclude that they are *not* mutually exclusive. But that would be wrong. The way you have to think in order to reach the correct conclusion is to go back to the definition of a walk. The walk is an *entire* sequence; the event in the walk probability is the complete sequence. So when you ask yourself if two walks are mutually exclusive, you are questioning whether they could both occur. If the two walks differ in any way—even if they have some parts in common—they could not both occur. By this logic we can conclude that *distinct walks are always mutually exclusive*. This is good news because it means we can add their walk probabilities without having to correct for intersections.

Consequently, the probability that we are interested in can be broken into the simple sum of four probabilities, one for each possible walk. Finding the corresponding probabilities for all four walks, and collecting results gives

$$P\{X_2 = 4 | X_0 = 1\} = p_{11}p_{14} + p_{12}p_{24} + p_{13}p_{34} + p_{14}p_{44}$$

This expresses, in terms of values already known, the probability that the truck will be at Detroit after the second trip, given that it started at Atlanta.

By similar logic, and keeping the same initial state, we may obtain the probabilities for being in states 2, 3 or 4 after the second trip:

$$P\{X_2 = 2 | X_0 = 1\} = p_{11}p_{12} + p_{12}p_{22} + p_{13}p_{32} + p_{14}p_{42}$$

$$P\{X_2 = 3 | X_0 = 1\} = p_{11}p_{13} + p_{12}p_{23} + p_{13}p_{33} + p_{14}p_{43}$$

$$P\{X_2 = 4 | X_0 = 1\} = p_{11}p_{14} + p_{12}p_{24} + p_{13}p_{34} + p_{14}p_{44}$$

The four expressions together give the complete probability distribution of the random variable X_2, given that $X_0 = 1$.

If the initial state is Baltimore, we get four different expressions for the distribution of X_2, another four for when it starts at Chicago, and four more for when it begins at Detroit. There are a total of sixteen expressions required to specify the distribution of X_2 under the various possible initial conditions. A general expression for an arbitrary term is

$$p_{ij}^{(2)} = \sum_{k} p_{ik} p_{kj}$$

where the sum index k runs over all states. Each term in this sum is a walk probability, and the sum runs over all of the possible states that the process could be in at epoch 1, so the sum lists all of the possible two-step walks from i to j.

Again, it is convenient to use a matrix arrangement for these sixteen terms. Let $\mathbf{P}^{(2)}$ denote the matrix of these probabilities, and let $p_{ij}^{(2)}$ represent the $(i,j)^{th}$ entry.

$$\mathbf{P}^{(2)} = \begin{bmatrix} p_{11}^{(2)} & p_{12}^{(2)} & p_{13}^{(2)} & p_{14}^{(2)} \\ p_{21}^{(2)} & p_{22}^{(2)} & p_{23}^{(2)} & p_{24}^{(2)} \\ p_{31}^{(2)} & p_{32}^{(2)} & p_{33}^{(2)} & p_{34}^{(2)} \\ p_{41}^{(2)} & p_{42}^{(2)} & p_{43}^{(2)} & p_{44}^{(2)} \end{bmatrix}$$

$\mathbf{P}^{(2)}$ is called the *two-step transition matrix*; its elements are the *two-step transition probabilities*. Of course, if there were S states, this matrix would have S rows and S columns, because every possible state provides both a row and a column.

The method of expressing n-step transition probabilities by adding the walk probabilities over all walks of length n in the transition diagram generalizes for any number of steps. However, if the transition diagram is large, complicated, or both; or if the number of steps is large, it is impractical to enumerate every such walk. Fortunately, there is a better way—one that it lends itself to easy computer implementation.

The method is based on the simple observation that the expressions previously derived for the elements of $\mathbf{P}^{(2)}$ are exactly what would be obtained if the matrix $\mathbf{P}$ were multiplied by itself using, of course, matrix multiplication. (If you do not see why, compare the result given above to the definition of matrix multiplication.) In symbols,

$$\mathbf{P}^{(2)} = \mathbf{PP} = \mathbf{P}^2$$

In words, the matrix of two-step transition probabilities is equal to the square of the matrix of one-step transition probabilities. This is not obvious; it is a derived relation, requiring both the Markov and stationarity assumptions. It is a most fortuitous relation because it means that a familiar, well-known matrix operation can be used to compute the probabilities for X_2.

Thus far we have described X_1 and X_2. To get the distribution of X_3 under varying possible conditions at time 0, we again have to consider the ways in which the states could occur. For example, suppose that $X_0 = 1$ and we are interested in the probability that $X_3 = 1$. Then at epoch 2, the state must have been 1, 2, 3, or 4, and the last trip must have been to Atlanta. Since these are mutually exclusive, collectively exhaustive possibilities,

$$p_{11}^{(3)} = p_{11}^{(2)}p_{11} + p_{12}^{(2)}p_{21} + p_{13}^{(2)}p_{31} + p_{14}^{(2)}p_{41}.$$

By this logic, we can write all three-step transition probabilities in terms of the two-step and one-step transition probabilities, which are of course already known. Furthermore, we may observe that the same results would be obtained by multiplying the two-step transition matrix by the one-step transition matrix. That is,

$$\mathbf{P}^{(3)} = \mathbf{P}^{(2)}\mathbf{P}$$

But since $\mathbf{P}^{(2)} = \mathbf{P}^2$,

$$\mathbf{P}^{(3)} = \mathbf{P}^2\mathbf{P} = \mathbf{P}^3$$

So the three-step transition matrix is just the one-step matrix cubed. The elements of this matrix completely describe X_3, the random variable describing the state at time 3.

By now it should be apparent how to continue. In general, the result is as shown below.

> **Result** The n-step transition matrix equals the one-step transition matrix raised to the nth power; that is,
>
> $$\mathbf{P}^{(n)} = \mathbf{P}^n$$

This is the single most important result for Markov chains. By raising **P** to the appropriate power, we can answer any question pertaining to the probability of being in a particular state at a particular time.

We can express an iterative form for the matrix multiplication in a few different ways. The most common (and probably most useful for computation) form is shown below.

> **Definition** The *Chapman-Kolmogorov forward equation* is the particular iterative form
> $$\mathbf{P}^n = \mathbf{P}^{n-1}\mathbf{P}$$

You can think of this as moving forward in time by post-multiplying (on the right) by one more transition matrix at a time.

It would be equally valid to state the reverse form, shown below.

> **Definition** The *Chapman-Kolmogorov backward equation* is the form
> $$\mathbf{P}^n = \mathbf{P}\mathbf{P}^{n-1}$$

This version is true not by reversal of the order of multiplication (because, in general, matrix multiplication is not commutative; that is $\mathbf{AB} \neq \mathbf{BA}$), but rather by regrouping terms, keeping the same order. Actually, we can accumulate the product in any grouping, so for any k between 1 and n,

$$\mathbf{P}^n = \mathbf{P}^k \mathbf{P}^{n-k}$$

The most natural software for these calculations is probably MATLAB. However, that either presumes familiarity with the program or requires a lot of explanation. If you have it and know how to use it, MATLAB is the way to go. A simpler, less elegant, but more accessible approach is to use spreadsheets. Modern spreadsheet programs provide operations that, in one way or another, accomplish matrix multiplication. Consequently, it is easy to set up a spreadsheet to perform the calculations automatically for any transition matrix of a given size. For example, the four-state example we have been using could be set up in Microsoft Excel as shown in Figure 3.1.

Figure 3.1 displays the formulas, not the data entries. It takes only a few moments and remarkably few keystrokes to set up such a spreadsheet. First, you define the range from B1 to E4 to have the name "P." If there were more than four states, this region would obviously have to be larger. This is the array where the elements of the four-by-four transition matrix will go. You can enter the values now or wait until later. Then, you highlight the range from B6 to E9, which is where $\mathbf{P}^{(2)}$ will go. Enter the formula "= MMULT(A1:D4, P)" in the single cell B6 and press the "enter" key while holding down the special key needed for matrix formulas. (See the spreadsheet manual or online help for specific instructions.)

When you enter the formula correctly, the entire matrix will fill in. The reason that you refer to the same region in two different ways is that you want a formula that you can copy. The first matrix reference (here shown as B1:E4) should be relative to the current position, that is, the region just above the current region. The second reference should be absolute, that is, always the same named region P. You can then define the formulas for all of the other ranges by simply copying the range B6 to E9. A quick way to do it is to highlight the entire region from B6 to E9, release the mouse button, move the cursor to the bottom-right corner of the highlighted region until the cursor changes to a solid cross, and click and drag down as many rows as you want. If you forget this shortcut, it works just as well to copy and paste the region multiple times. Once this is done for as many of the powers of **P** as you want, you can enter values for the one-step transition probabilities and all of the powers will be automatically computed.

78 Chapter 3 ■ Markov Chain Calculations

	A	B	C	D	E
1	P				
2					
3					
4					
5					
6	P2	=MMULT(B1:E4,P)	=MMULT(B1:E4,P)	=MMULT(B1:E4,P)	=MMULT(B1:E4,P)
7		=MMULT(B1:E4,P)	=MMULT(B1:E4,P)	=MMULT(B1:E4,P)	=MMULT(B1:E4,P)
8		=MMULT(B1:E4,P)	=MMULT(B1:E4,P)	=MMULT(B1:E4,P)	=MMULT(B1:E4,P)
9		=MMULT(B1:E4,P)	=MMULT(B1:E4,P)	=MMULT(B1:E4,P)	=MMULT(B1:E4,P)
10					
11	P3	=MMULT(B6,E9,P)	=MMULT(B6,E9,P)	=MMULT(B6,E9,P)	=MMULT(B6,E9,P)
12		=MMULT(B6,E9,P)	=MMULT(B6,E9,P)	=MMULT(B6,E9,P)	=MMULT(B6,E9,P)
13		=MMULT(B6,E9,P)	=MMULT(B6,E9,P)	=MMULT(B6,E9,P)	=MMULT(B6,E9,P)
14		=MMULT(B6,E9,P)	=MMULT(B6,E9,P)	=MMULT(B6,E9,P)	=MMULT(B6,E9,P)
15					
16	P4	=MMULT(B11,E14,P)	=MMULT(B11,E14,P)	=MMULT(B11,E14,P)	=MMULT(B11,E14,P)
17		=MMULT(B11,E14,P)	=MMULT(B11,E14,P)	=MMULT(B11,E14,P)	=MMULT(B11,E14,P)
18		=MMULT(B11,E14,P)	=MMULT(B11,E14,P)	=MMULT(B11,E14,P)	=MMULT(B11,E14,P)
19		=MMULT(B11,E14,P)	=MMULT(B11,E14,P)	=MMULT(B11,E14,P)	=MMULT(B11,E14,P)
20					
21	P5	=MMULT(B16,E19,P)	=MMULT(B16,E19,P)	=MMULT(B16,E19,P)	=MMULT(B16,E19,P)
22		=MMULT(B16,E19,P)	=MMULT(B16,E19,P)	=MMULT(B16,E19,P)	=MMULT(B16,E19,P)
23		=MMULT(B16,E19,P)	=MMULT(B16,E19,P)	=MMULT(B16,E19,P)	=MMULT(B16,E19,P)
24		=MMULT(B16,E19,P)	=MMULT(B16,E19,P)	=MMULT(B16,E19,P)	=MMULT(B16,E19,P)

■ **FIGURE 3.1** Spreadsheet calculation of powers of P

The borders and labels help in understanding the display but do not influence any of the results. Obviously, Markov chains with different numbers of states could be handled in a very similar way.

3.3 State Probabilities

There is one minor extension to be considered now. We have all of the random variables $\{X_0, X_1, X_2, X_3, \ldots\}$ characterized *provided* that the state at time 0 is known. Occasionally it happens that the value of X_0 is not known with certainty, but is given by a probability distribution. Suppose that such is the case, and let $p_i^{(0)}$ denote the probability that $X_0 = i$. For example, $p_1^{(0)}$ would be the probability that the state is 1 at time 0. Suppose we want to know the probability that the state will be 2 at some later time, say n = 5. Having the matrix $\mathbf{P}^{(5)}$, we can read off $p_{12}^{(5)}$, $p_{22}^{(5)}$, and $p_{32}^{(5)}$ from the second column. These are all probabilities of being in state 2 at time 5, but differ in the assumed state at time 0. The obvious thing to do is to take a weighted average of $p_{12}^{(5)}$, $p_{22}^{(5)}$, and $p_{32}^{(5)}$, weighting each according to the likelihood that the initial state it assumes is, in fact, the correct one. This logic would yield the expression,

$$P\{X_5 = 2\} = p_1^{(0)} p_{12}^{(5)} + p_2^{(0)} p_{22}^{(5)} + p_3^{(0)} p_{32}^{(5)} + p_4^{(0)} p_{42}^{(5)}$$

which is, indeed, the correct expression. Now consider the general case, stated below.

> **Definition** A *state probability* or an *absolute state probability*, denoted $p_j^{(n)}$ is the (unconditional) probability that the process is in a particular state j at a particular epoch n. That is, $p_j^{(n)} = P\{X_n = j\}$.

Such a probability is called a state probability to distinguish it from a transition probability. It would give the probability of being in a particular state at a particular time, *regardless of the state at time zero*. Since matrix notation has proved so convenient for the transition probabilities, let $\mathbf{p}^{(n)}$ be the row vector of state probabilities. (Note that this is a lowercase **p**; the uppercase is reserved for the transition matrix.)

Using the new notation, we may observe that the expressions we would obtain for the elements of $\mathbf{p}^{(5)}$—one of which was derived above—can also be obtained from

$$\mathbf{p}^{(5)} = \mathbf{p}^{(0)} \mathbf{P}^{(5)}$$

Furthermore, the logic will generalize to give the state probabilities at any time, as shown below.

> **Result** State probabilities are determined by the formula
> $$\mathbf{p}^{(n)} = \mathbf{p}^{(0)} \mathbf{P}^{(n)}$$

In words, the (row) vector of state probabilities for time n is given by the vector of initial state probabilities multiplied by the n-step transition matrix.

The right-hand side of this relation involves matrix multiplication, which is ordinarily not commutative (that is, $\mathbf{AB} \neq \mathbf{BA}$), so the order is important. Also it is important to remember that the $\mathbf{p}^{(n)}$ are *row* vectors. If you prefer column vectors, perhaps for reasons of convenience in writing the expressions on paper or in a spreadsheet, everything must be transposed. This would in turn reverse the order of multiplication on the right-hand side, giving the alternative form shown below.

> **Result** In conventional matrix form, state probabilities are determined by
> $$[\mathbf{p}^{(n)}]^T = [\mathbf{P}^{(n)}]^T [\mathbf{p}^{(0)}]^T$$

Setting up a spreadsheet to compute the state vectors is, if anything, even easier than doing one for the transition probabilities. Start the same way by defining a region for the transition matrix **P**. Then start a row for the initial vector $\mathbf{p}^{(0)}$. Say it is in the range A6 to D6. Immediately below it, highlight a region of the same size and input the formula "MMULT(A6:D6,P)" and hit return, remembering to hold down the special key for matrix formulas as you do. That formula computes the vector $\mathbf{p}^{(1)}$ in a way that can be copied downward for all the other row vectors.

3.4 A Numerical Example

EXAMPLE 3.1

Let us now put some specific numbers into our example and perform the calculations we have described. Suppose data from the field indicate that 10 percent of the trips starting from Atlanta are local; other trips are equally likely to go to the other three cities. Those that start from Baltimore divide equally among the three destinations. The ones that start from Chicago are local 40 percent the time, and split equally among the other three cities. Trips starting in Detroit have a 20 percent chance of going to Atlanta and are equally likely to go to Baltimore or Chicago.

From this description, the one-step transition matrix is

$$\mathbf{P} = \begin{bmatrix} 0.1 & 0.3 & 0.3 & 0.3 \\ 0.333 & 0 & 0.333 & 0.333 \\ 0.2 & 0.2 & 0.4 & 0.2 \\ 0.2 & 0.4 & 0.4 & 0 \end{bmatrix}$$

We get the matrix of two-step transition probabilities by squaring **P**

$$\mathbf{P}^{(2)} = \mathbf{P}^2 = \begin{bmatrix} 0.230 & 0.210 & 0.370 & 0.190 \\ 0.167 & 0.300 & 0.367 & 0.167 \\ 0.207 & 0.220 & 0.367 & 0.207 \\ 0.233 & 0.140 & 0.353 & 0.273 \end{bmatrix}$$

The interpretation of these numbers is related, of course, to their positions in the matrix. For example, the value in the first row and first column of $\mathbf{P}^{(2)}$ says that there is a probability of 0.23 that the truck will be back at Atlanta at the end of the second trip after leaving Atlanta.

Some of the transition matrices of higher order are

$$\mathbf{P}^{(3)} = \begin{bmatrix} 0.205 & 0.219 & 0.363 & 0.213 \\ 0.223 & 0.190 & 0.363 & 0.223 \\ 0.209 & 0.218 & 0.365 & 0.209 \\ 0.195 & 0.250 & 0.367 & 0.187 \end{bmatrix} \quad \mathbf{P}^{(4)} = \begin{bmatrix} 0.209 & 0.219 & 0.365 & 0.207 \\ 0.203 & 0.229 & 0.365 & 0.203 \\ 0.208 & 0.219 & 0.365 & 0.208 \\ 0.214 & 0.207 & 0.364 & 0.215 \end{bmatrix}$$

$$\mathbf{P}^{(5)} = \begin{bmatrix} 0.208 & 0.218 & 0.365 & 0.209 \\ 0.210 & 0.215 & 0.364 & 0.210 \\ 0.208 & 0.219 & 0.365 & 0.208 \\ 0.206 & 0.223 & 0.365 & 0.206 \end{bmatrix} \quad \mathbf{P}^{(6)} = \begin{bmatrix} 0.208 & 0.219 & 0.365 & 0.208 \\ 0.208 & 0.220 & 0.365 & 0.208 \\ 0.208 & 0.219 & 0.365 & 0.208 \\ 0.209 & 0.217 & 0.365 & 0.209 \end{bmatrix}$$

$$\mathbf{P}^{(7)} = \begin{bmatrix} 0.208 & 0.219 & 0.365 & 0.208 \\ 0.209 & 0.218 & 0.365 & 0.209 \\ 0.208 & 0.219 & 0.365 & 0.208 \\ 0.208 & 0.219 & 0.365 & 0.208 \end{bmatrix} \quad \mathbf{P}^{(8)} = \begin{bmatrix} 0.208 & 0.219 & 0.365 & 0.208 \\ 0.208 & 0.219 & 0.365 & 0.208 \\ 0.208 & 0.219 & 0.365 & 0.208 \\ 0.208 & 0.219 & 0.365 & 0.208 \end{bmatrix}$$

and so forth. These results are shown so that you can check your own calculations and also for later reference.

If the initial state of the truck is unknown, but is described by a probability distribution, we would use the formula

$$\mathbf{p}^{(2)} = \mathbf{p}^{(0)} \mathbf{P}^{(2)}$$

If, for example, the truck were equally likely to start from any of the four locations, then

$$\mathbf{p}^{(0)} = [0.25 \quad 0.25 \quad 0.25 \quad 0.25]$$

and

$$\mathbf{p}^{(2)} = [0.209 \quad 0.218 \quad 0.364 \quad 0.209]$$

Notice that rows of any power of **P** always sum to one. It is easy to prove that this must be so by purely algebraic means, but you can also deduce it from the fact that each row is a probability distribution for some X_n (given a particular initial state).

3.5 Expected Number of Visits

There is an easy transformation of the transition probabilities that can answer an entirely different class of questions. Suppose we wanted to know the expected number of times a state is visited over some time period—for example, the expected number of times that Detroit would be visited in eight trips. This should be a specific number, not a probability or a distribution. There is a related random variable, of course, which would count the number of times the state is visited, but we are now asking for the mean of that random variable. (We could define the random variable and look for its distribution, but the result is messy and rarely useful.) The notation for this mean is below.

Notation The expected number of times a state j is visited in the first n epochs is denoted by $e_{ij}^{(n)}$, where i is the initial state (which must be specified because it affects the result). The square matrix whose elements are $e_{ij}^{(n)}$ is denoted by $\mathbf{E}^{(n)}$.

For example, if we wanted to know the expected number of times Detroit is visited by the truck in the first eight trips, assuming that the process starts in Atlanta, we would want $e_{14}^{(8)}$.

A quick comment on the choice of notation here: It would be convenient to use v as the letter to represent "visits," but that would interfere with a use of that same letter later on when it appropriately stands for "value." It may help in remembering the meaning of $e_{ij}^{(n)}$ to think of it as counting "<u>en</u>tries" into state j. The only caution is to also remember to count the 0 epoch when using $e_{ii}^{(n)}$. Starting in a state may not seem like "entering" it, but it does count as a visit and is part of the definition of $e_{ii}^{(n)}$.

In order to develop an expression for computing $e_{ij}^{(n)}$, define a new random variable $I_{ij}^{(k)}$ that takes the value 1 if the state at epoch k is j and 0 otherwise. The subscript i is just specifying the initial state and is held constant throughout what follows. $I_{ij}^{(k)}$ is a Bernoulli indicator variable that, in effect, "counts" either zero or one depending upon whether its conditions are met. Its expectation is easily calculated as

$$E(I_{ij}^{(k)}) = 0 \times P(X_k > j | X_0 = i) + 1 \times P(X_k = j | X_0 = i) = P(X_k = j | X_0 = i) = p_{ij}^{(k)}$$

In words, the expected number of visits of the process to state j at the *k*th epoch is equal to the transition probability for that same state and epoch. Now the total number of visits in n steps—a random variable—is equal to the sum of the visits at each step, that is, the sum of the indicator variables over all the time steps. Consequently, the expectation of that sum is the sum of the expectations of the indicator variables, by one of the fundamental properties of expectations. Hence,

$$e_{ij}^{(n)} = \sum_{k=0}^{n} E(I_{ij}^{(k)})$$

Finally, by substitution, we find

$$e_{ij}^{(n)} = \sum_{k=0}^{n} p_{ij}^{(k)}$$

In words, we get the expected number of visits simply by adding up the transition probabilities for the corresponding subscript pair, over all of the time steps in the period of interest. Notice that the sum includes the zero epoch, so that it can include the possibility that the process starts in the state of interest. Of course, if the initial state i is different from j, the indicator variable would be 0, and if i is the same as j, it would be 1. So we could also write the basic equations as shown below.

> **Result** The expected number of visits to state j in n epochs is
>
> $$e_{ij}^{(n)} = \sum_{k=1}^{n} p_{ij}^{(k)} \quad \text{for } i \neq j$$
>
> $$e_{jj}^{(n)} = 1 + \sum_{k=1}^{n} p_{jj}^{(k)}$$

A matrix version of the same equations is

$$\mathbf{E}^{(n)} = \mathbf{I} + \mathbf{P} + \mathbf{P}^{(2)} + \cdots + \mathbf{P}^{(n)}$$

or, as shown below.

> **Result** The expected number of visits to state j in n epochs is (in matrix form)
>
> $$\mathbf{E}^{(n)} = \sum_{k=0}^{n} \mathbf{P}^{(n)}$$

Here it must be understood that $\mathbf{P}^{(0)}$ is the identity matrix. The identity matrix at the beginning of this series takes care of the difference in the cases of $i = j$ and $i \neq j$ at the initial epoch.

EXAMPLE 3.2

To answer the question we posed for the truck problem, the expected number of visits to Detroit in eight trips starting from Atlanta would be (The individual values in the sum were taken from the (1, 4) elements of the matrix powers shown earlier.)

$$e_{14}^{(8)} = 0.3 + 0.19 + 0.213 + 0.207 + 0.209 + 0.208 + 0.208 + 0.208$$
$$e_{14}^{(8)} = 1.744$$

By the way, when we add these probabilities, we are *not* computing another probability. We would have to have mutually exclusive events to add probabilities without correcting for the intersections, and these events are certainly not mutually exclusive. What we are doing in this case is adding *expectations* (which happen to equal the transition probabilities).

3.6 Sojourn Times

When a process has virtual transitions (represented by loops in the transition diagram), one can ask the question, "How long does the process stay in a state?" Staying in a state means looping so that the value of the state does not change even though the time index has advanced. Our truck example has a loop at state 1 (Atlanta), so we could ask, "How many local trips will take place in the Atlanta area before there is an out-of-town trip?" The answer, of course, will be a random variable, so we should answer with a probability distribution rather than a single number.

Definition A *sojourn time* is a random variable N_i that counts the number of consecutive epochs that the stochastic process is in state i, given that it starts there. We count the initial epoch, when it first enters the state, so the sojourn time is always at least 1.

It is easy to see what the distribution of N_i is if you focus on the loop transition and lump all others together. The only way the process can remain in the state i for k steps is to take the loop transition $k - 1$ times, followed by any one of the nonloop transitions. Since, by the Markov property, these are all independent events, we can multiply the probabilities to get the result shown below.

Result The distribution of the sojourn time N_i is

$$P(N_i = k) = (p_{ii})^{k-1}(1 - p_{ii})$$

This is a geometric distribution over the values k = 1, 2, 3, ..., with the loop value p_{ii} as the probability of failure. You can think of it as "failure to escape the state." The mean sojourn time—simply the expectation of the geometric random variable N_i—is shown below.

> **Result** The mean sojourn time in state i is
> $$E(N_i) = \frac{1}{1 - p_{ii}}$$

The fact that sojourn times have geometric distributions is always true for Markov chains; it is a fact derived from the Markov and stationarity assumptions. Sometimes, however, one might like to model a process in which the sojourn times follow some other distribution. For example, empirical data might suggest that a negative binomial distribution would fit the situation better. An ordinary Markov chain cannot accommodate that variation, but there is a way to model that situation. A more advanced class of stochastic processes called semi-Markov processes allow arbitrary sojourn time distributions.

3.7 First Passage and Return Probabilities

The computations treated so far answer questions of the general form, "What is the probability of being in a certain state at a certain time?" or "How many times will a certain state be visited?" Another type of question that is frequently of interest is, "How long will it take to reach a certain state?" The answer must involve probabilities, but the random variable is the number of transitions that occur before a specified state is reached rather than the state after a specified number of transitions.

> **Definition** The *first passage time* from state i to state j, denoted N_{ij}, is the (random variable) number of time steps necessary to reach state j for the first time, starting from state i. The *first return time* for any state i, denoted N_{ii}, is the number of time steps necessary to return to state i for the first time, starting from i.

The notation is similar to that used for sojourn times, but should offer no confusion because sojourn times have only one subscript. Although the words are different, the concept of first passage or first return is much the same, and the mathematics is identical, so we use the same letter N for both. You can tell which words to use by whether or not the two subscripts are the same.

It is possible that the distribution of N_{ij} will be defective (meaning that the probabilities sum to something less than 1). That will happen if there is some chance that j will never be reached, which can happen in certain kinds of Markov chains. All that follows in this section will be correct even if the distribution is defective.

One might be tempted to interpret $p_{ij}^{(n)}$ as the probability that n steps are required to reach state j given that the process starts in state i. If this were a correct interpretation, then the set $\{p_{ij}^{(1)}, p_{ij}^{(2)}, p_{ij}^{(3)}, \ldots\}$ would give the probability distribution of N_{ij}. However, this is not a correct interpretation of $p_{ij}^{(n)}$. To see what is wrong, consider $p_{ij}^{(3)}$ and suppose the event to which this probability refers does indeed happen. That is, suppose the process is in state j three steps after it was in state i. But this does not mean that you have waited three steps for state j to be reached. It might have been reached after only one or two steps, after which the process either stayed in state j or changed to another state and then returned to state j. Any of the possibilities could result in the event indicated by $X_3 = j$.

When we speak of the number of steps required to reach state j, we really mean the number of steps required to reach state j *for the first time*. So to get the distribution of this time, we must define a new term.

Definition The *first passage (return) probability*, denoted by $f_{ij}^{(n)}$, is the conditional probability

$$f_{ij}^{(n)} = P\{X_n = j, X_{n-1} \neq j, X_{n-2} \neq j, \ldots, X_1 \neq j | X_0 = i\}$$

The matrix of all $f_{ij}^{(n)}$ is $\mathbf{F}^{(n)}$.

In words, $f_{ij}^{(n)}$ is the probability that the process is in state j at time n and not before, given that it was in state i at time 0. This may be correctly reinterpreted as the probability that n steps are required to reach state j for the first time given that the process starts in state i, or one term from the distribution of N_{ij}. That is,

$$P(N_{ij} = n) = f_{ij}^{(n)}.$$

The whole set for $n = 1, 2, 3, \ldots$, to infinity gives the entire distribution of N_{ij}.

We now know what it means, but have not yet found a way to calculate $f_{ij}^{(n)}$. Clearly, $f_{ij}^{(1)} = p_{ij}$. By extending the logic of the explanation of $p_{ij}^{(n)}$, one can show the result shown below.

Result First passage time probabilities are determined by the formula

$$f_{ij}^{(n)} = p_{ij}^{(n)} - \sum_{k=1}^{n-1} f_{ij}^{(k)} p_{jj}^{(n-k)}$$

You can understand the logic of this expression if you observe that the first term on the right accumulates the probabilities of all walks of length n from i to j. From this you must take away the probabilities for the walks that arrive at j before time n, specifically at times 1, 2, 3, $\ldots$, $n - 1$. A single term in the sum of the form $f_{ij}^{(k)} p_{jj}^{(n-k)}$ gives the sum of all walk probabilities of length n that reach state j for the first time at epoch k and then complete the remaining $n - k$ steps in such a way that the walk ends up at state j.

Thus the $f_{ij}^{(n)}$ can be obtained iteratively if the $p_{ij}^{(n)}$ are known. There is no quick and easy matrix method for generating the $f_{ij}^{(n)}$; they must be generated one by one. For example, to obtain $f_{12}^{(4)}$, the formula would be

$$f_{12}^{(4)} = p_{12}^{(4)} - f_{12}^{(1)} p_{22}^{(3)} - f_{12}^{(2)} p_{22}^{(2)} - f_{12}^{(3)} p_{22}^{(1)}$$

which requires all of the intermediate $f_{12}^{(1)}$, $f_{12}^{(2)}$, and $f_{12}^{(3)}$, as well as the $p_{22}^{(1)}$, $p_{22}^{(2)}$, and $p_{22}^{(3)}$.

The above discussion tacitly assumed that i and j were distinct. If they are not, the formal definition and results would be exactly the same, but we would speak of first return rather than first passage.

EXAMPLE 3.3

For purposes of displaying some numerical results, here are a few of the $\mathbf{F}^{(n)}$ for the truck example.

$$\mathbf{F}^{(2)} = \begin{bmatrix} 0.220 & 0.210 & 0.250 & 0.190 \\ 0.133 & 0.300 & 0.233 & 0.167 \\ 0.187 & 0.220 & 0.207 & 0.207 \\ 0.213 & 0.140 & 0.193 & 0.273 \end{bmatrix} \quad \mathbf{F}^{(3)} = \begin{bmatrix} 0.160 & 0.129 & 0.153 & 0.131 \\ 0.123 & 0.190 & 0.148 & 0.132 \\ 0.144 & 0.158 & 0.135 & 0.154 \\ 0.128 & 0.130 & 0.143 & 0.187 \end{bmatrix}$$

$$\mathbf{F}^{(4)} = \begin{bmatrix} 0.122 & 0.099 & 0.103 & 0.099 \\ 0.092 & 0.139 & 0.099 & 0.095 \\ 0.110 & 0.115 & 0.089 & 0.114 \\ 0.111 & 0.089 & 0.090 & 0.141 \end{bmatrix} \quad \mathbf{F}^{(5)} = \begin{bmatrix} 0.093 & 0.071 & 0.067 & 0.073 \\ 0.076 & 0.101 & 0.064 & 0.071 \\ 0.084 & 0.084 & 0.058 & 0.084 \\ 0.080 & 0.066 & 0.060 & 0.103 \end{bmatrix}$$

These values are provided so that the reader can check his or her understanding of the formulas. Each individual value must be computed separately, as mentioned above; there are no easy shortcuts or matrix methods.

It would be very useful to have an easy way to calculate the full set of first passage probabilities, for then we could fully describe the passage time distributions. We could get the mean, the variance, and higher moments; we could find the probability of requiring longer than any given time or less than any given time to reach a state; we could even find out which of several states was most likely to occur first. Unfortunately, even if a spreadsheet is used, it will be tedious to enter the formulas, so these calculations are rarely used in practical applications. If you really want them, you would be advised to write a program to compute them. On the other hand, the formal definitions are commonly used in proving theorems related to, for example, mean first passage times. There are easy ways to obtain some of those results without having to calculate the first passage times explicitly.

We mentioned on page 83 that the set of $f_{ij}^{(n)}$ for all n might be a defective distribution. That will happen when it is less than certain that j will ever be reached (or returned to) starting from state i. It will be helpful later (when we discuss complications) to have notation for the probability of ever returning to a state.

Definition The *recurrence probability* for a state i, denoted f_{ii}, is the conditional probability that the state is ever visited again, given that the process starts from that state.

We can now express a way to calculate the recurrence probability, shown below.

Result The recurrence probability for state i is

$$f_{ii} = \sum_{n=1}^{\infty} f_{ii}^{(n)}$$

In words, the recurrence probability is equal to the sum of first return probabilities taken over all time steps from 1 to infinity. The reason we can sum these probabilities is that they refer to mutually exclusive events; the first return, if it occurs at all, will have to be at epoch 1 or 2 or 3 or . . . et cetera.

Once again, this last result is more for theoretical use (that is, in proving theorems) than for practical computation. Except in very special cases, it will not be practical to find the full distribution of first passage or return probabilities.

Before going on, be certain you are clear on the distinction between transition probabilities, $p_{ij}^{(n)}$, and first passage/return probabilities, $f_{ij}^{(n)}$. Both are probabilities and both refer to the same states and epochs in the same process, but they are answering different questions. For $p_{ij}^{(n)}$, we fix i and n as given values and ask about the probability of j. For $f_{ij}^{(n)}$, we fix i and j and ask about the probability of n. The former is uncertain about the state; the latter is uncertain about the time.

3.8 Computational Formulas for All Markov Chains

Name	Notation	Formula	Matrix Form
Walk Probability	$p_{j_0 j_1 j_2, \ldots, j_n}$	$p_{j_0 j_1} p_{j_1 j_2} p_{j_2 j_3}, \ldots, p_{j_{n-1} j_n}$	none
Transition Probability	$p_{ij}^{(n)}$	$p_{ij}^{(n)} = \sum_k p_{ik}^{(n-1)} p_{kj}$	$\mathbf{P}^n = \mathbf{P}^{n-1}\mathbf{P}$
State Probability	$p_j^{(n)}$	$p_j^{(n)} = \sum_i p_i^{(0)} p_{ij}^{(n)}$	$\mathbf{p}^{(n)} = \mathbf{p}^{(0)}\mathbf{P}^{(n)}$
Expected Number of Visits in n Epochs	$e_{ij}^{(n)}$	$e_{ij}^{(n)} = \sum_{k=1}^{n} p_{ij}^{(k)}$ for $i \neq 0$ $e_{ii}^{(n)} = 1 + \sum_{k=1}^{n} p_{ii}^{(k)}$	$\mathbf{E}^{(n)} = \sum_{k=0}^{n} \mathbf{P}^{(n)}$
First Passage (or Return) Probability	$f_{ij}^{(n)}$	$f_{ij}^{(n)} = p_{ij}^{(n)} - \sum_{k=1}^{n-1} f_{ij}^{(k)} p_{jj}^{(n-k)}$	none

3.9 Special Classes of Markov Chains

Everything about Markov chains up to this point would apply to any Markov chain. From this point on, however, we need to impose some extra conditions. The remaining calculations will work only if the Markov chain has certain properties. If you try to perform the calculations on a Markov chain of the wrong kind, you will fail. That is, you will not be able to get an answer, because the equations are inconsistent or perhaps have an infinite number of solutions. There is less danger here than there was in the formulation stage, because the conditions will enforce themselves. That is, if you make the mistake of trying to compute something when these new conditions are not satisfied, you will simply be unable to get an answer. That is safer, in some sense, than the danger of misformulation. If you set up a Markov chain model without realizing that the Markov or stationarity assumptions are wrong, you might be able to get numerical results that might even appear to be reasonable, but could be very misleading.

Instead of interrupting the presentation of computational results to delve into the complications, we will defer the details and simply mention that we will have to return to them. Barring these complications, most Markov chains commonly fall into one of two classes, called *ergodic* and *terminating*. (We will define these precisely later.) The truck example that we have been using is of the ergodic variety. The college progression example in Chapter 2 illustrates the terminating type. In simplest terms, ergodic Markov chains continue forever; terminating ones eventually end. Both types are very useful in applications, but they differ in how they behave over the long term. Consequently, the computational methods for limiting results (as time goes to infinity) differ significantly.

3.10 Steady-State Probabilities

The most common type of Markov chain model in practical applications is of the ergodic category. The word "ergodic" means that the results eventually become independent of initial conditions. The truck example is ergodic, so we can use it to examine what happens as time increases.

For small values of n, we find the probability of being in any state from a transition probability that depends, of course, upon the initial state. As more and more time accumulates, it seems logical to expect that the significance of the initial state diminishes. For example, after a hundred trips of the truck, it would not seem to matter where the truck started out. What this means mathematically is that, for large values of n, the rows of $\mathbf{P}^{(n)}$ should be identical. This, in fact, is exactly what happens. By the time $\mathbf{P}$ has been raised to the 34th power, we have

$$\mathbf{P}^{(34)} = \begin{bmatrix} 0.2083 & 0.2188 & 0.3646 & 0.2083 \\ 0.2083 & 0.2188 & 0.3646 & 0.2083 \\ 0.2083 & 0.2188 & 0.3646 & 0.2083 \\ 0.2083 & 0.2188 & 0.3646 & 0.2083 \end{bmatrix}$$

That is, every row is the same to four significant digits. That means that the probability of finding the truck in any particular city is the same regardless of which of the cities was the starting point. This should not be surprising; it is only saying that information loses value as it grows old. Notice, though, that the columns are different; that is, we cannot say that all cities become equally likely. One direct interpretation of this particular result is that the truck is more likely to be found in Chicago (state 3) than any of the other cities after the process has run for some time. Somehow, the transition mechanism "favors" trips to Chicago. Those different values that appear in the different columns are the steady-state probabilities, and they tell you a lot about the long-term behavior of the system you are modeling.

We might also expect that when you are predicting far into the future, it should not make much difference which particular value of n is specified. Mathematically, $\mathbf{P}^{(n)}$ and $\mathbf{P}^{(n+1)}$ are essentially the same for large n, or the transition matrix approaches stable limits as n grows large. For the truck problem, $\mathbf{P}^{(35)}$ is identical to $\mathbf{P}^{(34)}$ to four decimal places, and subsequent powers continue to give the same results. (There will be small differences in the fifth digit, but these will diminish as the powers grow, and so on.) Obviously, it is unnecessary to continue calculating once you reach the stage that the values have stabilized.

Another way to understand what happens is to visualize a graph of probability values over time, plotted for some particular transition pair. That is, plot $p_{ij}^{(n)}$ for fixed i and j, and $n = 0, 1, 2, 3, \ldots$. You will get a graph something like Figure 3.2. (This particular plot graphs the values of $p_{12}^{(n)}$ for the truck problem; you can find those values on page 80.) Typically, the value will jump around somewhat for small values of n (always between 0 and 1, of course, because they are probabilities), but eventually settle down to an asymptotic value. The same kind of behavior occurs for each of the other transition probability functions, though the specific values will of course be different.

After this informal discussion, let us move on to precise definitions and methods.

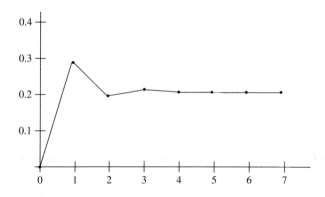

■ FIGURE 3.2 A particular transition probability

> **Definition** The *steady-state probabilities*, or *steady-state distribution*, for a Markov chain are the set of limit values
>
> $$\pi_j = \lim_{n \to \infty} p_{ij}^{(n)}$$
>
> The *steady-state vector* is the row vector $\boldsymbol{\pi} = [\pi_1 \ \pi_2 \ \ldots]$.

This notation already makes use of the fact that the limits are independent of the initial state i. That is, i does not appear in the notation π_j because it does not matter what i is. Then our comments above suggest that as n grows large,

$$\mathbf{P}^{(n)} \to \begin{bmatrix} \pi_1 & \pi_2 & \pi_3 & \cdots \\ \pi_1 & \pi_2 & \pi_3 & \cdots \\ \cdots & \cdots & \cdots & \cdots \\ \pi_1 & \pi_2 & \pi_3 & \cdots \end{bmatrix}$$

In other words, each row becomes the steady-state vector. As long as the process is ergodic, it can be proved that such limits exist and behave as we have suggested.

One obvious way to find the steady-state probabilities is to raise the transition matrix to higher and higher powers until the rows become equal to as many significant digits as you care about. But this method would be tedious, particularly when **P** is large. Fortunately we can obtain equations to determine them directly, as follows:

1. Write the Chapman-Kolmogorov equation: $\mathbf{P}^{(n)} = \mathbf{P}^{(n-1)}\mathbf{P}$
2. Take the limit on both sides: $\lim_{n \to \infty} \mathbf{P}^{(n)} = \lim_{n \to \infty} \mathbf{P}^{(n-1)}\mathbf{P}$.
3. Substitute the definition:

$$\begin{bmatrix} \pi_1 & \pi_2 & \pi_3 & \cdots & \cdots \\ \pi_1 & \pi_2 & \pi_3 & \cdots & \cdots \\ \pi_1 & \pi_2 & \pi_3 & \cdots & \cdots \\ \cdots & \cdots & \cdots & \cdots & \cdots \\ \pi_1 & \pi_2 & \pi_3 & \cdots & \cdots \end{bmatrix} = \begin{bmatrix} \pi_1 & \pi_2 & \pi_3 & \cdots & \cdots \\ \pi_1 & \pi_2 & \pi_3 & \cdots & \cdots \\ \pi_1 & \pi_2 & \pi_3 & \cdots & \cdots \\ \cdots & \cdots & \cdots & \cdots & \cdots \\ \pi_1 & \pi_2 & \pi_3 & \cdots & \cdots \end{bmatrix} \mathbf{P}$$

This represents many replications of the same set of linear equations

$$\boldsymbol{\pi} = \boldsymbol{\pi}\mathbf{P}$$

3.10 Steady-State Probabilities

or, if we want to express this equation in terms of the more conventional column vectors

$$\boldsymbol{\pi}^T = \mathbf{P}^T \boldsymbol{\pi}^T$$

This is what the individual equations look like in symbolic form:

$$\pi_1 = p_{11}\pi_1 + p_{21}\pi_2 + p_{31}\pi_3 + \ldots$$
$$\pi_2 = p_{12}\pi_1 + p_{22}\pi_2 + p_{32}\pi_3 + \ldots$$
$$\pi_3 = p_{13}\pi_1 + p_{23}\pi_2 + p_{33}\pi_3 + \ldots$$
$$\ldots$$

The π_j's are the unknowns and the transition matrix gives the known coefficients, but notice that the coefficient matrix gets transposed. You should practice a bit, but once you get used to the form, you will be able to write out these equations quickly.

It turns out that this set of linear equations, though it has as many equations as unknowns, is a dependent set and therefore possesses an infinite number of solutions. (The dependency derives from the fact that every row of $\mathbf{P}$ sums to 1.) Only one of the infinite number of solutions, however, will qualify as a probability distribution. This one solution can be "forced" by requiring that the π_j sum to 1. When we append that linear equation to those previously expressed, the resulting set of linear equations will possess a unique solution satisfying all the requirements of a probability distribution.

> **Definition** The *normalizing equation* is a linear equation the requires the steady-state probabilities to sum to one:
>
> $$\sum_{\text{all } j} \pi_j = 1 \quad \text{or} \quad \boldsymbol{\pi}\mathbf{1} = 1$$

The usual practice when solving by hand is first to obtain an unnormalized solution by manipulating the $\boldsymbol{\pi} = \boldsymbol{\pi}\mathbf{P}$ equations to express all of the π_j in terms of one of them; second, to use the normalizing equation to fix the value of this last one; and, third, to substitute this value into the expressions for the others.

> **Result** The steady-state probabilities are uniquely determined by the dependent set of linear equations
>
> $$\boldsymbol{\pi}^T = \mathbf{P}^T \boldsymbol{\pi}^T$$
>
> (from which any one equation can be deleted) and the additional normalizing equation
>
> $$\boldsymbol{\pi}\mathbf{1} = 1$$

As long as the process is ergodic, you can be certain that these equations have a unique solution that satisfies all of the probability requirements.

EXAMPLE 3.4

Referring once again to our truck example, the steady-state equations for that particular case would be given by

$$\pi_1 = 0.1\pi_1 + 0.333\pi_2 + 0.2\pi_3 + 0.2\pi_4$$
$$\pi_2 = 0.3\pi_1 + 0.0\pi_2 + 0.2\pi_3 + 0.4\pi_4$$
$$\pi_3 = 0.3\pi_1 + 0.333\pi_2 + 0.4\pi_3 + 0.4\pi_4$$
$$\pi_4 = 0.3\pi_1 + 0.333\pi_2 + 0.2 + 0.0\pi_4$$

Solving these by simple substitution methods will show the necessity of using the additional normalization equation, which in this case is

$$\pi_1 + \pi_2 + \pi_3 + \pi_4 = 1.$$

You may substitute this equation for any of the others to get four independent equations in four unknowns. The solution is

$$\pi_1 = 0.2083 \qquad \pi_2 = 0.2188$$
$$\pi_3 = 0.3646 \qquad \pi_4 = 0.2083$$

which, of course, matches the result we got by raising **P** to the 34th power. Notice that the ease with which these exact steady-state probability values were obtained using the linear equations contrasts rather sharply with the effort required to raise (by hand calculation) the transition matrix to higher and higher powers.

Computer methods are a great convenience, but you should solve a few small problems the old-fashioned way (with pencil and paper) just to be sure you understand the set-up procedure. Once you are confident with that, you will certainly prefer to resort to computer methods for larger problems (provided they are finite—you may have to use hand-solution methods when there are an infinite number of states). There are many programs for solving linear equations, any of which would do. Modern spreadsheets contain a matrix inversion function, which is enough to do the job. Some care is required, however, to arrange the calculations properly because the normalizing equation has to substitute for one of the equations from the transition matrix. You will have to manipulate the transition matrix through several steps before it is ready for inversion.

Specifically, you need to get into the form $Ax = b$ so that you can solve using the matrix inverse function for $x = A^{-1}b$. But you are starting from something in a different form and also have to deal with the dependency. To show you what that might look like in Microsoft Excel, start with the transition matrix in its natural form. In another space of the same size, input the matrix function "TRANSPOSE(P)" using the special key combination for matrix operations. That will give you the coefficients of the dependent equations in the right order. Then you have to subtract "1" from each diagonal element, which is equivalent to getting all of the unknowns on the same side of the equations. Next you have to replace one of the rows (say, the last one) with a row of "1"s to substitute the coefficients of the normalizing equation. Finally, you invert that matrix and read the last column for the results (which is equivalent to multiplying the inverted matrix by a column vector with 0s in all but the bottom element, which has a 1). It looks something like Figure 3.3 (spread out more than it has to be, just for clarity).

If the algebra seems a bit awkward, there are ways to relate all of this math directly to the transition diagram. To begin, we can visualize the correct steady-state equations using a kind of physical analog. The trick is to think of the points as small reservoirs and the arcs as connecting pipes through which liquid can flow, with valves to ensure that the flow goes only in the direction of the arrows. The probability p_{ij} associated with any arc is to be thought of as the fraction of the liquid in reservoir i that will pass to reservoir j in one transition time unit. One unit of liquid is poured into the system according to the initial distribution. If one particular state is known to be the initial state, all of the liquid is poured in at that point. If, for example, $p_1^{(0)} = 1/4$, then 1/4 of the liquid is poured in at point 1, and so on for the other points. After awhile, a dynamic equilibrium is attained; the liquid continues to flow, but the amount in every reservoir remains level. When this happens, the amount in each reservoir gives the steady-state

3.10 Steady-State Probabilities

	A	B	C	D	E
1	original P	0.100	0.300	0.300	0.300
2		0.333	0.000	0.333	0.333
3		0.200	0.200	0.400	0.200
4		0.200	0.400	0.400	0.000
5					
6	transpose of P	0.100	0.333	0.200	0.200
7		0.300	0.000	0.200	0.400
8		0.300	0.333	0.400	0.400
9		0.300	0.333	0.200	0.000
10					
11	adjust diag.	−0.900	0.333	0.200	0.200
12		0.300	−1.000	0.200	0.400
13		0.300	0.333	−0.600	0.400
14		0.300	0.333	0.200	−1.000
15					
16	replace row	−0.900	0.333	0.200	0.200
17		0.300	−1.000	0.200	0.400
18		0.300	0.333	−0.600	0.400
19		1	1	1	1
20					
21	invert	−0.903	−0.087	0.017	0.208
22		0.052	−0.716	0.143	0.219
23		0.087	0.056	−1.001	0.365
24		0.764	0.747	0.851	0.208

■ **FIGURE 3.3** Spreadsheet solution of steady-state probabilities

probability for the corresponding state. They are proper probabilities because they are nonnegative and sum to 1.

The analogy can be made more exact by thinking of the liquid in terms of its molecules. The trajectory of an individual molecule describes a realization of the stochastic process. The effect of pouring in many molecules is to consider many realizations simultaneously. Hence, we are using, in effect, a statistical mechanics approach. This technique is used in chemical diffusion models, in electronics, and elsewhere.

Now when is flow equilibrium reached? Certainly a necessary condition is that the flow rate into any reservoir must equal the flow rate out, because if the two were not equal, the amount in the reservoir would be changing. If this condition is met for *all* reservoirs, it is sufficient. For any point j, the flow rate out is

$$\sum_k \pi_j p_{jk} = \pi_j \sum_k p_{jk} = \pi_j \quad \left(\text{since } \sum_k p_{jk} = 1\right)$$

Another way to see that this relation is true is to realize that all of the liquid in reservoir j must flow out each time unit, so the total flow out per unit time is just π_j. During the same time period, the flow rate in from other reservoirs is

$$\sum_k \pi_k p_{kj}$$

Hence, in equilibrium,

$$\pi_j = \sum_k \pi_k p_{kj}$$

for all j; or, in matrix form

$$\pi = \pi P$$

If you are starting with the transition diagram, an easy way to write down the steady-state equations is to look at each point one by one and apply the flow balance principle to write down one equation for each point. For example, for point 1 of the example, we would write down π_1 as the flow rate out, then set that equal to the flow rate in from each of the points, giving

$$\pi_1 = 0.1\pi_1 + 0.333\pi_2 + 0.2\pi_3 + 0.2\pi_4$$

which matches the first equation shown earlier. You then move on to the second point, use the same logic to write the second equation, and so forth. You will still need to remember to use the normalizing equation, which does not correspond to anything visible in the transition diagram.

Although it is not at all obvious, there is actually a way to use the transition diagram to obtain the steady-state probabilities directly without writing out the equations. If you know how to do it, you can look at the diagram and write down the answers. The method uses graph theory instead of linear algebra. In fact, there are similar graph theoretic methods for virtually all of the computations involved in Markov chains. Although they are quite handy to know, they are unconventional—indeed, unique to this book—so you should learn the standard algebraic methods first. The graph theoretic techniques are explained in Chapter 8.

3.11   The Uses of Steady-State Results

Before learning more about interpretation, the reader should be warned about a common confusion. The term *steady-state* is used somewhat differently in the context of stochastic processes from the way it is used in deterministic mathematical models. For many models or mechanical or electrical systems, steady-state refers to a condition of unchanging output that occurs after the effect of some initial disturbance or impulse wears off. That is, after some initial period of variation, some physical measurement of the system becomes constant. In the case of these stochastic models, the systems do not converge to some stable configuration, but continue to change state in accordance with the random mechanism. The term steady-state really refers to what happens to probabilities, not to the process itself.

Another possibly confusing factor is that different authors use different terms for the same thing. Steady-state probabilities are sometimes called the stationary distribution, limit probabilities, or equilibrium probabilities. None of these terms is really ideal, so we will stick with the most commonly used term, steady-state.

The steady-state probabilities have several useful interpretations. If you fix a point in time in the distant future, π_j is the probability that you will find the process in state j at that time. This is the most obvious interpretation. For example, in the truck problem, if the question were, "What is the probability that the truck will end its next trip at Atlanta?" when it has been quite some time since you had any information about where the truck is, the answer would be given by π_1.

A steady-state probability can also be viewed as a time average; π_j would be the fraction of time, over the long run, that the process spends in state j. Remember that time is measured discretely, so we are really talking about the fraction of epochs or transitions. For the truck example, if you wanted to know what fraction of the trips, over the long run, terminate in Atlanta, the answer would again be π_1.

A third interpretation is as an ensemble average; if you ran many identical processes simultaneously, π_j would be the fraction of processes that you would find in state j (after a long period of time). For example, if you had many trucks operating simultaneously, all according to the same transition probabilities, π_1 could be interpreted as the fraction of all those trucks that can be expected to be involved in trips to Atlanta at any given time.

Finally, a steady-state probability can be viewed as the reciprocal of the mean number of transitions between recurrences of the state. That is, the average time between the recurrences of the event that the process is in state j is $1/\pi_j$. So the question, "How many trips, on the average, will occur between visits to Atlanta?" would be answered by $1/\pi_1$. We will have more to say about recurrence times shortly.

The full set of steady-state probabilities is a complete probability distribution, so you can get the mean, the variance, or any other quantity associated with distributions. In the truck example, the numbers of the states have no particular meaning except as a way to label cities, but in many cases they measure something. For example, the state might be the level of inventory, or a cost, or a size. In such cases, the steady-state distribution would provide a complete description of how that quantity behaves over the long run, after the influence of initial conditions diminish. For many engineering and business decisions, the long run is what really matters, so the steady-state probabilities provide essential information.

3.12 Mean First Passage Times

Another commonly required quantity that is easily computed when the model is ergodic is the mean first passage time from one state to another, denoted m_{ij}. It has already been shown that it is possible to define and calculate the full probability distribution of the random variable N_{ij}, which is the number of transitions to reach j for the first time starting from state i—that is, the $f_{ij}^{(n)}$—but the calculations are tedious and rarely used. If all that is desired is the *mean* first passage times, a rather simple calculation will suffice.

> **Definition** The *mean first passage* time from state i to state j (where $i \neq j$), denoted m_{ij}, is the expected value of the first passage time random variable N_{ij}.

Formally, we can express m_{ij} using the distribution of N_{ij} and the definition of a mean,

$$m_{ij} = \sum_{k=1}^{\infty} k f_{ij}^{(k)}$$

but that will not be of much help because the $f_{ij}^{(n)}$ probabilities are too difficult to obtain. Fortunately, there is a much better way.

We may obtain a formula for m_{ij} by conditioning on the state at step 1. Given that the process is in state i at time 0, either the next state is j, in which case the passage time is exactly 1, or it is some other state k, in which case the passage time will be 1 plus the passage time from that state k to j. Weighting the possibilities by the appropriate probabilities gives

$$m_{ij} = 1 p_{ij} + \sum_{k \neq j} (1 + m_{kj}) p_{ik}$$

which then manipulates

$$m_{ij} = p_{ij} + \sum_{k \neq j} p_{ik} + \sum_{k \neq j} p_{ik} m_{kj} \quad \text{(multiplying and separating the sum)}$$

$$m_{ij} = \sum_{\text{all } k} p_{ik} + \sum_{k \neq j} p_{ik} m_{kj} \quad \text{(collapsing the first two terms)}.$$

Finally, since the first sum equals 1, we obtain the formula below.

> **Result** A particular mean first passage time m_{ij} satisfies
> $$m_{ij} = 1 + \sum_{k \neq j} p_{ik} m_{kj}$$

This equation expresses the desired m_{ij} as a linear function of the m_{kj} (itself and other mean first passage times with the same second subscript but different first subscripts). By using the same relation for other m_{ij}'s, a full set of independent linear equations may be expressed. The solution to the set gives the mean first passage times from any state into state j.

EXAMPLE 3.5

For the truck problem, the equations we would need to solve to find the mean number of trips to reach state 2 (Baltimore) from state 1 (Atlanta) would be

$$m_{12} = 1 + p_{11}m_{12} + p_{13}m_{32} + p_{14}m_{42}$$
$$m_{32} = 1 + p_{31}m_{12} + p_{33}m_{32} + p_{34}m_{42}$$
$$m_{42} = 1 + p_{41}m_{12} + p_{43}m_{32} + p_{44}m_{42}.$$

Substituting the numerical values for the transition probabilities,

$$m_{12} = 1 + 0.1m_{12} + 0.3m_{32} + 0.3m_{42}$$
$$m_{32} = 1 + 0.2m_{12} + 0.4m_{32} + 0.2m_{42}$$
$$m_{42} = 1 + 0.2m_{12} + 0.4m_{32} + 0.0m_{42},$$

which solves by easy substitution to give $m_{12} = 3.51$, $m_{32} = 3.93$, and $m_{42} = 3.27$.

Notice that, even though there are four states in the example, there are only three equations in three unknowns. In particular, there is no equation for m_{22}. The definition for mean first passage time did not allow for the case of i = j, so we do not yet even recognize such a term. However, it does make perfect sense to speak of a mean recurrence time.

> **Definition** The *mean recurrence time* for a state i, denoted m_{ii}, is the expectation of the random variable N_{ii} (the first return time).

This is defined, and can therefore be obtained, just like the mean first passage times. It is really only the words that change. Conditioning on the state after one transition, we can develop the equation below.

> **Result** A particular mean recurrence time m_{ii} satisfies
> $$m_{ii} = 1 + \sum_{k \neq i} p_{ik} m_{ki}$$

This really looks just like the mean first passage time equations, except that it expresses the mean recurrence time of state i in terms of the mean first passage times from other initial states to state i. This is an "extra" equation, in the sense that it is not needed to determine the mean first passage times. But after having determined those, it tells us one way to get the mean recurrence time. For our example, and using the values $m_{12} = 3.51$, $m_{32} = 3.93$, and $m_{42} = 3.27$, which we obtained earlier, we would get $m_{22} = 4.57$.

The equations for the mean first passage and recurrence times contain a sum that looks almost like matrix multiplication. The only problem is the sum runs over all k except for one term. If we are willing to modify our matrix, we can express the equations and their solution in matrix form.

> **Notation** For any matrix $\mathbf{A}$, let $\mathbf{A}_j^*$ represent the same matrix with the entries in column j replaced by zeros.

Applying this modification to the transition matrix $\mathbf{P}$ and defining, we may express the mean first passage and recurrence time equations all at once with the matrix form

$$\mathbf{m}_j = \mathbf{1} + \mathbf{P}_j^* \mathbf{m}_j$$

where $\mathbf{m}_j$ is a column vector of the mean times to reach state j and $\mathbf{1}$ is a column vector of 1's. That is,

$$\mathbf{m}_j = \begin{bmatrix} m_{1j} \\ m_{2j} \\ m_{2j} \\ \ldots \\ \ldots \end{bmatrix} \text{ and } \mathbf{1} = \begin{bmatrix} 1 \\ 1 \\ 1 \\ \ldots \\ \ldots \end{bmatrix}$$

From that expression, we can manipulate using matrix operation to obtain the solution below.

> **Result** The mean first passage times to state j and mean recurrence time to j are given by
>
> $$\mathbf{m}_j = (\mathbf{I} - \mathbf{P}_j^*)^{-1} \mathbf{1}$$

(Recall that multiplying any matrix by $\mathbf{1}$ on the right has the effect of adding across rows, so you can skip the multiplication if you know that.) In words, to generate all of the mean first passage times to a state j, substitute a column of zeros for the *j*th column of $\mathbf{P}$; then subtract that matrix from an identity matrix of the same dimension and invert the result. Finally, add the elements in each row. The resulting column vector is $\mathbf{m}_j$, the vector of mean first passage times from each state to state j. This method will give you the mean recurrence time as part of the same calculation.

If you need only mean recurrence times, and have no particular interest in the mean first passage times, you may find it easier to calculate them as reciprocals of steady-state probabilities. This is shown in symbols below.

> **Result** Any mean recurrence time m_{ii} in an ergodic Markov chain satisfies
>
> $$m_{ii} = 1/\pi_i$$

Although you could prove this result by working from one set of equations to the other, it will make complete sense to you if you just think about what the terms mean. If the fraction of time the process spends in state i over the long run is π_i, then one would expect an average of $1/\pi_i$ transitions between visits to i. In the case of the example, π_2 was found to be 0.219—meaning that the truck makes a little more than 20 percent of its trips to Baltimore. Thus it must also be true that 1/0.219 or 4.57 is the average number of trips between visits to Baltimore.

3.13 Computational Formulas for Ergodic Markov Chains

Name	Notation	Formula	Matrix Form
Steady-State Probability	π_j	$\pi_j = \sum_k \pi_k p_{kj}$ $\sum_j \pi_j = 1$	$\boldsymbol{\pi} = \boldsymbol{\pi}\mathbf{P}$ $\boldsymbol{\pi}\mathbf{1} = 1$
Mean First Passage Time	m_{ij}	$m_{ij} = 1 + \sum_{k \neq j} p_{ik} m_{kj}$	$\mathbf{m}_j = (\mathbf{I} - \mathbf{P}_j^*)\mathbf{1}$
Mean Recurrence Time	m_{ii}	$m_{ii} = \dfrac{1}{\pi_i}$	none

3.14 Terminating Markov Chains

Processes that end after some period of time are most conveniently modeled as terminating chains. Sometimes it is useful to model continuing processes as terminating Markov chains. For example, one might be interested in which of two states is entered first and would therefore artificially make them both absorbing states. More commonly, the absorbing states represent natural terminating conditions. There are many opportunities to use such models. Any process representing the progression through a series of stages, such as students moving through grades or employees moving through job categories or any growth or aging process, suggest applications where a terminating processes model may be useful.

EXAMPLE 3.6

For one specific example, let us imagine a production process having the structure indicated in Figure 3.4. This example is similar, but just a bit smaller, than the college progression example that was shown in Chapter 2.

Notice that we have added loops to states 4 and 5 to indicate that, once the process reaches either of these states, it remains there. One might first imagine that these loops are not needed because the process simply stops as soon as it reaches one of those states. However, in order to preserve the desirable property that each row of the transition matrix sums to 1, we must include those loops.

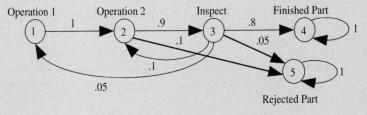

■ FIGURE 3.4   A terminating process

3.14 Terminating Markov Chains

> **Definition** An *absorbing state* j is any state for which $p_{jj} = 1$.

Absorbing states represent the end conditions for the Markov chain. They are recognizable in the transition diagram by the loops with weight 1 (and no other outgoing arcs) and in the transition matrix by 1s on the diagonal (and 0s everywhere else in the same row). The terminology makes sense because once the process enters an absorbing state, it cannot escape to any other state.

> **Definition** A *transient state* j is any state for which the recurrence probability f_{jj} is less than 1.

With a less-than-certain recurrence probability, a transient state might be visited a number of times. Eventually, however, a time is reached after which the transient state is never visited again. States 1, 2, and 3 in Figure 3.4 are transient states. Usually you can tell without any calculation whether a state is transient; you just think about whether it is going to continue to be visited forever or not.

> **Definition** A *terminating* or *absorbing* Markov chain is one that contains only transient and absorbing states.

There are Markov chains that contain states that are neither transient nor absorbing, so these terms are not quite opposites of one another. But this is a complication that does not have to be addressed until later. Within a terminating process, every state must be one of these two types.

Because any terminating Markov chain has a mix of absorbing states and transient states, it will make things easier to sort them into the two types and rearrange the transition matrix into the following form. (A terminating process can always be arranged in this manner, though it may require ordering states in a way that is different from the original or natural order.)

$$\mathbf{P} = \begin{bmatrix} p_{11} & p_{12} & \cdots & p_{1K} & p_{1K+1} & \cdots & \cdots \\ p_{21} & p_{22} & \cdots & & p_{2K+1} & \cdots & \cdots \\ \cdots & \cdots & \cdots & & \cdots & \cdots & \cdots \\ p_{K1} & p_{K2} & \cdots & p_{KK} & p_{KK+1} & \cdots & \cdots \\ 0 & \cdots & & 0 & 1 & 0 & \cdots \\ 0 & \cdots & & 0 & 0 & 1 & \\ 0 & \cdots & & 0 & \cdots & & \cdots \end{bmatrix}$$

The lower right portion of the **P** matrix will be an identity matrix of dimension equal to the number of absorbing states. The lower left will be all zeros. There are no constraints on the elements in the first rows (the transition probabilities out of the transient states), other than that each row sum to 1. However, it is useful to partition the matrix and name the portions for later reference.

$$\mathbf{P} = \begin{bmatrix} [\mathbf{Q}] & [\mathbf{R}] \\ [\mathbf{0}] & [\mathbf{I}] \end{bmatrix}$$

The square matrix **Q** contains the probabilities for the transitions among the transient states. The submatrix **R**, which usually will be rectangular rather than square, contains the probabilities for the transitions from transient to absorbing states. We will also use K to represent the number of transient states and speak of them as the first K states.

EXAMPLE 3.7

In the case of our production system example, we have already numbered the states appropriately, so the transition matrix is in the order necessary for partitioning. In general, you may have to rearrange the order of states to put the matrix in this form:

$$P = \begin{bmatrix} 0 & 1 & 0 & 0 & 0 \\ 0 & 0 & 0.9 & 0 & 0.1 \\ 0.05 & 0.10 & 0 & 0.80 & 0.05 \\ 0 & 0 & 0 & 1 & 0 \\ 0 & 0 & 0 & 0 & 1 \end{bmatrix}$$

Hence, the designated submatrices for this problem would be

$$Q = \begin{bmatrix} 0 & 1 & 0 \\ 0 & 0 & 0.9 \\ 0.05 & 0.10 & 0 \end{bmatrix}$$

$$R = \begin{bmatrix} 0 & 0 \\ 0 & 0.1 \\ 0.80 & 0.05 \end{bmatrix}$$

$$\mathbf{0} = \begin{bmatrix} 0 & 0 & 0 \\ 0 & 0 & 0 \end{bmatrix}$$

$$I = \begin{bmatrix} 1 & 0 \\ 0 & 1 \end{bmatrix}$$

Notice that **Q** and **I** are square matrices and the others are not. We will make use of these particular matrices to illustrate several computations that follow.

3.15   Expected Number of Visits

The first questions we address have to do with how long the process stays among the transient states. Suppose that you want to know how many times a particular transient state will be visited before absorption occurs. This would, of course, be a random variable that depended on the initial state. There is a probability distribution associated with the random variable, but we will be content here to find the expectation.

> **Definition** The *mean number of times a transient state* j *is visited* given that the initial state is i, denoted e_{ij}, is the expectation of the (random variable) number of times that j is visited prior to termination in some absorbing state. The matrix of all e_{ij} is demoted **E**.

The definition requires that j be transient because the question does not make any sense if it is absorbing. Also, if i were absorbing, the possibility of ever visiting any transient state would be zero. So, for sensible results, both i and j must be transient. Since i and j each range over the transient states, **E** is a $(K \times K)$ square matrix.

This notation is similar, and in fact related to, the notation we used earlier for the expected number of visits in the first n epochs. That earlier use was for an arbitrary Markov chain, including both ergodic and terminating cases. However, we had to specify how many time units had passed because the answer would be infinite in some cases if we did not. For a terminating process, we can be sure that a transient state will be visited only a finite number of times, so we can be sure that the mean will be finite. In fact, we could say

$$e_{ij} = \lim_{n \to \infty} e_{ij}^{(n)}$$

which is true, but that would not be very useful for computing.

To develop an expression for the e_{ij}, we may use the same logic already used several times—that of conditioning on the state at time 1. There are two cases. If $i \neq j$,

$$e_{ij} = \sum_{k=1}^{K} p_{ik} e_{kj}$$

because j will be occupied a mean of e_{kj} times if the first transition is to any transient state k, and 0 times if the first transition is to any of the absorbing states. If $i = j$, the equation is slightly different:

$$e_{ii} = 1 + \sum_{k=1}^{K} p_{ik} e_{ki}$$

because i will be occupied once (it spends the first epoch in i) plus a mean of e_{ki} times if the first transition is to any transient state k. Putting these relations into matrix form,

$$\mathbf{E} = \mathbf{I} + \mathbf{QE},$$

and solving for the matrix $\mathbf{E}$,

$$\mathbf{E} - \mathbf{QE} = \mathbf{I}$$

$$(\mathbf{I} - \mathbf{Q})\mathbf{E} = \mathbf{I}$$

Finally, using matrix inversion and the fact that multiplying by the identity matrix changes nothing, we obtain the result below.

Result The expected number of times a transient site is visited is given by (in matrix form)

$$\mathbf{E} = (\mathbf{I} - \mathbf{Q})^{-1}$$

It can be shown that the matrix $(\mathbf{I} - \mathbf{Q})$ is always nonsingular, so the inverse always exists. For finite problems of reasonable size, a spreadsheet can easily handle the inversion.

EXAMPLE 3.8

For the production example, the matrix $\mathbf{I} - \mathbf{Q}$ is

$$\mathbf{I} - \mathbf{Q} = \begin{bmatrix} 1 & -1 & 0 \\ 0 & 1 & -0.9 \\ -0.05 & -0.10 & 1 \end{bmatrix},$$

and its inverse is

$$(\mathbf{I} - \mathbf{Q})^{-1} = \begin{bmatrix} 1.052 & 1.156 & 1.040 \\ 0.052 & 1.156 & 1.040 \\ 0.058 & 0.173 & 1.156 \end{bmatrix}.$$

In words, then, the (i, j)-th entry of the matrix $(\mathbf{I} - \mathbf{Q})^{-1}$ gives the mean number of times that the state j will be visited before absorption occurs, given that the process starts in state i. For our example, if the process begins in state 1, it will visit state 2 an average of 1.156 times before absorption occurs. Or, to say it in terms more directly related to the application, a product just entering the production process can expect to visit the second stage of production an average of 1.156 times.

3.16 Expected Duration of a Terminating Process

The previous section asked and answered a question about how many times a particular transient state is visited. A related question asks about how long a terminating process stays among the transient states, which could also be phrased as, "How long will the terminating process last?" since the process essentially ends as soon as any absorbing state is reached. A full answer would provide a probability distribution for a random variable, but we will be content with the expectation.

> **Definition** The *expected duration* of a terminating process that begins in transient state i, denoted d_i, is the expected number of transitions until the first visit to any absorbing state.

The expected duration is easy to calculate once you realize that you can get it by adding up all the expected number of times that each transient state is visited. That is, the process will visit state 1 some number of times, state 2 some number of times, and so forth. Without knowing or having to know anything about the order in which states are visited, we can still say that the expected total number of visits to all transient states is the sum of the individual state visits. Hence, we find the result below.

> **Result** The expected duration of a terminating process that starts in transient state i is given by
> $$d_i = \sum_{k=1}^{K} e_{ik}$$

This is just the sum of elements across row i of **E**. In the numerical example, if the process begins in state 1, it makes an average of 3.249 transitions before being absorbed in one of the two absorbing states. Or, in production terms, a job will take an average of 3.249 operations to reach completion.

3.17 Absorption Probabilities

The next question we address is *which* absorbing state will be entered. Of course, if a process has only one, the answer is trivial. If there is more than one possibility, the question is answered by a probability distribution over all of the possible absorbing states. For our production example, if chance determines that state 4 is entered, then state 5 will never occur, and vice versa. What we need is a method to determine the probability that state 4 is *ever* entered (as opposed to state 5, since one of the two must eventually occur).

> **Definition** An *absorption probability* for an absorbing state j, given an initial transient state i, denoted a_{ij}, is the probability that state j is ever visited. The matrix of all a_{ij} is denoted by **A**.

Absorption probabilities, like steady-state probabilities, are limit values of transition probabilities as n goes to infinity. You can, in fact, read them from the appropriate cells in a sufficiently high power of **P**. However, unlike steady-state probabilities in an ergodic chain, they *do* depend upon the initial state. Intuitively, one can well imagine that a product that has already reached the last stage of production may have a better chance of finishing as a good product than one which is just starting out. In addition, the definition makes sense only if i is transient and j is absorbing, so A will be a rectangular matrix with as many rows as there are transient states and as many columns as there are absorbing states.

3.17 Absorption Probabilities

There are some other interpretations of the absorption probabilities that are useful. If you were to run the process many times, or were to run many copies of the same process in parallel, the absorption probabilities would indicate the fraction of them that would terminate in one final state versus any other. For example, we could interpret a_{14} as the fraction of parts that survive the production process to emerge as good finished products versus being scrapped somewhere along the way.

We are now looking for a probability, rather than an expected value of a random variable. Nevertheless, we can still exploit the same idea of conditioning on the state at epoch 1 to get an equation relating a_{ij} to other absorption probabilities. There are two immediate possibilities leading to the ultimate occurrence of state j. Either the first transition is to state j (immediate absorption), or the first transition is to some *transient* state k and the process ultimately enters state j from that state. (There is, of course, another possibility for the first transition, namely that it is to an absorbing state other than j, but if that happens, j will never occur.) The probability of the first event is just p_{ij}, of the second is

$$\sum_{k=1}^{K} p_{ik} a_{kj}$$

where the sum is over all transient states k. Since the two possibilities are mutually exclusive, we find the result below.

Result A particular absorption probability a_{ij} satisfies the equation

$$a_{ij} = p_{ij} + \sum_{k=1}^{K} p_{ik} a_{kj}$$

This is one linear equation in several unknown absorption probabilities. A complete set of independent linear equations will be obtained if the same formula is reapplied to all possible transient states, regarding each in turn as the initial state.

EXAMPLE 3.9

By way of example, if we want a_{14} for the process illustrated in Figure 3.4, our first equation would be

$$a_{14} = p_{14} + p_{11}a_{14} + p_{12}a_{24} + p_{13}a_{34}$$

This is an equation in three unknowns, a_{14}, a_{24}, and a_{34}. If we now write the equations for a_{24} and a_{34}

$$a_{24} = p_{24} + p_{21}a_{14} + p_{22}a_{24} + p_{23}a_{34}$$
$$a_{34} = p_{34} + p_{31}a_{14} + p_{32}a_{24} + p_{33}a_{34}$$

we get two additional equations in the same three unknowns. Substituting numerical values, our equations are

$$a_{14} = 0 + 0a_{14} + 1a_{24} + 0a_{34}$$
$$a_{24} = 0 + 0a_{14} + 0a_{24} + 0.9a_{34}$$
$$a_{34} = 0.8 + 0.05a_{14} + 0.1a_{24} + 0a_{34}$$

Solving these, we obtain

$$a_{14} = 0.832$$
$$a_{24} = 0.832$$
$$a_{34} = 0.925$$

Notice that even if we have no direct interest in a_{24} or a_{34}, it was necessary to obtain them in order to find a_{14}. Notice also that a_{14}, a_{24}, and a_{34} do not sum to 1; there is no reason that they should. On the other hand, if the initial state is fixed, then the process must ultimately enter one of the two absorbing states, so each of the following would hold

$$a_{14} + a_{15} = 1$$
$$a_{24} + a_{25} = 1$$
$$a_{34} + a_{35} = 1$$

From this, we can deduce that

$$a_{15} = 0.168$$
$$a_{25} = 0.168$$
$$a_{35} = 0.075$$

Alternatively, we could have used the formula to get three linear equations in a_{15}, a_{25}, and a_{35} and solved for them directly.

As an alternative method, it is convenient to express the equations for the absorption probabilities in matrix form and to solve for all of them at once. Recall that **A** is the matrix of the a_{ij}. **A** will not necessarily be square, but will have as many rows as there are transient states, and as many columns as there are absorbing states. Thus, it has the same dimension as **R** does. In fact, examination of the equations for the a_{ij} reveals that

$$\mathbf{A} = \mathbf{R} + \mathbf{QA}$$

Manipulation of this matrix equation gives **A** in terms of **R** and **Q**, which are just portions of the transition matrix **P**:

$$\mathbf{A} - \mathbf{QA} = \mathbf{R}$$
$$(\mathbf{I} - \mathbf{Q})\mathbf{A} = \mathbf{R}$$

Finally, inversion gives us the result below.

Result The full matrix of absorption probabilities A is given by

$$\mathbf{A} = (\mathbf{I} - \mathbf{Q})^{-1}\mathbf{R}$$

If we have already calculated $(\mathbf{I} - \mathbf{Q})^{-1}$ in order to find **E**, then only one more matrix multiplication is necessary to find **A**, as shown below.

Result The full matrix of absorption probabilities A is given by

$$\mathbf{A} = \mathbf{ER}$$

In the case of the numerical example, the matrix of absorption probabilities is

$$\mathbf{A} = (\mathbf{I} - \mathbf{Q})^{-1}\mathbf{R} = \begin{bmatrix} 0.832 & 0.168 \\ 0.832 & 0.168 \\ 0.925 & 0.075 \end{bmatrix}$$

which agrees with our previous calculation.

3.18 Hit Probabilities

Sometimes, depending upon the structure, a terminating process will have transient states that might never occur. For example, in a model of a production process, the item being worked on might be scrapped before ever reaching the final stages. Or, in a model of a student progressing through college, the student might drop out before ever becoming a senior. In such cases, it can be useful to compute the probability that a particular transient state is ever visited.

Definition The *hit probability* for a transient state j, denoted f_{ij}, is the conditional probability that j is ever visited, given that the initial state is i. The matrix of all hit probabilities is denoted by **F**.

Recall that in the section on classification of states we used the notation f_{jj} as the probability of eventual recurrence of a state j. This reuse of the same notation is entirely consistent with that. The probability of recurrence is just a special case. Again the definition makes sense only if i is transient.

Formally, it is true that

$$f_{ij} = \sum_{n=1}^{\infty} f_{ij}^{(n)}$$

but that is not a useful expression for calculation purposes.

To develop an equation for f_{ij}, we employ the now-familiar device of conditioning on the state at time 1. On the first transition, we either visit state j, which would mean that we definitely reached j, or we visit some other transient state k, from which the probability of ever reaching j is f_{kj}, or we pass to an absorbing state, which would mean that we never reach j. Weighting these conditional probabilities by the respective probabilities that the various states occur, we get the result below.

> **Result** A particular hit probability f_{ij} satisfies
> $$f_{ij} = p_{ij} + \sum_{\text{transient } k \neq j} p_{ik} f_{kj}$$

This same equation holds true even when $i = j$. If you construct a full set of such equations for each possible initial state in the transient set, you will have a set of independent linear equations, the solution to which gives the f_{ij}. If you want to find similar results for a different j, you will have to construct a different set of equations.

The matrix form of these equations requires a modification of $\mathbf{Q}$, and we must express one column at a time. The column vector $\mathbf{f}_j$ will contain the hit probabilities for all of the possible transient initial states and for reaching state j. Let $\mathbf{Q}_j^*$ be the matrix $\mathbf{Q}$ with a column of zeros replacing the original jth column. That adjustment is necessary to account for the fact that $k = j$ is skipped in the sum. Then

$$\mathbf{f}_j = \mathbf{q}_j + \mathbf{Q}_j^* \mathbf{f}_j$$

When we solve this by matrix manipulation, we get the solution below.

> **Result** A vector of hit probabilities (for different initial states) is given by
> $$\mathbf{f}_j = (\mathbf{I} - \mathbf{Q}_j^*)^{-1} \mathbf{q}_j$$

You can be certain that these equations will have a unique solution, and that it will satisfy all of the requirements to be interpreted as hit probabilities. If you want hit probabilities for a different target state j, you will need to solve a different set each time.

3.19 Conditional Mean First Passage Times to Absorbing States

We have already dealt with the calculation of the total number of transitions that a terminating process lasts, namely d_i. It is quite a different question if one specifies a particular absorbing state and asks how many transitions will occur before it is entered. This question would be answered by something like a mean first passage time to a particular absorbing state. In fact, if there is only one absorbing state, so that ultimate passage to that state is certain, then an ordinary mean first passage time will do. In other cases, when there is some possibility that passage to the specified final state will never occur, the ordinary mean first passage time would be infinite. In these cases, what is really desired is the *conditional* mean first passage time, given that passage does occur. In other words, we have to adjust for the fact that the probability of *ever* reaching the state (given by a_{ij}) is less than 1.

We have previously used the notation m_{ij} for the mean first passage time from i to j in an ergodic process. We will use almost the same notation for the *conditional* mean first passage time in a terminating process, only adding a parenthetical lowercase superscript "c" (for conditional) to remind you of the distinction. When you see this, remember that the superscript is just part of the notion; it does not signify a power or a time epoch.

> **Definition** The *conditional mean first passage time* to an absorbing state j from an initial transient state i, denoted $m_{ij}^{(c)}$ is the expected value of the number of epochs required to reach state j for the first time, given that that event does occur.

Then the result below it can be shown (by arguments that are too complex for this presentation).

> **Result** A particular conditional mean first passage time $m_{ij}^{(c)}$ satisfies
> $$a_{ij}m_{ij}^{(c)} = a_{ij} + \sum_{k=1}^{K} p_{ik}a_{kj}m_{kj}^{(c)}$$

Here the sum is over all transient states. Assuming that the absorption probabilities a_{kj} are already known, the above equation is linear in the unknowns $m_{kj}^{(c)}$. By writing a complete set, one such equation for each transient state i, a sufficient number of independent equations will be available to determine the $m_{kj}^{(c)}$.

EXAMPLE 3.10

To illustrate, suppose that we want to know the mean number of transitions that would occur in the process we used to illustrate absorption probabilities, if we know that it started in state 1 and ultimately reached state 4. We want $m_{14}^{(c)}$, so the associated equation would be,

$$a_{14}m_{14}^{(c)} = a_{14} + p_{11}a_{14}m_{14}^{(c)} + p_{12}a_{24}m_{24}^{(c)} + p_{13}a_{34}m_{34}^{(c)}$$

Since this equation has the three unknowns $m_{14}^{(c)}$, $m_{24}^{(c)}$, and $m_{34}^{(c)}$, we need two other equations. Writing the one for $m_{24}^{(c)}$ and the one for $m_{34}^{(c)}$, we get

$$a_{24}m_{24}^{(c)} = a_{24} + p_{21}a_{14}m_{14}^{(c)} + p_{22}a_{24}m_{24}^{(c)} + p_{23}a_{34}m_{34}^{(c)}$$
$$a_{34}m_{34}^{(c)} = a_{34} + p_{31}a_{14}m_{14}^{(c)} + p_{32}a_{24}m_{24}^{(c)} + p_{33}a_{34}m_{34}^{(c)}$$

Substituting the values of the p_{ij} and the a_{ij}, which were previously calculated, we have the three equations

$$0.832m_{14}^{(c)} = 0.832 + (0)(0.832)m_{14}^{(c)} + (1)(0.832)m_{24}^{(c)}$$
$$+ (0)(0.925)m_{34}^{(c)}$$

$$0.832m_{24}^{(c)} = 0.832 + (0)(0.832)m_{14}^{(c)} + (0)(0.832)m_{24}^{(c)}$$
$$+ (0.9)(0.925)m_{34}^{(c)}$$

$$0.925m_{34}^{(c)} = 0.925 + (0.05)(0.832)m_{14}^{(c)} + (0.1)$$
$$\times (0.832)m_{24}^{(c)} + (0)(0.925)m_{34}^{(c)}$$

These can be solved to yield

$$m_{14}^{(c)} = 3.365 \quad m_{24}^{(c)} = 2.365 \quad m_{34}^{(c)} = 1.364$$

To answer our original question, then, it will take a mean of 3.365 transitions to get from state 1 to state 4, assuming that state 4 does occur. In terms of the original application, products that make it through the production process as good parts take an average of 3.365 steps to do so.

By the way, if a Markov chain has only one absorbing state, then it is certain to occur, so $a_{kj} = 1$ for all k and they all just drop out of the equation. In that case, the equation reduces to the one derived earlier for ergodic processes. Furthermore, when the Markov chain has only one absorbing state, we can even express the entire distribution of first passage time probabilities to the absorbing state j. The reason we can do so is that we know $p_{jj}^{(n)} = 1$ for all numbers of time steps, so the equation for the first passage time probabilities simplifies considerably. In fact, the cumulative distribution function of the passage time (which would also be the duration of the process) is just $p_{ij}^{(n)}$ for any initial state i. You can, of course, obtain that result mathematically, or you can just recognize that if the process is in the only absorbing state at epoch n, it must have arrived in n or less steps.

3.20 Computational Formulas for Terminating Processes

Name	Notation	Formula	Matrix Form
Expected Number of Visits (to transient state j over the full duration of the process)	e_{ij}	$e_{ii} = \sum_{k=1}^{K} p_{ik} e_{kj}$ $e_{ij} = 1 + \sum_{k=1}^{K} p_{ik} e_{ki}$	$\mathbf{E} = (\mathbf{I} - \mathbf{Q})^{-1}$
Expected Duration	d_i	$d_i = \sum_{k=1}^{K} e_{ik}$	Sum across ith row of $\mathbf{E}$
Absorption Probability	a_{ij}	$a_{ij} = p_{ij} + \sum_{k=1}^{K} p_{ik} a_{kj}$	$\mathbf{A} = \mathbf{E}\mathbf{R}$
Hit Probability	f_{ij}	$f_{ij} = p_{ij} + \sum_{\text{transient } k \neq j} p_{ik} f_{kj}$	$\mathbf{f}_j = (\mathbf{I} - \mathbf{Q}_j^*)^{-1} \mathbf{q}_j$
Conditional Mean First Passage Time to an Absorbing State j	$m_{ij}^{(c)}$	$a_{ij} m_{ij}^{(c)} = a_{ij} + \sum_{k=1}^{K} p_{ik} a_{kj} m_{kj}^{(c)}$	none

3.21 Call Center Calculations

EXAMPLE 3.11

In Chapter 2, Section 2.8, we gave careful attention to the formulation of a Markov chain model to assist in the evaluation of a proposed call center. We developed a terminating model with seven states, two of which were absorbing. The transition matrix was

$$\mathbf{P} = \begin{bmatrix} 0 & 0.6 & 0.2 & 0.2 & 0 & 0 & 0 \\ 0 & 0 & 0 & 0.04 & 0.16 & 0.8 & 0 \\ 0 & 0.1 & 0 & 0 & 0.4 & 0.5 & 0 \\ 0 & 0.3 & 0 & 0 & 0.1 & 0.4 & 0.2 \\ 0 & 0 & 0 & 0 & 0 & 0.75 & 0.25 \\ 0 & 0 & 0 & 0 & 0 & 1 & 0 \\ 0 & 0 & 0 & 0 & 0 & 0 & 1 \end{bmatrix}$$

where the states were as follows: (1) Initial, (2) Sales, (3) Billing, (4) Support, (5) Supervisor, (6) Satisfied, and (7) Unresolved.

Now that we have computational formulas, let us explore how we would answer some specific questions.

1. What percentage of customers will be satisfied?
2. Of all the calls coming in, what fraction will have to be dealt with by a supervisor?
3. Between billing or support, which will receive more calls?
4. How many people, on average, will a customer speak to?

In general, you cannot expect a question to be phrased in a way that translates directly to the terminology of Markov chain theory. That is, you will have to think clearly to identify what you have to calculate to answer the question. That, in turn, will require a clear understanding of the exact meanings of terms such as "hit probability" and "absorption probability." Many of them look similar, but of course they produce very different results. Obviously, you cannot answer the question correctly if you calculate the wrong thing, even if you set up the equations and solve them perfectly.

Here are some hints to help you keep the notation straight as you seek a match between a question and a computable term. First, the letter "i" in a subscript consistently refers to initial state. It has to be there, but it does not offer any clue to what you are calculating, so just ignore it when you are looking for a match. Second, you should distinguish easily between something that is a probability and something that is a time or count. If a question is of the form, "What fraction ... ?" or "What percentage ... ?" it might be answered by one of the probability terms, but if it takes the form, "How long ... ?" or "How many times ... ?" you would want to calculate a mean value. Those basic differences should narrow your search immediately. Third, you have to think about how the question translates to particular states and the classification of those states. For example, if the question relates to the probability that a state is ever visited and that state is transient, it would involve a hit probability, whereas if the state is absorbing, it would be an absorption probability.

Now let us apply that kind of thinking to the questions posed. The first question asks for the percentage (think probability) of satisfied (think state 6) customers. Since state 6 is an absorbing state, you want to compute an absorption probability, specifically, a_{16}. The model represents what happens to an individual customer, but if you imagine many realizations, the probability can be interpreted as the fraction of those realizations that will terminate in state 6, or the fraction of satisfied customers. Just multiply by 100 to translate the fraction to a percentage after you find the value of a_{16}.

The second question, "Of all the calls coming in, what fraction will have to be dealt with by a supervisor?" is quite similar, except that it asks about state 5, which is a transient state. Hence, it relates to a hit probability. Out of many realizations, the fraction that contain a visit to state 5 will correspond to the probability that state 5 is ever visited, or f_{15}.

The third question, "Between billing or support, which will receive more calls?" is a little more indirect. Two transient states, 3 and 4, are mentioned. One might consider comparing hit probabilities for the two states, reasoning that the one with the higher value will be visited more times. However, that logic is a little imprecise, because the hit probability involves only the first visit to a state. A better approach is to compare the expected number of visits, or e_{ij}'s, for the two states.

The fourth question, "How many people, on average, will a customer speak to?" translates roughly to how many states will be visited before the process terminates, or d_1. One should be careful, however, about what is counted. In this case, the initial state in our model uses an automated answering machine, not a person. What the question actually asks for is the expected total number of visits to states 2, 3, 4, and 5, which we would calculate as $e_{12} + e_{13} + e_{14} + e_{15}$.

Actually, that answer interprets the question in such a way that each time a person answers adds to the count. If the question really meant to count the number of *distinct* people spoken to (not counting return visits), we might want the expected number of *first* visits to the appropriate states. We can also answer that question, but it takes more advanced thinking. Define, for each state, a Bernoulli random variable X_{ij} as 1 if the state j is visited (starting from initial state i) and 0 if it is not. That random variable "counts" the first visit, so in this case we would want the expected value of the sum $X_{12} + X_{13} + X_{14} + X_{15}$. But the expected value of a Bernoulli random variable is just the probability of "success," which in this case is the same as the hit probability. Hence the expected number of distinct people spoken to would be $f_{12} + f_{13} + f_{14} + f_{15}$. The point of this last example was to emphasize that you have to be very clear about what the question really means. Often verbal statements are somewhat ambiguous, but the mathematical definitions are not. When you address a question that might have multiple meanings, you should take pains to restate the question more precisely.

If any of these explanations seem unclear, be sure to work out the logic in your own mind. One of the problems asks you to complete the calculations, and the final answers are provided in the back of the book.

3.22 Classification Terminology

We now return to previous deferred discussion of the complications that could interfere with routine calculations. Many students have difficulty with this section, so let's take a moment to explain why it is necessary. The first calculations we have seen will work for any Markov chain, regardless of any special properties or structure. You may use those methods without fear of anything going wrong. Then we considered ergodic and terminating processes, which are defined by certain requirements. Those calculations depend upon things being "right" or "normal." In other words, you might run into trouble if you try to perform the calculations for a Markov chain that fails to meet certain requirements. For example, you might set up some linear equations, in complete accordance with the formulas, but then discover that there is no solution or an infinite number of solutions. We need to identify those "abnormal" situations, establish terminology to separate the workable cases from the unworkable ones, and make sure we can tell the difference. Some of these abnormal cases are rare in practical situations, but others are common. These complications should not alarm you. When we eliminate the exceptional cases, we will be left with only the two common classes of Markov chains, for which clean solutions will always be obtainable.

EXAMPLE 3.12

We will start with an easy case. Suppose we had a transition matrix like this:

$$\mathbf{P} = \begin{bmatrix} 0.5 & 0.0 & 0.3 & 0.0 & 0.2 \\ 0.0 & 0.3 & 0.0 & 0.7 & 0.0 \\ 0.4 & 0.0 & 0.2 & 0.0 & 0.4 \\ 0.0 & 0.6 & 0.0 & 0.4 & 0.0 \\ 0.7 & 0.0 & 0.2 & 0.0 & 0.1 \end{bmatrix}$$

Seeing nothing remarkable, we might start raising that matrix to higher powers, which would be a perfectly correct way to obtain transition probabilities. But then we might be surprised to find that the rows do not seem to be converging to the same numbers. For example, the 20th power of this matrix is

$$\mathbf{P}^{20} = \begin{bmatrix} 0.520 & 0.000 & 0.252 & 0.000 & 0.228 \\ 0.000 & 0.462 & 0.000 & 0.538 & 0.000 \\ 0.520 & 0.000 & 0.252 & 0.000 & 0.228 \\ 0.000 & 0.462 & 0.000 & 0.538 & 0.000 \\ 0.520 & 0.000 & 0.252 & 0.000 & 0.228 \end{bmatrix}$$

Even worse, we might assume that we could obtain the steady-state probabilities in the usual way, and then discover that we cannot get a solution. In this case, the transition matrix would provide five equations, but only three independent equations, so the normalizing equation is not enough to compensate.

The entire process is a legitimate Markov chain, but it behaves somewhat abnormally. What is happening? The mystery is unraveled if we look at the transition diagram shown in Figure 3.5. We can see that the Markov chain really consists of two separate pieces. If we start in state 1, 3, or 5, we will never be able to visit state 2 or 4, and vice versa. That may not have been obvious from just a casual glance at the transition matrix. We might be assisted in recognizing the separated structure in the matrix if we rearranged rows and columns to put the states in the order {1, 3, 5, 2, 4}, giving

$$\mathbf{P} = \begin{bmatrix} 0.5 & 0.3 & 0.2 & 0.0 & 0.0 \\ 0.4 & 0.2 & 0.4 & 0.0 & 0.0 \\ 0.7 & 0.2 & 0.1 & 0.0 & 0.0 \\ 0.0 & 0.0 & 0.0 & 0.3 & 0.7 \\ 0.0 & 0.0 & 0.0 & 0.6 & 0.4 \end{bmatrix}$$

Now we can see from the blocks of zeros that there is no link between the first three and last two states.

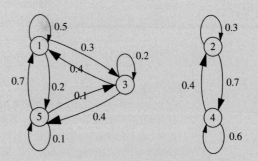

■ **FIGURE 3.5** An abnormal case

What should you do if you encounter a situation like this? Since the two subsets of states really have nothing to do with one another, you can simply break the Markov chain into separate, smaller Markov chains and do whatever analysis you want on each of them independently. The same approach would work even if there were many pieces, as long as they do not connect in any way.

EXAMPLE 3.13

A slightly more difficult situation is reflected by the transition diagram shown in Figure 3.6. and the matrix.

This Markov chain is neither ergodic nor terminating, but has some aspects of both. It has some one-way connections: The process can move from the set {1, 2} to the set {3, 4}, but can never get back. If you look at Figure 3.6 and think about what the process will do, you can see that it may spend some time bouncing back and forth between states 1 and 2 (provided that it starts in one of them), but will eventually take one of the two arcs to the right, after which it will spend the rest of eternity bouncing back and forth between states 3 and 4. If we raise the transition matrix for a process like this to a high power, the probability of finding the process in either state 1 or state 2 would go to zero. States 1 and 2 fit the definition of transient states, because the recurrence probability for each is less than 1. We do not yet have terms to describe states 3 and 4.

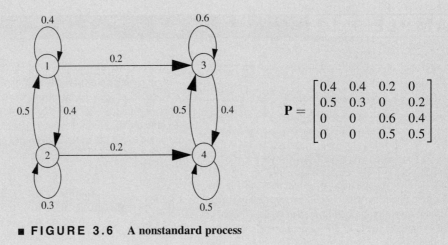

■ **FIGURE 3.6** A nonstandard process

Recall the definition of transient states from page 97: they are states whose recurrence probability is less than 1. The term that describes any nontransient state is recurrent. The formal definition is below.

Definition A *recurrent* state j is any state for which the probability of recurrence f_{jj} is 1.

An individual state must be either *transient* or *recurrent*. The property of being either recurrent or transient is permanent; that is, a state could not be recurrent for some time and then switch to transient or back the other way. Informally, a transient state is one that may occur only a finite number of times and eventually "drops out of" the process. As time goes to infinity, the probability of being in such a state goes to zero. A recurrent state, on the other hand, is one that will continue to occur (at least occasionally) no matter how long the process runs and therefore has some probability of occurring even as time goes to infinity. Notice that the definition does not say that a recurrent probability is certain to occur. The requirement $f_{ii} = 1$ says that it is certain is that state i will occur again *given* that it occurs once. Any absorbing state will automatically be recurrent. But, as the example shows, a state does not have to be an absorbing state to be recurrent.

How does one know the difference? In most cases, the answer will be obvious from the way you formulate the model—that is, what you intend and how you set it up. But if you were handed a model that was formulated by someone else, you should have a way to determine which states are transient and which are recurrent, and that method should cover every conceivable situation. The formal definitions that use the recurrence probability are useful in theorems, but are not very helpful for practical classification because the f_{jj} are too difficult to calculate. In most cases (the only exception being an oddity that can occur only when there are an infinite number of states), we can tell whether states are recurrent or transient by how they are connected to one another. So now we want some terminology to formalize notions of connectedness.

> **Definition** A particular state j is *reachable* from another state i if there exists some sequence of transitions that lead from i to j. (A state is always reachable from itself.)

Less formally, one state j is reachable from another i if you can trace a walk from i to j in the transition diagram. This is a one-way relationship; we also need a word to describe being able to get back again.

> **Definition** Two states *communicate* if each is reachable from the other.

The forward and backward routes do not have to involve the same intermediate states. This is a symmetric relationship between two states. In fact, it is an equivalence relation, because it is reflexive (a state always communicates with itself), symmetric (if i communicates with j then j communicates with i) and transitive (if i communicates with j and j communicates with k then i communicates with k). If we identify all of the states that communicate with one another, we get a set which might be the whole state space or only a subset.

> **Definition** A *communicating class* is a maximal collection of states that all communicate with one another.

The word "maximal" in the definition is important; we must include *all* of the states that communicate with one another. Obviously, any subset of a communicating class will also be a set of states that communicate with one another.

These terms (unlike some that follow) are quite intuitive and are unlikely to cause confusion. In a transition diagram, it is usually easy to see in a glance which states are reachable from one another and therefore to identify the communicating classes. In a very complex diagram, you might have to trace out walks to see which states can be reached, but that would be rare.

A more mechanical way to see which states can be reached that works for finite chains is to raise the matrix to a power equal to one less than the number of states. For example, if there are ten states, raise **P** to the 9th power. Every nonzero term means that a state can be reached in 9 or less transitions. (If it can be reached at all, it can be reached within that many steps. Otherwise, you would have to repeat states on a walk, and would be going around in circles.) Of course, if the process has an infinite number of states, that procedure will not work. But in that case, you will certainly have some repetitive structure that will tell you what you need to know about how states are connected.

The terms transient and recurrent refer to individual states. The terms reachable and communicating refer to pairs of states, or by extension to groups of states. We can now extend the concept of connectedness to an entire Markov chain.

Definition A Markov chain is said to be *irreducible* if all states in the state space S communicate; that is, every state can be reached from every other state, not necessarily in a single transition but through some sequence of allowable transitions. A Markov chain is *reducible* if it is not irreducible.

Another way to phrase it is that the states in an irreducible Markov chain belong to a single communicating class. This property assures that the process cannot get "caught" within a subset of states, as it did in the example in Figure 3.6.

EXAMPLE 3.14

Figure 3.7 illustrates another reducible process, just to illustrate a number of terms simultaneously. Transition probability values are omitted, because the structure alone is sufficient to determine connectedness. In this process, states 1, 3, and 4 communicate (but are not closed); states 6 and 7 communicate and form a closed communicating class; and states 2 and 5 are reachable from states 1, 3, and 4, but not from 6 or 7.

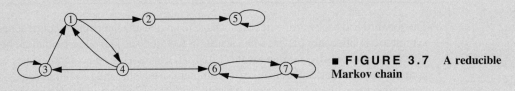

■ **FIGURE 3.7** A reducible Markov chain

You should be able to look at this diagram and conclude that states 1, 2, 3, and 4 are transient, and states 5, 6, and 7 are recurrent. If you cannot, you should review the definitions.

The fact that states communicate, or belong to the same communicating class, is not sufficient to ensure that they are recurrent, as the last example illustrates. (States 1, 3, and 4 communicate, but are transient, because the process could exit the class without a way to return.) So we need to add another restriction before we can conclude anything about whether they are recurrent or transient.

Definition A communicating class of states is *closed* if there are no transitions from any member of the class to any state outside of the class.

A communicating class is a generalization of an absorbing state, in the sense that if the process even enters the set it will be trapped there forever. It may circulate freely among the members of the set (because they all communicate), but it cannot escape from the set. Of course, an absorbing state is a special case of a communicating class.

Finally, we now have enough to relate how the states connect to whether they are transient or recurrent.

Result In a finite Markov chain (that is, one with a finite number of states), all states that are members of closed communicating classes are recurrent; all the others are transient.

This result leads immediately to some other conclusions. An absorbing state, which is a special case of a closed communicating class, is therefore always recurrent. If a process has a finite number of states and is irreducible (so they all communicate), then all states are necessarily recurrent. More generally, a finite Markov chain that is reducible must contain at least one recurrent state (because if all states were transient, the process would run out of places to go), but could contain one or more subsets of recurrent states, as Figure 3.7 does.

When an arbitrary finite Markov chain is given, we can classify states as recurrent or transient by the following method.

1. Starting from any state, determine all of the states that communicate with it. That set will form a communicating class. (If the state does not communicate with any other state, it must be absorbing.)
2. If the communicating class is closed, all of the states in it are recurrent; if not, they are all transient.
3. Take any state that is not yet classified, and repeat steps 1 and 2.
4. Continue until all states are classified.

3.23 Additional Complications in Infinite Chains

The last result in Section 3.22 applies cleanly to finite Markov chains. In a process with an infinite number of states, several oddities can occur. It is possible for all states to be transient.

EXAMPLE 3.15

An example transition diagram showing this possibility is shown in Figure 3.8. In this diagram, we obviously cannot see all of the infinite number of states, but the regular structure enables us to infer all of the properties.

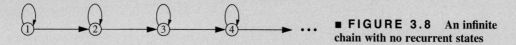

■ **FIGURE 3.8** An infinite chain with no recurrent states

This process may loop around a state a few times, but eventually always moves one state to the right. In infinite time, the state drifts to the right and never runs out of new states to move to. None of the states are recurrent; they are all transient. In fact, you can see that the probability of recurrence, f_{jj}, is less than 1 for every state.

You may have noticed that these states do not communicate (each state is alone in its communicating class) so the process is reducible. One might be tempted to assume that if the process is irreducible (so all the states communicate), then all states would have to be recurrent. We have already said that statement is true for finite processes, but it is not necessarily true when there are an infinite number of states. Depending on the probability values, it is possible for the process to drift off to infinity, so that the recurrence of any state is less than certain.

EXAMPLE 3.16

An example of an infinite process where the structure alone is not enough to indicate whether the states are recurrent or transient is shown in Figure 3.9.

If $p < q$, there is a bias that tends to keep the state number small. Even if the process occasionally grows large (moves far to the right in the diagram), it is certain that it will return to the lower states. That is, $f_{jj} = 1$ for every state. In that case, all states are recurrent. However, if $p > q$, it is possible for the process to move infinitely far to the right and never return to the lower states. In this case, all states are transient. Again, such behavior could occur only in a process with an infinite number of states; when the number of states is finite, an irreducible process will have only recurrent states.

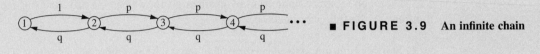

■ **FIGURE 3.9** An infinite chain

The case $p = q$ illustrates another interesting possibility that can occur only in processes with an infinite number of states. It turns out that in this case, all states are recurrent ($f_{jj} = 1$, for all j), but that the mean recurrence time (m_{jj}, the expected time for a state to occur again once it has occurred) is infinite. This apparent paradox is simply one of the bizarre things that can happen when dealing with infinity.

> **Definition** A *null* state is a recurrent state whose mean recurrence time is infinite. A *non-null*, or *positive*, recurrent state is one whose recurrence time is finite.

Although they may seem bizarre, null states do occasionally appear in practical applications. We will see some in Chapter 6 on queueing models. The way they would show up is as follows. You would have an infinite set of steady-state equations, any one of which appears reasonable. Taking advantage of whatever repetitive structure there is in the equations, you solve for all of the probabilities in terms of one unknown. That remaining unknown has to be determined by the normalizing equation. However, when you attempt to find the sum of all steady-state probabilities to set the sum equal to 1, you realize that you have a divergent infinite series. Hence, the only possible value for all of the steady-state probabilities is zero. Intuitively, what happens is that the probability spreads out evenly over the infinite number of states, leaving only an infinitesimal amount for each. Hence as time goes to infinity, the probability of being in any particular state goes to zero.

> **Result** If a state j is null, the limit probability $\lim_{x \to \infty} p_{ij}^{(n)}$ is zero.

This result tells you both that you do not have to bother solving steady-state equations once you have identified that the states are null, and that you may run into difficulties if you attempt to solve them.

3.24   Dealing with a Reducible Process

We know from earlier sections of this chapter how to compute various quantities for ergodic processes (which are necessarily irreducible) and for terminating processes (which are reducible, but have only transient and absorbing states). As we have seen, however, sometimes Markov chains are neither ergodic nor terminating. So what should you do with a reducible model like the one in Figure 3.7? You should reduce it by breaking it apart into two models, as shown in Figure 3.10. That is, we replace each closed communicating class that is not already an absorbing state by a single absorbing state that represents the entire communicating class. That gives us a terminating process (because it contains only transient and absorbing states) which models everything that happens in the original process up until the time it reaches one of the closed communicating classes. Then we separately model each closed communicating

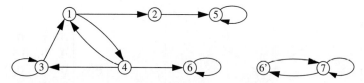

FIGURE 3.10   A reducible Markov chain

class. Each of those will be an ergodic model (provided that certain additional requirements are met, which we have yet to discuss.)

Although the transformation is not precisely equivalent, we can answer any question about the original model by judiciously combining results from the two submodels. For example, the conditional mean first passage time $m_{17}^{(c)}$ in the original model is equal to $m_{16}^{(c)}$ in the terminating submodel plus m_{67} in the ergodic submodel. In general, you use the terminating submodel to analyze what happens until one of the closed sets of states is entered, then use separate ergodic models to analyze what happens subsequently.

3.25   Periodic Chains

We now have all that we need to identify and use terminating processes, but there is one more condition to be met before we can be sure that we have an ergodic Markov chain. We will assume that we have already reduced the process if necessary, so we will be dealing with a single communicating class. All of the states can be reached from one another.

EXAMPLE 3.17

Here is the smallest, simplest example of the next problem you might encounter. Consider a two-state Markov chain that does nothing but alternate between the two states. The transition diagram (Figure 3.11) and matrix would be as shown below:

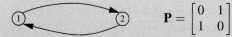 $\mathbf{P} = \begin{bmatrix} 0 & 1 \\ 1 & 0 \end{bmatrix}$

FIGURE 3.11   A two-state chain

If we start taking powers of P to see what happens over time, we quickly find that successive powers simply alternate between

$$\begin{bmatrix} 1 & 0 \\ 0 & 1 \end{bmatrix} \text{ and } \begin{bmatrix} 0 & 1 \\ 1 & 0 \end{bmatrix}$$

Specifically, every even-numbered power gives the matrix on the left, and every odd-numbered power gives the matrix on the right. That behavior is not surprising if you look at the diagram; all the process does is to alternate between the two states. But it presents a dilemma when we ask about steady-state probabilities. The problem is that no ordinary limit exists for $p_{ij}^{(n)}$. No matter how high we increase n, the probability is 0 or 1 depending only on whether n is odd or even.

EXAMPLE 3.18

Here is a second example (Figure 3.12 and matrix) where the same problem appears in a slightly more subtle form.

This time the first few powers of **P** are different from one another, but when you reach $\mathbf{P}^5$, you realize that you have a matrix that is the same as $\mathbf{P}^2$. Then $\mathbf{P}^6$ is the same as $\mathbf{P}^3$, $\mathbf{P}^7$ is

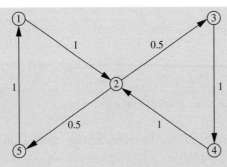

■ **FIGURE 3.12** A five-state chain

the same as $\mathbf{P}^4$, and $\mathbf{P}^8$ is back to the same as $\mathbf{P}^2$. The cycle continues from there, rotating among the same three matrices. Clearly, these values will never reach a limit, at least not in the conventional manner.

In both of these examples, the states all communicate. Both models are finite, so all states are recurrent. The problem is that they are periodic.

> **Definition** A state j is *periodic* if $p_{jj}^{(n)} = 0$ except for values of n that are multiples of some integer r > 1. The *period* is the value of r. A state is *aperiodic* if it is not periodic.

Periodic states can recur only at certain epochs—epochs that are multiples of some integer larger than 1. For example, if a state could recur only on even-numbered epochs, it would be periodic of period two, as in Figure 3.11. Figure 3.12 illustrates a process in which every state is of period three. Notice that the definition does not assert that the state is *certain* to recur every r epochs, but only that it cannot recur *except* at those times. All of the states of a communicating class will share the same periodicity property. That is, if any one is periodic, they all are and must have the same period. Or if any one is aperiodic (not periodic), they all are. The problem with periodic states is that they do not possess limiting probabilities in the normal sense. The probability of being in the state will alternate between zero (at nonmultiples of the period) and positive values. The issue does not appear very often in applications and, even then, is more of a nuisance than a serious obstacle. If you should encounter a problem with periodic states, they are treated fully in W. Feller's *An Introduction to Probability Theory and Its Applications* (full citation, page 115).

How can you tell if a state is periodic? First, separate the state space into communicating subclasses (if the process is reducible), then find out whether any single state in each class is periodic. In visual terms, if all cycles connecting the communicating states in the transition diagram are of lengths that are multiples of some number greater than 1 (such as three in Figure 3.12), then all of the states of that communicating class would be periodic. If there is a loop (a virtual transition), then that state and hence all states communicating must be aperiodic. Or, if you can find two cycles that are relatively prime in length (that is, have no common divisor greater than 1), then all of the states in the communicating class are aperiodic.

3.26 Ergodic Chains

Now that we have words to describe the exceptions, we may consider the classes of Markov chains that are well behaved for purposes of limit calculations. When the process is irreducible, recurrent, positive, and aperiodic, the entire Markov chain and every state in it are called *ergodic*.

> **Definition** An *ergodic* Markov chain is one that is irreducible, and every state is recurrent, nonnull, and aperiodic.

Although the word may seem strange (it comes from deeper theorems than we will cover later), ergodic Markov chains are the "ordinary" kind, and it is the exceptions that are rare. Almost always, when you are modeling a real-life process that does not terminate, the Markov chain will be ergodic. The truck problem is a small, but otherwise typical, example.

How can we be sure that a Markov chain is ergodic? If it is finite, it is sufficient to verify that it is irreducible (every state can be reached from every other) and aperiodic. Usually, that is easy to do. If it is infinite, you also have to establish that the states are recurrent and positive, which could take some work. Fortunately, you can also plunge ahead with any of the calculations you want to perform. If you encounter a problem in obtaining a solution, such as not enough independent equations to determine a unique solution, it is because you have one of complications in your model. If the methods give you a solution, you need not worry about those possible complications.

It is often possible to model the same physical process as either an ergodic or terminating Markov chain, depending on the point of view taken and the questions you want answered. For example, you could look at one of the work stations in a production process and model the randomly varying sequence of parts passing through it. Most likely, you would view this as a continuing process and end up with an ergodic chain. Alternatively, you could focus on a single workpiece, as in the model above, and end up with a terminating chain.

3.27 Recommended Reading

For many years, most of the literature of random processes was so mathematically advanced that a beginner would have difficulty in gaining access to the methods. This problem, however, was remedied in the 1970s; there are now many good sources for the mathematics available (though most are weak on applications). Bhat (2), Cinlar (3), Clark and Disney (4), Cox and Miller (5), Heyman and Sobel (7), Parzen (14), and Ross (15) all offer the basic fundamentals of random processes as used in operations research. The styles of treatment and the depths to which they probe vary considerably; still, you could begin with any of these books. Recently published introductory textbooks include Berger (1), Higgins (8), Kao (10), Kulkarni (12), Minh (13), and Tijms (16).

1. Berger, Marc A., *An Introduction to Probability and Stochastic Processes*. Springer-Verlag, New York, 1993.
2. Bhat, U. N., *Elements of Applied Stochastic Processes*, 3rd ed. Wiley, New York, 2002.
3. Cinlar, E. *Introduction to Stochastic Processes*. Prentice-Hall, Englewood Cliffs, NJ, 1975.
4. Clarke, A. B., and R. Disney, *Probability and Random Processes for Engineers and Scientists*, 2nd ed. Wiley, New York, 1985.
5. Cox, D. R., and H. D. Miller, *Theory of Stochastic Processes*. Wiley, New York, 1965.
6. Feller, W., *An Introduction to Probability Theory and Its Applications*, vol. I, 2nd ed. Wiley, New York, 1957.
7. Heyman, D. P., and M. J. Sobel, *Stochastic Models*. North-Holland, New York, 1990.
8. Higgins, James J., and Sallie Keller-McNulty, *Concepts in Probability and Stochastic Modeling*. Duxbury Press, Belmont, CA, 1995.
9. Howard, R., *Dynamic Probabilistic Systems*, vols. I and II. Wiley, New York, 1971.
10. Kao, Edward, *An Introduction to Stochastic Processes*. Brooks/Cole, New York, 1996.
11. Kemeny, J. G., and L. Snell, *Finite Markov Chains*. Van Nostrand, New York, 1960 (republished by Springer-Verlag, New York, 1976).
12. Kulkarni, V. G., *Modeling, Analysis, Design, and Control of Stochastic Systems*. Springer-Verlag, New York, 1999.

13. Minh, D. L., *Applied Probability Models*. Duxbury, New York, 2001.
14. Parzen, E., *Stochastic Processes*. Holden-Day, New York, 1962.
15. Ross, S., *Introduction to Probability Models*. Academic, New York, 1972.
16. Tijms, H., *A First Course in Stochastic Models*. Wiley, New York, 2003.

Chapter 3 Problems

1. Referring to problem 1 in Chapter 2 (the fishing problem), calculate what you need to answer the following questions.
   a. If the fishing is medium on Monday, what is the probability that it will be medium on Thursday?
   b. If yesterday's fishing was bad, what is the expected number of days of good fishing over the next week (7 days)?
   c. What percentage of days over the long run are good fishing days?
   d. If the fishing is bad today, what is the expected time (in days) until it is good?
   e. If the fishing is bad today, how long (in expected number of days) will it remain bad before it gets better?

2. Referring to problem 2 in Chapter 2 (the elevator problem), use the methods of this chapter to calculate answers to the following questions.
   a. If you arrive at the building and find the elevator on the ground floor, what is the probability that it goes to the second floor and then immediately returns?
   b. If the elevator starts at the ground floor, what is the expected number of times it will go to the top floor in the next five trips?
   c. If you walk into the building with no knowledge of where the elevator is, what is the probability that you will find the elevator at the ground floor (assuming that any observation of the state was long ago).
   d. If it is not at the ground floor, but at the second floor, how many stops will it make (on the average) before coming to the ground floor?

3. Referring to problem 3 in Chapter 2 (the air quality problem), answer the following questions.
   a. If the air quality is currently good (green), what is the probability that it will go from bad to worse (that is, to yellow, then orange, then red) in the next three days?
   b. Over the long run, what fraction of the time will the air quality be in the best (green) category?
   c. If the air quality reaches the worst level (red), what is the expected number of days until it reaches the best level (green)?
   d. If the air quality is now at the best level (green), what is the expected number of consecutive days that it will remain at that level?

4. Referring to problem 4 in Chapter 2 (the wine problem), answer the following questions.
   a. Out of 1,000 bottles of wine sampled randomly from many years, what is the expected number that would be excellent?
   b. If this was a "good" year, how many of the next five years can be expected to be good or excellent?
   c. If this year was a "good" year, what is the expected number of years until the next "excellent" year?
   d. If this year was good and it turns out that next year is bad, what is the probability that the following year will be excellent?
   e. What is the mean number of years between occurrences of "excellent" years?

5. Referring to problem 5 in Chapter 2 (the crop rotation problem), answer the following questions.
   a. What is the amount of land devoted to each category at some distant future time?
   b. If a particular acre chosen at random happens to be currently planted in soybeans, how many of the next five years can that same acre be expected to grow corn?
   c. What will be the acreages devoted to each use two years beyond the most recent data?
   d. If you were to pick out an acre of farmland at random three years from now, what is the probability that it will be fallow?
   e. What is the average number of years between fallow years?

6. Referring to problem 6 in Chapter 2 (the housing market problem), use the methods of this chapter to calculate answers to the following questions.
   a. What is the probability that someone who is now in an apartment will own his or her own home after two moves?
   b. What is the probability that the same person will own his or her own home after two moves but not sooner?
   c. How many moves will occur, on the average, before a homeowner will live in a condominium?
   d. What, if anything, can you say about the long-term demand for condominiums, relative to the other residence types?

7. Referring to problem 7 in Chapter 2 (the robot problem), use the methods of this chapter to calculate answers to the following questions.
   a. If the arm is now turning, what is the most likely action to follow the next action (without specifying the intermediate action)?
   b. If the arm is turning, how many action steps will occur on the average before the grip action is taken?

c. If the most recently observed move was a grip, what is the expected number of times the arm will turn in the next six moves?
d. Over the long run, how does the use of the lift motor compare to the use of the extension motor?

8. Referring to problem 8 in Chapter 2 (the automobile problem), use the methods of this chapter to calculate answers to the following questions.
   a. Selecting a car owner at random, what is the probability that he or she will buy a Chrysler product for either his or her next car or the one after that?
   b. Assuming that the estimated transition probabilities are correct, that they will remain stable, and that Chrysler stays in business, what share of the market in terms of a percentage would you expect Chrysler to achieve in the distant future?
   c. If a buyer chosen at random is currently an owner of a Ford, what is the expected number of consecutive purchases of Fords that he or she will make before switching to another brand?
   d. If a person now last purchased a GM product, how many purchases will occur before he or she buys a Ford?

9. Referring to problem 9 in Chapter 2 (the jumping frog problem), use the methods of this chapter to calculate answers to the following questions.
   a. If the frog starts on pad 2, what is the probability that he is on pad 3 two jumps later?
   b. Interpret the steady-state probabilities in terms of the behavior of the frog.
   c. Explain in terms of the frog, his motivation, his behavior, and so forth, what the Markov and stationarity assumptions mean.

10. Referring to problem 10 in Chapter 2 (the paint mix problem), use the methods of this chapter to calculate answers to the following questions.
    a. Out of a production run of 1,000 batches, what is the expected number of white batches?
    b. If the most recent batch is white, what is the expected number of cleanings (or returns to white) out of the next six batches?
    c. What is the expected number of batches produced between thorough cleanings of the vat?
    d. If red is being mixed now, how long will it be on the average (in terms of number of batches) before the vat must be cleaned?

11. Referring to problem 11 in Chapter 2 (the accounting problem), use the methods of this chapter to calculate answers to the following questions.
    a. What is the probability that a new sale will ultimately be paid for?
    b. What is the probability that a bill that is not paid within the thirty-day period will never be paid?
    c. What is the probability that a bill will reach the point where it has gone three months without being paid?
    d. Calculate the mean time that a bill that is ultimately paid is outstanding.

12. Referring to problem 12 in Chapter 2 (the highway problem), use the methods of this chapter to calculate answers to the following questions.
    a. Explain how to obtain the probability distribution of the lifetime. You do not have to provide numerical results, but indicate exactly what has to be computed.
    b. What is the probability that the highway will need repair within five years?
    c. Comment on the appropriateness of the Markov and stationarity assumptions in this context.
    d. Suppose that when the highway reaches state D, it can be resurfaced with asphalt to restore it to state B, but that this can be done only once. Show how the process model should be changed.

13. Referring to problem 13 in Chapter 2 (the hospital problem), use the methods of this chapter to calculate answers to the following questions.
    a. Classify the states.
    b. Suppose that two-thirds of the heart patients entering the hospital are emergency cases, and the rest are for planned surgery. Out of 1,000 cases, how many are discharged as survivors?
    c. What is the expected number of operations required once a heart patient enters the hospital?
    d. What is the probability that an emergency patient will undergo surgery at least once?
    e. What is the expected number of transfers that a patient will go through during a hospital stay, if he or she is admitted as an emergency?

14. Referring to problem 14 in Chapter 2 (the salt usage problem), use the methods of this chapter to calculate answers to the following questions.
    a. What is the expected number of storms for which the initial inventory of salt can be expected to last?
    b. What is the probability that the fifth storm uses up the last of the initial inventory?

15. Referring to problem 15 in Chapter 2 (the solar heating problem), use the methods of this chapter to calculate answers to the following questions.
    a. If today is sunny, how long will it be until the heating system requires back-up fuel?
    b. If yesterday was sunny and today is cloudy, what is the expected number of consecutive days that the house can go until requiring back-up fuel?
    c. If the system is running on back-up fuel, how many more consecutive days will it continue to operate on back-up before the solar heating can take over?
    d. If the back-up heating is needed today, how long will it be (on average) until it is needed again?
    e. If a heating season is 120 days long, what is the expected number of days per year that will require back-up fuel?

16. Referring to problem 16 in Chapter 2 (the print shop problem), use the methods of this chapter to calculate answers to the following questions.
   a. Classify the states (transient versus recurrent, and so forth).
   b. What fraction of the time, over the long run, will all of the machines be operable?
   c. What fraction of the time, over the long run, will there be a machine out of service that the repair person cannot work on because he or she is already occupied in servicing another?
   d. Explain in detail how to obtain the long-term average production rate of the print shop.

17. Referring to problem 17 in Chapter 2 (the game of craps), use the methods of this chapter to calculate answers to the following questions.
   a. For a simple bet on a single round of play, what is the probability of winning?
   b. What is the expected number of rolls of the dice needed to determine whether the player wins or loses?

18. Referring to problem 18 in Chapter 2 (the tournament problem), use the methods of this chapter to calculate answers to the following questions.
   a. The probability that the better team, A, will win the series, as calculated before any games are played.
   b. The probability that A will win the series, given that B wins the first game.
   c. The probability that A will win the series, given that B has won two games and A has won one (in any order).
   d. The expected number of games that will be played to determine the winner of the series.

19. In Section 2.8 of Chapter 2, a case study model of a call center was developed. The motive for building the model was to evaluate customer service. Time was measured in transfers, as a customer was (potentially) passed from one service operator to another to get the problem resolved. Too many transfers would leave the customer with "getting the run-around." (In Chapter 4, this problem will be extended further.)
   a. Using the Markov chain that was developed in that section, calculate the fraction of incoming calls that achieve satisfactory resolution.
   b. What is the expected number of transfers that a customer would experience? [Hint: You should not count the entry into the system as a transfer.]

CHAPTER 4

Rewards on Markov Chains

CHAPTER CONTENTS

4.1. Formulation 120
4.2. A Numerical Example 120
4.3. Expected Total Reward 121
4.4. Random Variable Rewards 124
4.5. Semi-Markov Processes 126
4.6. Limiting Results for Ergodic Processes 126

4.7. Total Reward for Terminating Processes 130
4.8. Case Study 132
4.9. Discounting 133
4.10. Case Study 135
4.11. Recommended Reading 137

Although Markov chains are very useful in modeling the stochastic behavior of systems and processes, they do not by themselves provide a means to evaluate that behavior, that is, attach a profit or loss to it. The objective of this chapter is to add a reward structure to Markov chains and to consider some of the mathematical expressions arising naturally from economic concerns. We do not yet allow any decisions so there is no opportunity for influencing the rewards received; we can only observe the process and collect the rewards it generates. Eventually, of course, we would like to extend the model to include decision variables and to choose these so as to produce optimal rewards. This is possible to do, using the theory of Markov decision processes. The mathematics developed here will provide the basis for formulating appropriate objective functions. Even without optimization, this extension provides many opportunities for useful applications.

We begin with a treatment that presupposes that rewards are fixed quantities and the only uncertainties stem from the transitions. Later, it will be shown that one could almost as easily treat the case in which the rewards are themselves random variables.

It should be noted that although we use the term "reward" consistently, we could just as well have used the term "cost." For that matter, we could have chosen a term completely devoid of economic connotations, but the feeling here is that it is better to risk prejudicing the interpretation than to handicap the student by failing to use natural, suggestive terminology. In any case, these rewards are unrestricted in sign, so costs can simply be thought of as negative rewards. It is even possible to interpret the rewards as time intervals, which leads to a new class of stochastic processes known as semi-Markov processes. We will have more to say about this extension later; for now, just think of rewards in the obvious way.

4.1 Formulation

Let r_{ij} denote the reward received in making a transition from state i to state j, and let **R** be the matrix of such values.[1] That is, whenever a transition from state i to state j occurs, someone collects the reward r_{ij}, which could be positive, negative, or zero. Whether the reward is collected all at the beginning, all at the end, or continuously during the transition is immaterial. We choose to associate the rewards with making transitions, rather than with being in states, for the simple reason that it is more general. We can cover the variation in which the reward is associated with being in a state just by setting all of the rewards for transitions out of that state to the same value. (Think about this if it is not immediately obvious to you.) We assume that the rewards accumulate additively over time; that is, for each transition, we collect the reward and add it to the total received so far. Negative or zero rewards are acceptable variations. Of course, a negative reward would subtract from the total.

We are interested in the expected total reward collected over n transitions. It is an expected value, rather than a completely predictable quantity because of the uncertain transitions, not because of any uncertainty about the rewards for specific transitions. Since the total reward collected will be dependent on the initial state, which we shall consistently denote by i, let this quantity be denoted by $v_i^{(n)}$. (As a mnemonic aid, think of "v for value.") Notice that we are placing the time index as a superscript to ensure that it is not interpreted as an event or the value of a random variable. It is convenient and natural to assume that we start with nothing; that is, $v_i^{(0)} = 0$.

Result The expected reward from one transition is expressible as

$$v_i^{(1)} = \sum_j p_{ij} r_{ij}$$

This result follows because the reward is just r_{ij} with probability p_{ij}. (Note that this is not matrix multiplication, but term-by-term multiplication across corresponding rows of **P** and **R**.) We call this quantity the expected immediate reward. To simplify the notation just a little, we will drop the time index when it is equal to 1; that is, v_i means the same as $v_i^{(1)}$.

4.2 A Numerical Example

EXAMPLE 4.1

Before going any further, let's introduce a numerical example to use as we develop the theory. We will return to the trucking example that we used to introduce Markov chains in Chapter 2. (Once again, we are using a small example just for clarity; larger problems are easily handled.) The rewards in this case could be the profits collected for each load

---

[1] Note that we are using the same notation (the matrix **R**) that we previously used for a portion of the transition matrix of a terminating process. Although there is a small risk of confusion, the prior use conforms with standard notation, and the current use is so logical that any other choice of notation would seem forced. We will accept the risk, with the assurance that we will not make any use of the prior meaning in this chapter.

delivered to its destination. Or, they could be the distances traveled, the gasoline consumed, or any of numerous other possibilities. We will think of them as dollar profits, after costs have been covered.

Recall that the trucking example had four states, corresponding to the cities Atlanta, Baltimore, Chicago, and Detroit. Suppose the transition matrix is:

$$\mathbf{P} = \begin{bmatrix} 0.2 & 0.3 & 0.5 & 0 \\ 0.1 & 0 & 0.3 & 0.6 \\ 0.1 & 0.2 & 0.3 & 0.4 \\ 0.1 & 0.4 & 0.5 & 0 \end{bmatrix}$$

We will add rewards to the process by assuming fixed profits for every possible trip based on distance, giving us the reward matrix

$$\mathbf{R} = \begin{bmatrix} -20 & 70 & 60 & 120 \\ 50 & 0 & 30 & 50 \\ 100 & 80 & 30 & 50 \\ -10 & 30 & 20 & 0 \end{bmatrix}$$

For example, the profit associated with delivering a load from Baltimore to Atlanta (r_{21}) is $50. The numbers in this example are made up for the purpose of demonstration; they do not represent any realistic situation. There are some negative rewards, or losses, just to show that they can be handled.

Before calculating anything, you may find it interesting to see if you can guess which city would be the best one to start from, just from looking at **R** and **P**. You can immediately see from **R** that the most profitable individual trip is the one from Atlanta to Detroit ($r_{14} = \$120$). However, when you look at **P**, you see that that particular trip will never occur ($p_{14} = 0$). That means that the trip from Chicago to Atlanta ($r_{31} = \$100$) is the most profitable one that can actually occur. From just scanning across the rows of **R**, it looks like Chicago is the best city to start from, but that conclusion does not take the transition probabilities into account.

For this problem, the expected immediate rewards for the four states are

$$v_1 = 47$$
$$v_2 = 44$$
$$v_3 = 55$$
$$v_4 = 21$$

These can be directly interpreted; for example, if the truck driver is in Atlanta not yet knowing where the next load will go, the expected value of the profit for that next trip is $47. So, based on just that first trip, the most profitable starting state would be Chicago with an expected profit of $55, but that does not necessarily tell you which would be best over the long run.

This same example will be used to illustrate further computations, after we have developed the formulas. Our first interest is in the total reward collected over n transitions, or $v_i^{(n)}$ for arbitrary initial state i and number of transitions n.

4.3 Expected Total Reward

When n is greater than 1, a recurrence relation may be used to determine $v_i(n)$. We employ the usual device of conditioning on the state occurring after the first transition and obtain

$$v_i^{(n)} = \sum_j \left\{ r_{ij} + v_j^{(n-1)} \right\} p_{ij}$$

In words, given that the first transition is from i to j, the total reward will be r_{ij}, collected immediately, plus the expected reward collected thereafter in $n - 1$ transitions. So the expected total reward under the condition that j is the state after the first transition is $r_{ij} + v_j^{(n-1)}$. Then you weight each of these by the probability that the condition is satisfied and add up.

Once you have the expression, it can be simplified as

$$v_i^{(n)} = \sum_j p_{ij} r_{ij} + \sum_j p_{ij} v_j^{(n-1)}$$

which upon substitution gives the final iterative form.

Result The total reward after n transitions, starting from state i, is

$$v_i^{(n)} = v_i + \sum_j p_{ij} v_j^{(n-1)}$$

EXAMPLE 4.2

For small values of n, it is practical to use this recurrence relation directly. For example, we can get $v_i^{(2)}$ for the numerical example as follows

$$v_1^{(2)} = v_1 + p_{11}v_1^{(1)} + p_{12}v_2^{(1)} + p_{13}v_3^{(1)} + p_{14}v_4^{(1)}$$
$$= 47 + (0.2)(47) + (0.3)(44) + (0.5)(55) + 0(21)$$
$$= 97.10$$
$$v_2^{(2)} = v_2 + p_{21}v_1^{(1)} + p_{22}v_2^{(1)} + p_{23}v_3^{(1)} + p_{24}v_4^{(1)}$$
$$= 44 + (0.1)(47) + (0)(44) + (0.3)(55) + (0.6)(21)$$
$$= 77.80$$
$$v_3^{(2)} = v_3 + p_{31}v_1^{(1)} + p_{32}v_2^{(1)} + p_{33}v_3^{(1)} + p_{34}v_4^{(1)}$$
$$= 55 + (0.1)(47) + (0.2)(44) + (0.3)(55) + (0.4)(21)$$
$$= 93.40$$
$$v_4^{(2)} = v_4 + p_{41}v_1^{(1)} + p_{42}v_2^{(1)} + p_{43}v_3^{(1)} + p_{44}v_4^{(1)}$$
$$= 21 + (0.1)(47) + (0.4)(44) + (0.5)(55) + (0)(21)$$
$$= 70.80$$

This tells us, perhaps surprisingly, that when the expected profit for two trips is found, the profit starting from Atlanta ($97.10) is higher than from Chicago ($93.40). That is a reversal of the order when only one trip was considered. We still do not know which starting city will prove to best over the long run.

It is convenient to use matrix notation at this point, so let $\mathbf{v}^{(n)}$ denote the column vector whose ith element is $v_i^{(n)}$. Again, we will drop the time index n when it is 1, so $\mathbf{v}$ means the same as $\mathbf{v}^{(1)}$.

> **Result** The total reward after n transitions in vector form is
> $$\mathbf{v}^{(n)} = \mathbf{v} + \mathbf{P}\mathbf{v}^{(n-1)}$$

This equation written out looks like the following for a four-state case such as our numerical example.

$$\begin{bmatrix} v_1^{(n)} \\ v_2^{(n)} \\ v_3^{(n)} \\ v_4^{(n)} \end{bmatrix} = \begin{bmatrix} v_1 \\ v_2 \\ v_3 \\ v_4 \end{bmatrix} + \begin{bmatrix} p_{11} & p_{12} & p_{13} & p_{14} \\ p_{21} & p_{22} & p_{23} & p_{24} \\ p_{31} & p_{32} & p_{33} & p_{34} \\ p_{41} & p_{42} & p_{43} & p_{44} \end{bmatrix} \begin{bmatrix} v_1^{(n-1)} \\ v_2^{(n-1)} \\ v_3^{(n-1)} \\ v_4^{(n-1)} \end{bmatrix}$$

In words, we get the vector of expected total reward for n transitions by multiplying the corresponding vector for n − 1 transitions by the transition matrix, **P**, and then adding the expected immediate gain vector. An alternative, closed-form expression for **v**(n) may be obtained by "solving" the recurrence relation:

$$\mathbf{v}^{(2)} = \mathbf{v} + \mathbf{P}\mathbf{v}$$
$$\mathbf{v}^{(3)} = \mathbf{v} + \mathbf{P}\mathbf{v}^{(2)}$$
$$= \mathbf{v} + \mathbf{P}(\mathbf{v} + \mathbf{P}\mathbf{v})$$
$$= \mathbf{v} + \mathbf{P}\mathbf{v} + \mathbf{P}^2\mathbf{v}$$
$$\mathbf{v}^{(4)} = \mathbf{v} + \mathbf{P}\mathbf{v}^{(3)}$$
$$= \mathbf{v} + \mathbf{P}\mathbf{v} + \mathbf{P}^2\mathbf{v} + \mathbf{P}^3\mathbf{v}$$

It is apparent from this simple substitution method that the general solution is given by the following result.

Result The total reward after n transitions in vector form is
$$\mathbf{v}^{(n)} = (\mathbf{I} + \mathbf{P} + \cdots + \mathbf{P}^{n-1})\,\mathbf{v}$$

This form suggests a little more clearly how the rewards "collect" over time. One more transition would add $\mathbf{P}^n\mathbf{v}$ to the total. Another variation that can be obtained by collecting terms differently would be:
$$\mathbf{v}^{(n)} = \mathbf{v}^{(n-1)} + \mathbf{P}^{n-1}\mathbf{v}$$

For purposes of iterative calculation in a spreadsheet, the original form is the simplest to implement.

For our numerical example, we may iteratively determine the $\mathbf{v}^{(n)}$ from any of these relations. These calculations are easily carried out in a spreadsheet. Set up an area to contain $\mathbf{P}$, name the region, then put the first values—the $v_1^{(1)}$—in a column named $\mathbf{v}$. The next column can be expressed as the formula "=MMULT(**P**,(previous column)) + **v**" whereby (previous column) we mean that you select the column to the left. Remember to hold down the special key as you hit enter to ensure that the matrix formula is used. Then you can just copy that formula across to additional columns by dragging. It takes only a few keystrokes to set up the whole table if you know how. Results for the first few n are shown in Table 4.1.

A table such as Table 4.1 can be generated quite easily for other cases as long as there are not too many states (columns) or too many transitions (rows). To give a verbal interpretation of the result, we would say, for example, that the expected total reward to be gained over four transitions, starting in state 1, is 179.86. Remember that these are *expected* total rewards; the *actual* total rewards are random variables.

Before proceeding to a more detailed study of $\mathbf{v}(n)$, let us take note of a few points that may be of use in applications. First, observe that as we have used n, it simply refers to the number of transitions. If one wanted to count off the transitions, one could use either the order $\{1, 2, \ldots, n\}$, or the order $\{n, n-1, \ldots, 1\}$. The former method is more natural when observing a process evolving over time, but the latter is also useful, particularly when we introduce decisions. In other words, we may think of n as the number of transitions that have occurred so far, with the idea that they are to continue, or we may think of it as the number left to go, in a process that is destined to terminate after a fixed number of transitions.

Second, it is worth noting that the transition rewards r_{ij} do not appear explicitly in the expressions for $v_i^{(n)}$. They are only used once, to construct v_i. If you think about it, you will realize that v_i, the expected reward from one transition, beginning in state i, can be thought of as a reward for being in state i. Hence, the generality we thought we would achieve by associating rewards with transitions rather than states was somewhat illusory. On the other

■ **TABLE 4.1**
Total Expected Rewards Over Time

Starting City	Trips	1	2	3	4	5	6	7	8
Atlanta	$v_1^{(n)}$	47.00	97.10	136.46	179.86	221.04	263.36	305.09	347.13
Baltimore	$v_2^{(n)}$	44.00	77.80	124.21	163.75	206.93	248.21	290.48	332.23
Chicago	$v_3^{(n)}$	55.00	93.40	136.61	177.88	220.15	261.91	303.93	345.81
Detroit	$v_4^{(n)}$	21.00	70.80	108.53	152.64	193.43	235.95	277.57	319.67

hand, one has to be more careful with questions of timing if one elects to view rewards as associated with states. Is the reward collected upon entering or upon leaving the state? Does one get a reward for being in the initial state? The final state? Both? Any of these variations can be handled; you merely have to be sure that you have the appropriate expressions for whichever conventions are adopted.

4.4 Random Variable Rewards

We now consider the effect of relaxing the assumption that the rewards are fixed quantities. We want to repeat the development we just did, only now using the assumption that the rewards associated with individual transitions are random variables. This whole section will end up saying that we could just as well have assumed that rewards were random in the first place. That is, the iterative equation for the expected total reward is that same; however, the argument that leads to this conclusion is more involved. For a first reading, you may prefer to accept this statement and go on to the next section, avoiding the digression into the complications we must deal with in developing the theory.

Let R_{ij} be a random variable denoting the reward received in making a transition from state i to state j. Note that the reward distribution depends upon both the origination and destination state. If one were to think in terms of simulating the process, one would first have to sample from the transition matrix to determine the next state, then sample from a reward distribution which is selected accordingly. We do assume that the R_{ij} are independent of one another and independent of the transition number.

Again, it is immaterial whether the reward R_{ij} is collected upon leaving i, upon entering j, or continuously during the transition. Also, special cases, such as when $R_{ij} = R_{ik}$ for all j and k, will permit the treatment of variations such as rewards associated with being in a state rather than with making a transition.

We are interested in the total reward collected over n transitions (a random variable) and its expected value. Accordingly, let $V_i^{(n)}$ represent the total reward, when i is the initial state, and let $v_i^{(n)}$ be its expected value. Similarly, let r_{ij} denote $E(R_{ij})$. We have, of course, used the lowercase letters before. One of the objectives of this section is to show that the present use of the notation (for the mean of a random variable) is consistent with the previous use (as a fixed value).

For the random variable $V_i^{(1)}$, one may be tempted to write

$$V_i^{(1)} = \sum_j p_{ij} R_{ij}$$

but this would be wrong. The error is subtle, very commonly made, and of no great consequence if only the expected value is to be calculated. It would, however, affect the variance. To see what is wrong, imagine how one would obtain a value for $V_i^{(1)}$ using this equation. One would sample each of the R_{ij}, multiply the values obtained by the corresponding p_{ij}'s and add. In other words, we would take a weighted average of values of the R_{ij}. What is actually desired is a single value, of one of the R_{ij}, the particular one being selected according to the p_{ij}. Consequently, the best expression for $V_i^{(1)}$ would be

$$V_i^{(1)} = R_{ij} \quad \text{with probability } p_{ij}, \quad \text{for } j = 1, 2, \ldots.$$

In developing expressions involving this kind of "selection" of random variables, it is usually best to think of what the random variable is, given the transition.

In dealing with the distribution of $V_i^{(1)}$, it is fortunate that the second equation applies; for if the first were correct, one would have to convolve the distributions of the random variables $p_{ij}R_{ij}$. Instead, we have only to take the linear combination of the distributions of the R_{ij}. That is,

$$P\left\{V_i^{(1)} < x\right\} = \sum_j p_{ij} P\{R_{ij} < x\}$$

One may use this expression to prove rigorously that

$$E(V_i^{(1)}) = \sum_j p_{ij} E(R_{ij})$$

or

$$v_i^{(1)} = \sum_j p_{ij} r_{ij}$$

which is consistent with the case in which r_{ij} is deterministic. That is, we do not have to know anything more about the random variable rewards than their means in order to write $v_i^{(1)}$. That is not the case if we need to know about the entire distribution of the total reward.

We can develop a "sort of" recursion relation for $V_i^{(n)}$ by conditioning on the first transition.

$$V_i^{(n)} = R_{ij} + V_j^{(n-1)} \text{ with probability } p_{ij},$$

$$\text{for } j = 1, 2, \ldots.$$

This time, there is no escaping the fact that random variables are being summed, and therefore a convolution of distributions is involved. If the reward distributions are continuous,

$$P\left\{V_i^{(n)} < x\right\} = \sum_j p_{ij} \int_0^x P\left\{V_i^{(n-1)} < x - u\right\} dP\{R_{ij} < u\}$$

Although this expression does not appear to hold much promise, there are ways (more advanced than we will deal with here) to compute such distributions conveniently. For our immediate purposes, we return to the calculation of the expected value of the random variables $V_i^{(n)}$. We may use this equation directly to prove

$$E\left(V_i^{(n)}\right) = \sum_j p_{ij} E\left(R_{ij} + V_j^{(n-1)}\right)$$

$$= \sum_j p_{ij} E(R_{ij}) + \sum_j p_{ij} E\left(V_j^{(n-1)}\right)$$

or

$$v_i^{(n)} = \sum_j p_{ij} r_{ij} + \sum_j p_{ij} v_j^{(n-1)}$$

Again, this is consistent with the deterministic reward case and requires only the expected immediate rewards as data.

In summary, then, if it is only the *expected* total reward in n transitions that concerns us—not the distribution—then one may use all of the expressions developed for the deterministic case, using for r_{ij} the expected reward for going from i to j. In other words, if all we care about is the expected total reward, it makes no difference whether the r_{ij} are fixed values or means of random variables. Once we know that, we do not have to worry about any of the details of the derivation, what the distributions are, or any other complication. Since, in most applications involving optimization, it is an expected value that is used for the objective function, this result is both appealing and useful.

It is worth a reminder at this point that the values we have used as rewards could have been something other than profits. They could have been costs or distances or anything else that adds up over the transitions. We now know that they could be fixed quantities, random variables, or any combination.

4.5 Semi-Markov Processes

There is another useful interpretation of a random variable reward that extends the theory of Markov chains to a considerably more general class called semi-Markov processes. The trick is to think of the reward as a (clock) *time* to make the transition. In the case of the truck example, we used transitions or trips of the truck as our time counter. If we want to know how much actual clock time is consumed in these trips, we could associate a random variable R_{ij} with the time to get from i to j. This should include any idle time waiting for a load plus the travel time from i to j. Then, by accumulating these rewards over trips, just as we accumulated fares or distance, we will be adding up the total time consumed. All of the results we have obtained so far and those yet to come would apply.

Semi-Markov processes provide a powerful modeling tool. Notice that we do not have to make any particular assumptions about the forms of the R_{ij} distributions; they could be anything. We do not even have to identify what family of distribution they are, much less their distribution functions; all we need are their means, the r_{ij}. It *is* necessary to assume that the R_{ij} are independent of one another and that the underlying transition mechanism is that of a Markov chain. However, these are conditions that are commonly met in many applications.

The conventional treatment of semi-Markov processes uses much more advanced mathematics than we have here. Usually, the development works through expressions of the distributions and eventually derives the simple results for the means. Our approach, which was based upon the conceptual trick of thinking of times as rewards, is limited to the expected values.

4.6 Limiting Results for Ergodic Processes

As n grows large, it becomes less feasible to obtain exact values for $v_i^{(n)}$ by direct calculation. However, we do want to know what kinds of rewards to expect over long intervals of time. That is, we want to know something about the limiting behavior of $v_i^{(n)}$. Now it is apparent both from the numerical example and from common sense, that we cannot normally expect $v_i^{(n)}$ to converge to some value. Except possibly for special cases, the total reward accumulated over infinite time will be unbounded. On the other hand, we know that the Markov chain generating the rewards eventually stabilizes (in a probabilistic sense), which suggests that some sort of regularity ought to be expected in the $v_i^{(n)}$. We will first cover the case of an ergodic Markov chain, in which the process continues to make transitions for infinite time. Later, we will find that it is actually somewhat easier to deal with the terminating process case.

Referring to the following form of the equation for $\mathbf{v}^{(n)}$ developed earlier,

$$\mathbf{v}^{(n)} = \mathbf{v}^{(n-1)} + \mathbf{P}^{n-1}\mathbf{v}$$

we can see intuitively what must happen. The difference between $\mathbf{v}^{(n+1)}$ and $\mathbf{v}^{(n)}$ is $\mathbf{P}^n\mathbf{v}^{(1)}$. In the ergodic process case, as n becomes large, $\mathbf{P}^n$ approaches its asymptotic value

$$\lim_{n \to \infty} \mathbf{P}^n = \begin{bmatrix} \pi_1 & \pi_2 & \pi_3 & \cdots \\ \pi_1 & \pi_2 & \pi_3 & \cdots \\ \pi_1 & \pi_2 & \pi_3 & \cdots \\ \cdots & \cdots & \cdot & \cdot \end{bmatrix}$$

Therefore,

$$\lim_{n \to \infty} \mathbf{P}^n \mathbf{v} = \begin{bmatrix} \pi_1 & \pi_2 & \pi_3 & \cdots \\ \pi_1 & \pi_2 & \pi_3 & \cdots \\ \pi_1 & \pi_2 & \pi_3 & \cdots \\ \cdots & \cdots & \cdot & \cdot \end{bmatrix} \begin{bmatrix} v_1 \\ v_2 \\ v_3 \\ \cdot \\ \cdot \end{bmatrix}$$

where the π_j are steady-state probabilities. These differences, then, between successive expected total rewards become constant. Furthermore, because the rows of π are identical, the differences are the same for each state. That is

$$\lim_{n \to \infty} \left\{ v_i^{(n+1)} - v_i^{(n)} \right\} = \sum_j \pi_j v_j \qquad \text{for all } i$$

We call this common difference the *gain rate* for the reward process (without, however, intending to imply that it is necessarily positive), and denote it by g.

> **Definition** The long-term gain rate is defined by
>
> $$g = \sum_j \pi_j v_j$$

This is just a single number, not a vector. In vector notation, however, it can be calculated as $g = \pi^T \mathbf{v}$. It could be a negative number if some or all of the v_j are negative.

EXAMPLE 4.3

For the numerical example used earlier, we find $\pi_1 = 0.111$, $\pi_2 = 0.224$, $\pi_3 = 0.379$, and $\pi_4 = 0.286$, so the gain rate is

$g = (0.111)(47) + (0.224)(44) + (0.379)(55) + (0.286)(21)$
$= 41.93$

Examination of the later entries of Table 4.1 confirm that the expected total rewards are increasing at just about that rate (n has to be larger to establish a precise value).

Another way of saying that the differences eventually become constant is to say that the $v_i^{(n)}$, viewed as functions of n, are asymptotically linear and have the same slope. To see this visually, let's graph the first few entries of Table 4.1 and observe that fact. Figure 4.1 shows the four functions; they all, after a few transitions, acquire the constant slope of g.

It appears that the expected total reward should be easy to calculate approximately, even for large n, by simply knowing the linear functions that the $v_i^{(n)}$ asymptotically approach. In fact, it appears from the graph that the linear approximation would be quite accurate for n larger than 3 or 4. Although one has to be careful about presuming too much from one example, it is in fact known that the error in this approximation approaches zero geometrically fast, which means that it will not take many transitions before the exact curve is essentially the same as the linear approximation.

A linear function of n would take the form $f(n) = f(0) + gn$, where g is the slope of the line. We already know g. Therefore, in order to determine these linear functions, we require only the y-intercepts, where the lines intercept the y axis. Unlike the slope, which is the same for

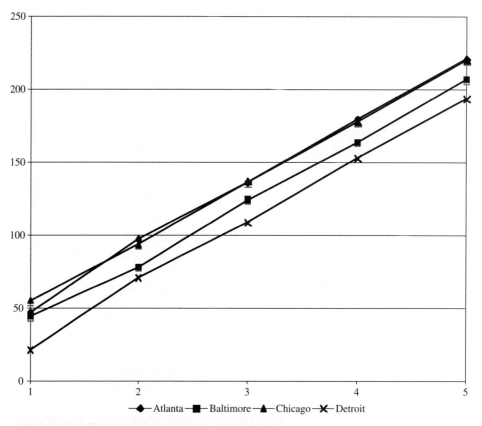

■ **FIGURE 4.1** Expected total reward over time

all of the lines, the y-intercepts will depend upon the initial state. Let y_i denote the y-intercept for the linear approximation for $v_i^{(n)}$. Then for large n,

$$v_i^{(n)} \cong y_i + gn$$

If we substitute these approximations for the $v_i^{(n)}$ into the recurrence relation (4), we obtain equations that determine the y_i,

$$y_i + gn = v_i + \sum_j p_{ij}\{y_j + g[n-1]\}$$

$$y_i + gn = v_i + \sum_j p_{ij}y_j + \sum_j p_{ij}g[n-1]$$

After cancelling out common terms, we obtain the final result.

> **Result** The y-intercepts are related by the equation
> $$y_i = v_i - g + \sum_j p_{ij}y_j$$

Since an equation of this form is valid for every i, we have a set of as many linear equations as there are unknowns. In vector form, they would be

$$\mathbf{y} = \mathbf{v} - g\,\mathbf{1} + \mathbf{P}\mathbf{y}$$

EXAMPLE 4.4

Our numerical example would produce the equations

$$y_1 = 47 - 41.93 + 0.2y_1 + 0.3y_2 + 0.5y_3 + 0.0\,y_4$$
$$y_2 = 44 - 41.93 + 0.1y_1 + 0.0y_2 + 0.3y_3 + 0.6y_4$$
$$y_3 = 55 - 41.93 + 0.1y_1 + 0.2y_2 + 0.3y_3 + 0.4y_4$$
$$y_4 = 21 - 41.93 + 0.1y_1 + 0.4y_2 + 0.5y_3 + 0.0\,y_4$$

Unfortunately, these equations contain a dependency, so they do not determine the y_i uniquely. For some purposes (notably, optimization), it is sufficient to fix relative values for the y_i. In these cases, any one of the y_i can be arbitrarily set equal to some convenient value (usually zero), and the above equations can be used to solve for the others. This method is often used in Markov decision processes. There are also several methods for calculating the exact y_i's using advanced matrix analysis techniques, such as spectral decomposition. However, these methods are practical for only very small cases, perhaps up to 4 or 5 states.

A much more practical method to use if you really want to determine these functions for applications is to just plot the curves as we did in Figure 4.1. If you ignore the first few terms, which may deviate from the linear approximation, and use the later values from the table to fit straight lines, you can easily determine the functions. In fact, since the slope of the function is already known precisely, you can fit each line with a single point.

EXAMPLE 4.5

To illustrate this observation using our numerical example, we can take a calculated value from the table for, say, n = 15. (This can be any value that is large enough to be sure that the long-term behavior of the Markov chain is established.) Then we want the line that passes through this point with a slope of g = 25.17. We find

$$v_1^{(6)} = y_1 + (g)(15)$$
$$640.61 = y_1 + (41.93)(15)$$

This is one equation in one unknown, which is easily solved:

$$y_1 = 11.81$$

Similarly,

$$v_2^{(6)} = y_2 + (g)(15)$$
$$625.81 = y_2 + (41.93)(15)$$
$$y_2 = -2.99$$

and

$$v_3^{(6)} = y_2 + (g)(15)$$
$$639.35 = y_2 + (41.93)(15)$$
$$y_2 = 10.55$$

and

$$v_4^{(6)} = y_2 + (g)(15)$$
$$613.13 = y_2 + (41.93)(15)$$
$$y_2 = -15.67$$

The ultimate interpretation, then, is that the expected total reward accumulated over n transitions is approximately (for large n)

11.81 + 41.93n	if the initial state is 1
−2.99 + 41.93n	if the initial state is 2
10.55 + 41.93n	if the initial state is 3
−15.67 + 41.93n	if the initial state is 4

One could also say that the eventual difference in value between starting in state 1 as opposed to state 3 is

$$(11.81 + 41.93n) - (10.55 + 41.93n) = 1.26$$

That is, one should be willing to pay that much more to begin in state 1 as opposed to state 3. Table 4.1 confirms that this figure is right.

This means that it is slightly better to start in Atlanta rather than in Chicago, over the long run. Over the short run of one or two or three trips, the expected profits cross, as shown in Figure 4.1. Eventually, as the Markov chain reaches steady-state, the "bounce" effects of the initial condition lose their influence and the long term straight-line trends show up. The gain rate g is the long-term profit per trip, taking into account all of the states and probabilities. That is the same regardless of the initial state. But the initial state affects the early profits, and the differences established in those early transitions produce the long-term spacing of the straight lines.

4.7 Total Reward for Terminating Processes

It has been assumed to this point that the underlying Markov chain is ergodic. Of the exceptions to this case—null states, periodic states, and so forth—the one of greatest interest is that in which the process consists of a number of transient states and one or more absorbing states. Because this case provides a natural model for terminating processes, which occur often in applications, it warrants special treatment.

Typically, one would want the rewards to stop once absorption occurs, so we assume that $r_{ij} = 0$ unless i is a transient state. Let K denote the set of indices of the transient states. Repeating the argument used to develop the expected total reward over n transitions,

$$v_i^{(n)} = \sum_j \{r_{ij} + v_j^{(n-1)}\} p_{ij}$$

$$v_i^{(n)} = v_i + \sum_j p_{ij} v_j^{(n-1)}.$$

But now, since there is no further reward to be gained once an absorbing state is reached,

$$v_j^{(n)} = 0 \quad \text{for any n, any } j > K$$

and we can restrict the sum to those indices j which represent transient states

$$v_i^{(n)} = v_i + \sum_{j \in K} p_{ij} v_j^{(n-1)} \quad \text{for } i \in K$$

As is usual for terminating process models, we may partition the transition matrix **P** according to

$$\mathbf{P} = \begin{bmatrix} \mathbf{Q} & \mathbf{R} \\ \mathbf{0} & \mathbf{I} \end{bmatrix}$$

where the square matrix **Q** refers to transitions within the transient set, the rectangular matrix **R** refers to transitions from the transient set to the absorbing set, **0** is a rectangular matrix of zero elements, and **I** is the identity matrix referring to the absorbing set. (Note that this **R** is not the same as the reward matrix **R**; there should be no danger of confusion in anything that follows.) Then in matrix form, we have

$$\mathbf{v}^{(n)} = \mathbf{v} + \mathbf{Q}\mathbf{v}^{(n-1)}$$

Here, the dimensions of the column vectors $\mathbf{v}^{(n)}$ and **v** have been reduced by eliminating elements corresponding to absorbing states, since these are known to be zero. This iterative form may be "solved" to obtain the closed form expression.

> **Result** The total expected reward in n steps for a terminating Markov chain is given by
>
> $$\mathbf{v}^{(n)} = (\mathbf{I} + \mathbf{Q} + \mathbf{Q}^2 + \ldots \mathbf{Q}^{n-1})\mathbf{v}$$

4.7 Total Reward for Terminating Processes

$\mathbf{Q}$ is a nonsingular matrix. Provided that it is finite, the inverse $(\mathbf{I} - \mathbf{Q})^{-1}$ will exist, and we may express this equation in a form that is notationally (though not necessarily computationally) more convenient.

Result The total expected reward in n steps for a terminating Markov chain with a finite number of states is given by

$$\mathbf{v}^{(n)} = (\mathbf{I} - \mathbf{Q})^{-1}(\mathbf{I} - \mathbf{Q}^n)\mathbf{v}$$

Either form will permit us to see what happens to $\mathbf{v}^{(n)}$ as n increases. Since $\mathbf{Q}^n$ refers to probabilities of being in transient states after n transitions, all of the elements of this matrix approach zero as n increases. In fact, it can be shown that they approach zero geometrically fast, so the infinite series

$$\mathbf{E} = \sum_{n=0}^{\infty} \mathbf{Q}^n$$

is convergent. When $\mathbf{Q}$ is finite the infinite series is represented by

$$\mathbf{E} = (\mathbf{I} - \mathbf{Q})^{-1}$$

We observe that as $n \to \infty$,

$$\lim_{n \to \infty} \mathbf{v}^{(n)} = \mathbf{E}\mathbf{v}$$

or in element form,

$$\lim_{n \to \infty} v_i^{(n)} = \sum_j e_{ij} v_j$$

These are finite quantities, so the expected total reward starting from any transient state i approaches a bounded value.

Result The total expected reward over infinite time for a terminating Markov chain with a finite number of states is given by

$$\mathbf{E}\mathbf{v}$$

By recalling an earlier interpretation of the matrix $\mathbf{E}$, we may show that this equation for the limiting values makes intuitive sense. The element e_{ij} was obtained as the expected number of visits to transient state j before absorption occurs, given that the process started in transient state i. On each visit, we collect the expected reward v_j, so multiplying these and summing over all transient j accumulates the expected total reward.

In some applications, it may be useful to associate a reward with first entry into an absorbing state. That is, one gets the reward for reaching the state but does not continue to collect the reward just for staying there. Provided that we compute v_i appropriately, this variation is already covered by the theory given. In the derivation above, the sum used in calculating v_i included the reward for the "last" transition, that is, the one that takes the process from the set of transient states to one of the absorbing states. Because such a transition occurs only once, we can associate the reward for entering a particular absorbing state with each of the transitions into that state, and thereby accomplish our objective.

4.8 Case Study

EXAMPLE 4.6

For a numerical example, let us extend the case study developed in Section 2.8 of Chapter 2 of a proposed call center providing support to customers of a software company. Recall that the transition matrix was

$$P = \begin{bmatrix} 0 & 0.6 & 0.2 & 0.2 & 0 & 0 & 0 \\ 0 & 0 & 0 & 0.04 & 0.16 & 0.8 & 0 \\ 0 & 0.1 & 0 & 0 & 0.4 & 0.5 & 0 \\ 0 & 0.3 & 0 & 0 & 0.1 & 0.4 & 0.2 \\ 0 & 0 & 0 & 0 & 0 & 0.75 & 0.25 \\ 0 & 0 & 0 & 0 & 0 & 1 & 0 \\ 0 & 0 & 0 & 0 & 0 & 0 & 1 \end{bmatrix}$$

where the states were: (1) Initial, (2) Sales, (3) Billing, (4) Support, (5) Supervisor, (6) Satisfied, and (7) Unresolved. The primary concern initially was for customer satisfaction, but now we will think about the cost of providing the service.

Some of the operators require more knowledge and experience than others. For example, providing technical support requires more knowledge than taking a sales order. It is therefore logical that the average cost per minute would vary from one function to another. Moreover, the average length of call can be expected to vary by function. Although both are random variables, we can assume that they are independent, so that the expected cost per call to any department is just the product of the mean cost per minute and mean call duration for that department.

Let us assume the following data values in Table 4.2 for purposes of the example. Now in this case, the rewards (really costs) are associated with visits to the states, as opposed to transitions between them. We do not have to calculate the vector $\mathbf{v}$ of immediate rewards; we already have it in the right-hand column of Table 4.2 (except for the cost of the initial state, which is zero). Hence the $\mathbf{v}$ vector is

$$\mathbf{v} = \begin{bmatrix} 0 \\ 80 \\ 96 \\ 300 \\ 140 \end{bmatrix}$$

Now, to get the expected total reward/cost for each initial state, we have to premultiply $\mathbf{v}$ by $\mathbf{E}$, the matrix of the expected number of times each transient state is visited. You were asked to calculate $\mathbf{E}$ in one of the problems in Chapter 3, but you can find in it the answers in the back of the book. The final result is

$$E\mathbf{v} = \begin{bmatrix} 172 \\ 116 \\ 164 \\ 349 \\ 140 \end{bmatrix}$$

These values are in dollars, so the expected cost of a call originating from the outside is $172. A call that has reached the supervisor, in contrast, has an expected *remaining* cost of $349.

It may have occurred to you that this way of modeling the costs is not very realistic. For one thing, the real costs are not incurred minute by minute during the calls but depend on the number and kind of people that you have to hire, whether they are busy or not. That is, they are paid for being there, even when they are not talking to customers. Also, the model does not take into account the cost of delaying customers because all operators are busy. We will improve the model further in Chapter 7, where we deal with queueing networks. At that time, we will be able to evaluate the trade-off between improving customer service and the cost of more operators, including the effects upon delay. However, this basic Markov chain model will model the routing of customer calls, which is something we will need even for the more sophisticated model.

■ **TABLE 4.2**
Call Center Costs

	Cost/Min	Expected Time	Cost per Visit
Sales	$8.00	10 min.	$80
Billing	$12.00	8 min.	$96
Support	$15.00	20 min.	$300
Supervisor	$28.00	5 min.	$140

4.9 Discounting

There are two good reasons to raise the subject of discounting. One is the obvious point that economic considerations frequently call for discounting, particularly when we want to deal with streams of costs or rewards over time. The other reason is that adding a discount factor actually simplifies some of the calculations. In the case of ergodic chains, the total accumulated reward will continue to grow (or decline) as time progresses; however, with discounting, the contributions from the distant future shrink to insignificance so that the discounted total will be finite. That would give a concrete way to compare alternatives over the long run.

There is an issue to be careful about when applying discounted rewards. Normally, discounting assumes calendar time units (days, weeks, months, or years). If we define the discrete time units of our model as steps, counting off the transitions independently of real time, we may not have an easy translation between the time units. Of course, this will not be a problem if the time units of the model are simple time increments.

Before developing the formal expressions, let us be sure that the meaning of discounting is clear. The idea, of course, is to take account of the economic reality that rewards in the future are not really worth the same as the identical dollar amount in the present. If one were to deposit $100 in a bank today and withdraw it, say, ten years from now, the interest added and compounded would have increased it considerably. (As a rule of thumb, a 6 percent interest rate, compounded annually, will double your money every twelve years.) Consequently, the promise of receiving $100 ten years from now is not worth nearly as much as having $100 today—if one had it today one could deposit it in the bank and be much better off ten years from now. The opportunity to use the money and to make it grow through compounded interest payments marks the difference between having a fixed amount of money now and getting that same amount sometime in the future.

Given a certain amount, D, now and assuming a fixed interest rate per period of x, one should receive after one period the amount

$$D + xD = (1 + x)D$$

After one more period, the amount should be this amount multiplied by $(1 + x)$,

$$(1 + x)[(1 + x)D] = (1 + x)^2 D$$

and so forth.

If, on the other hand, one required the amount F after n periods and were uncertain how much to deposit initially so as to accumulate F, one can solve for D in the expression

$$(1 + x)^n D = F$$

obtaining,

$$D = (1 + x)^{-n} F$$

It is this relation that indicates the equivalent value now of D dollars n periods from now. That is, one would be indifferent between receiving D dollars now (and depositing it in a bank paying interest at the rate x) or F dollars n periods from now.

It should be noted that x refers to the per-period rate. Banks and other lending institutions customarily speak of the annual interest rate, but then in the same breath mention quarterly or monthly compounding. If one is to compound quarterly, one must not use the annual interest rate, but one-fourth of that rate. Table 4.3 shows the effect of different compounding periods on the yield after one year from an initial deposit of $1 using an annual interest rate of 6 percent.

The last row of Table 4.3, referring to continuous compounding, is obtained by taking the limit and recognizing the identity

$$\lim_{n \to \infty} \left(1 + \frac{x}{n}\right)^n = e^x$$

■ **TABLE 4.3**
Comparing Different Compounding Periods

Compounding Period	Yield after One Year	Value for x = 0.06
Annual	$1 + x$	1.06
Quarterly	$\left(1 + \dfrac{x}{4}\right)^4$	1.061364
Monthly	$\left(1 + \dfrac{x}{12}\right)^{12}$	1.061678
Continuously	e^x	1.061836

One can see from Table 4.3 that, although it makes a definite difference how often interest is compounded, this difference is relatively minor compared to a change of 1 percent in the interest rate. Thus, there is little cause for concern over the issue of frequency of compounding. The important thing to remember is to use the appropriate interest rate for the period you have elected to use.

Although it is easier to think in terms of interest rates, it turns out to be more convenient in some computations to use a slightly transformed variation called the discount factor.

Definition The discount factor, α, is defined in terms of the interest rate i by

$$\alpha = \frac{1}{1+x}$$

Because x typically ranges from 0.01 to 0.10, α will typically range from 0.90 to 0.99. The smaller the interest rate, the closer α will be to 1. Sometimes it is convenient to use e^{-x} for the discount factor instead of α. This would, of course, correspond to the continuous compounding case. Again, the effect upon the result would be minor.

Returning our attention now to the rewards accumulated from an ergodic Markov chain, let $v_i(\alpha)$ denote the expected discounted total reward, over an infinite horizon, beginning in state i. The value of α we use is determined by the time units of our process; that is, if our time units were days, we would want to use an α that corresponds to a daily interest rate. By conditioning as usual on the first transition, we obtain

$$v_i(\alpha) = \sum_j p_{ij}(r_{ij} + \alpha v_j(\alpha))$$

Upon further simplification, this yields the following result.

Result

$$v_i(\alpha) = v_i + \alpha \sum_j p_{ij} v_j(\alpha)$$

This is a set of linear equations (not a recurrence relation), which in vector form would be:

$$\mathbf{v}(\alpha) = \mathbf{v} + \alpha \mathbf{P} \mathbf{v}(\alpha)$$

The general solution is given by

$$\mathbf{v}(\alpha) = \mathbf{v}\left[\mathbf{I} + \sum_{n=1}^{\infty}(\alpha \mathbf{P})^n\right]$$

or, if **P** is finite, by the equivalent form

$$\mathbf{v}(\alpha) = (\mathbf{I} - \alpha\mathbf{P})^{-1}\mathbf{v}$$

Because α is strictly less than 1 and the elements of $\mathbf{P}^n$ never exceed 1, we may be sure that the infinite series in the above equation converges. Given some particular value of α, we may even be able to predetermine when to safely truncate the series in computational problems. If α is very close to 1, it may be necessary to take quite a few terms to have an adequate approximation to the infinite sum. Similarly, because α is strictly less than one, we may be certain that the matrix $(\mathbf{I} - \alpha\mathbf{P})$ is nonsingular; though if α is close to one, we may encounter computational problems.

4.10 Case Study

EXAMPLE 4.7

For our case study, we will model a small portion of a small retail business, specifically, an appliance store. The focus will be on just one particular model of a refrigerator—a basic model that is representative of the business. We will be making a lot of simplifying assumptions. These are just to keep the illustration clear, not because we are forced to make them. Once you understand the flavor of the application, you could develop a much more realistic model.

The business owner's policy is to keep a maximum of four units of this model in stock. In any given week, the number of units sold might be zero, one, or two, with probabilities 0.5, 0.3, and 0.2, respectively. When the number in stock falls to one or zero at the end of the week (Sunday), the owner personally drives a truck to the supplier's warehouse on Monday to pick up three or four units, the number depending upon how many he needs to bring the stock up to four. Monday is the only day of the week when the store is closed.

From this description, we can formulate a Markov chain model, where time is measured in weeks and the states are the defined as the number of units in stock at the end of the sales week (Sunday). It will be convenient to arrange the states in reverse numerical order, from 4 to 0, so that sales move the state forward 0, 1, or 2 steps. With that convention, the transition matrix will be

$$\mathbf{P} = \begin{bmatrix} 0.5 & 0.3 & 0.2 & 0 & 0 \\ 0 & 0.5 & 0.3 & 0.2 & 0 \\ 0 & 0 & 0.5 & 0.3 & 0.2 \\ 0.5 & 0.3 & 0.2 & 0 & 0 \\ 0.5 & 0.3 & 0.2 & 0 & 0 \end{bmatrix}$$

The first three rows are obvious translations of the sales distribution. For the last two rows, you have to think about what happens. If the number of refrigerators at the end of the week is one or zero (because the sales in the previous week took the inventory to that level), then the owner will replenish to the level of four by Tuesday, after which the sales could take the state by the end of the week to four, three, or two.

We have the Markov chain completely formulated, and we could immediately compute a number of descriptive measures such as steady-state probabilities and mean first passage times. However, we will continue with the formulation stage and incorporate the rewards. Assume that the wholesale cost to the owner for each refrigerator is $178 and he sells them for $299. Further assume that the holding cost is $10 per refrigerator per week. Also, there is an order cost of $50 that occurs each time the owner has to pick up a new supply. This is basically the cost of the paperwork and the delivery charges, and is the same whether three or four units are involved. With these assumptions, the reward matrix is

$$\mathbf{R} = \begin{bmatrix} -40 & 259 & 558 & 0 & 0 \\ 0 & -30 & 269 & 568 & 0 \\ 0 & 0 & -20 & 279 & 578 \\ -594 & -295 & 4 & 0 & 0 \\ -762 & -463 & -164 & 0 & 0 \end{bmatrix}.$$

Again, the entries in the first three rows are fairly obvious; they consist of the revenues from the number of units sold minus the holding charges for the number of units held at the end of the previous week. For the last two rows, we have to subtract the fixed order cost and the costs of the refrigerators purchases. The zeros in the matrix are irrelevant, because there is no corresponding transition possible.

Notice that it is not immediately obvious from the reward matrix whether selling this particular refrigerator is even profitable over the long run. The owner will make money some weeks, but lose on others. We cannot tell whether the profitable weeks outnumber the loss weeks until we combine the rewards with the Markov chain.

The first step is to calculate the expected immediate rewards using the equation

$$v_i^{(1)} = \sum_j p_{ij} r_{ij}$$

The resulting values, in vector form, are

$$\mathbf{v} = \begin{bmatrix} 169.30 \\ 179.30 \\ 189.30 \\ -384.70 \\ -552.70 \end{bmatrix}$$

The differences among these values occur because of the value of the inventory on hand. For example, if the owner has no inventory at the end of a week (that is, the process is in state 0), he will have to buy four units. In such a week, the immediate effect will be an expected loss even if he has sales.

For the longer term, we weight the immediate rewards with the steady-state probabilities to find the expected gain rate, g. The calculated steady-state probabilities are:

$$\pi_4 = 0.212 \quad \pi_3 = 0.254 \quad \pi_2 = 0.322$$
$$\pi_1 = 0.147 \quad \pi_0 = 0.064$$

From these the gain is $50.09 per week. That means that on average, over the long run, the owner will obtain profits at that rate. Of course, in any given week the profit or loss will vary, so even though the business can be expected to be profitable over the long run, the owner will need a cash reserve to deal with the variations.

Now imagine that the owner has an opportunity to invest in a one-time "sales booster" that promises to alter the demand distribution permanently. This might be a more prominent display or a billboard extolling the virtues of this particular refrigerator. It will cost $10,000 (once), and it is expected to modify the probability of selling one unit only slightly. Specifically, the daily sales distribution will change from $P(0) = 0.5$ and $P(1) = 0.3$ to $P(0) = 0.4$ and $P(1) = 0.4$. The probability of selling two units in a single day will remain the same. The question is whether the investment would be worthwhile.

Under the given assumptions, the revised transition matrix is

$$\mathbf{P} = \begin{bmatrix} 0.4 & 0.4 & 0.2 & 0 & 0 \\ 0 & 0.4 & 0.4 & 0.2 & 0 \\ 0 & 0 & 0.4 & 0.4 & 0.2 \\ 0.4 & 0.4 & 0.2 & 0 & 0 \\ 0.4 & 0.4 & 0.2 & 0 & 0 \end{bmatrix}$$

We do not have to revise the reward matrix because the allowable transitions and the associated rewards are the same as before. When we recalculate the gain, it turns out to be $62.18 per week, which is certainly higher than the $50.09 that we had previously. There is no question that the owner would eventually earn back the investment, but the real issue is whether that would happen fast enough to justify the investment.

If the investment were a recurring cost (like, for example, a weekly advertising fee), we would just include it in the reward matrix. However, this is a case where we need to discount all the future earnings in order to compare to the one-time immediate cost. The results will be very sensitive to the discount rate, so we must choose this number carefully. Also, we must understand that the discounted total reward will not be a guaranteed value; it is the expected value of a random variable. Although the calculation incorporates a representation of risk, the risk is not eliminated. Hence, when we compare the uncertain gain to the certain cost, we should allow an extra margin to compensate for the risk. If the expected total discounted total reward resulting from the investment turns out to be less than or only slightly greater than the cost of the investment, it is not very attractive.

Let us assume for this case study that the owner can obtain a "certain" return of 12 percent on his normal business investments, and would therefore require more than that for this risky opportunity. We will use an annual interest rate of 12 percent, which translates to 0.23077 percent per week. That value in turn translates to a discount factor of $\alpha = 0.9977$.

Now, for the discounted total return vector, we use the equation

$$\mathbf{v}(\alpha) = (\mathbf{I} - \alpha \mathbf{P})^{-1} \mathbf{v}$$

which gives us, after all of the calculations, the vector

$$\mathbf{v}(\alpha) = \begin{bmatrix} 27,066 \\ 26,896 \\ 26,719 \\ 26,512 \\ 26,344 \end{bmatrix}$$

This is a vector of dollar values that represent the discounted expected total reward that will result from the investment of $10,000, differing only in what the inventory level is at the time of the investment. For every initial state, however, the discounted return is much higher than the cost, so the investment is clearly attractive.

4.11 Recommended Reading

This chapter explained how to formulate objective functions that depend upon an underlying process that is modeled as a Markov chain. It stopped short of explaining how to optimize those objective functions (typically subject to constraints) because to do so would require a good deal of additional theory that is extraneous to the major purpose of the book. However, the reader should be aware that such optimization is quite feasible, using the methods of dynamic programming and/or linear programming. If you are interested and have the necessary background in optimization theory, there are plenty of resources to learn from. The key phrase to look for is "Markov decision process."

If you can find it, one of the best introductions to the topic is Ronald Howard's doctoral dissertation published as a book (2), in which he essentially created the field. His two-volume treatment (3) first appeared about ten years later, and has been in print ever since. The other books below provide treatments of varying depth.

1. Bhat, U. N. *Elements of Applied Stochastic Processes*, 3rd. ed. Wiley, New York, 2002.
2. Howard, R., *Dynamic Programming and Markov Processes*. MIT Press, Cambridge, 1960.
3. Howard, R., *Dynamic Probabilistic Systems*, vols. I and II. Wiley, New York, 1971 (reprinted by Dover in paperback, 2007).
4. Kao, Edward, *An Introduction to Stochastic Processes*. Brooks/Cole, New York, 1996.
5. Kulkarni, V. G., *Modeling, Analysis, Design, and Control of Stochastic Systems*. Springer-Verlag, New York, 1999.
6. Mine, H. and S. Osaki, *Markovian Decision Processes*. Elsevier, New York, 1970.
7. Puterman, Martin L., *Markov Decision Processes: Discrete Stochastic Dynamic Programming*. Wiley-Interscience, New York, 2005.
8. Tijms, H., *A First Course in Stochastic Models*. Wiley, New York, 2003.
9. White, D. J., *Markov Decision Processes*. Wiley, New York, 1993.

Chapter 4 Problems

1. Return to the first example presented for the formulation of Markov chains, in which a trucker carried loads among four cities. Suppose that we wanted to calculate the expected total profit he will realize over his next three trips. The profit for each trip (revenues minus costs) can be figured separately, but the uncertainty in which routes will be taken implies that the total profit over several trips will be a random variable. Suppose that individual trip profits are given by the reward matrix (this is different from the one in the text at the beginning of the chapter):

$$\mathbf{R} = \begin{bmatrix} 0 & 100 & 150 & 150 \\ 200 & 0 & 180 & 220 \\ 80 & 90 & 20 & 30 \\ 190 & 200 & 60 & 0 \end{bmatrix}$$

a. If the driver is currently in Baltimore, what is his or her expected profit over the next three trips?

b. What is his or her long-term average profit per trip?

2. Using the same example, suppose that the driver has another option. Instead of waiting for the dispatcher to assign a load in Chicago (which has a relatively low expected immediate payoff), he or she could choose to a load to Detroit (which has a significantly better expected immediate payoff). Assume that the driver can elect to take a load from Chicago to Detroit (at the same profit as before) without waiting for a dispatcher's decision. That assumption would alter the transition matrix in one row, but would not change the reward matrix. Would the driver be wise in the long run to take that option? (Hint: Repeat the analysis with the altered transition matrix and compare the long-term expected gain rate.)

3. Referring to the commercial fishing problem that you were asked to formulate in Chapter 2, problem 1, and calculate some things in Chapter 3, problem 1, now let us suppose that the expected profits for a good, medium, and bad fishing day are $10,000, $3,000, and −$1,000 per day, respectively. (Note that these rewards are associated with the states, not the transitions.)

a. Assuming that the first day is good, what is the expected profit for the next week?

b. What is the expected profit per day over the long run?

4. Referring to the air quality problem that you were asked to formulate in Chapter 2, problem 3, and for which you calculated some things in Chapter 3, problem 4, now suppose that the number of people appearing at emergency rooms with critical breathing problems is strongly correlated with the air quality. In particular, the number of asthma cases increases dramatically as the air quality worsens. Suppose that the data

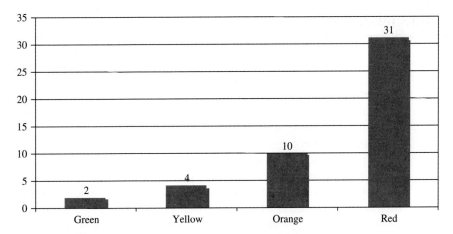

■ **FIGURE 4.2** Asthma cases as a function of air quality

shows the average number of asthma cases per day versus the air quality index is as shown in the Figure 4.2.

    a. If the air quality starts out in the orange category, what is the expected number of asthma cases in the city over the next five days?

    b. Over the long run, what will be the expected total number of asthma cases per week in this city?

5. Referring to the crop rotation problem that you were asked to formulate in Chapter 2, problem 5, and for which you calculated some things in Chapter 3, problem 5, there are different profits that can be expected from each usage, but there are also costs associated with changing usage. For example, if you want to convert from fallow to corn, you have to plow and harrow less thoroughly than if you are going to plant soybeans. There is also a penalty paid, in terms of yield, if the same crop is grown in successive years because the soil suffers a kind of fatigue. The expected profits per acre for the various crops, ignoring the transition costs, are corn, $180; soybeans, $150; other, $120; fallow, $0. The transition costs per acre are shown in the following table:

	Corn	Soybeans	Other	Fallow
Corn	−40	−25	−30	−5
Soybeans	−20	−30	−20	−5
Other	−15	−20	−25	−5
Fallow	−20	−10	−15	0

    a. Construct an appropriate reward matrix to include both the profits and transition costs.

    b. What is the vector of values expressing the expected immediate profit (that is, in one year) for each type of crop?

    c. The gross statistical values that were used to estimate the transition probabilities for the Markov chain model could not be used to predict anything about any particular acre or farm (which, after all, would be dependent upon the decisions of the farmer), but our interest is in predicting agricultural profits for the state as whole, containing twelve million acres of farmland devoted to crops. Assuming that the prices and costs remain stable over time (we could adjust for inflation later), what is the long-term total expected profits from cropland agriculture in this state?

6. Referring to the robot control problem that you were asked to formulate in Chapter 2, problem 7, and calculate some things in Chapter 3, problem 6, suppose that we are interested in knowing something about the usage of the motors, in order to predict reliability. The failure time of motors, like lightbulbs and automobiles, is affected as much by the number of times it is started up as it is by the hours of use. Receiving the surge of power when turned on and going from a cold to a warm state place more of a load on the electronics than continuous operation. Hence, the wear is affected by the sequence of moves, as well as the relative use of each motor. For purposes of this problem, assume that switching from one motor to another instantaneously causes as much wear as one minute of use, and if the same motor is used for successive moves, no additional wear beyond the time of actual use is accumulated. For each move, the average duration of motor use is: Turn = 1 minute; Lift = 1.2 minutes; Extend = 0.6 minutes; Grip = 0.2 minutes.

    a. Construct an appropriate reward matrix to include both the direct times for the moves and the extra penalties for switching motors.

    b. Find the expected time (including penalties) for each move.

    c. What is the long-term expected time for each robot motion, in minutes per move?

7. Returning again to the case study of the call center first developed in Chapter 2, Section 2.8, let us use the methods of this chapter to estimate the cost of servicing a customer. First, we have to calculate the expected number of minutes of each operator's time will be consumed by an average call, taking full account of the fact that the call may return to the same operator multiple times. Assume that data collected from similar call centers suggests that the typical length of a single conversation involving sales is nine minutes, billing is twelve minutes, support is twenty minutes, and the supervisor is five minutes. Transfers are essentially instantaneous.

   a. What are the values for the vector v (expected immediate cost) in the context of this example?
   b. What is the expected total duration of an incoming call, including all of the operators who have to deal with it?

8. Continuing the previous problem, it is logical that the operators would be paid differently, based upon their level of expertise and experience. Assume that sales operators earn $24 per hour, billing operators earn $30, technical support operators earn $54, and supervisors earn $72. For purposes of this problem, assume that a call that consumes an operator's time will cost the company the wage rate per minute times the number of minutes (which is admittedly simplistic, but will suffice for this example). What is the expected total cost of serving one customer's call to the center?

9. Consider a model of a production process having the structure indicated in Figure 4.3.

There are two production operations, followed by an inspection. Assume that the cost incurred in putting the part through operation 1 is $20, whether or not it has been through before. The operation 2 cost is $10, regardless of what happens after. The cost of inspecting a part is $1 plus an additional cost of $3 and $2 if it is sent back to operation 1 or operation 2.

   a. Construct an appropriate reward matrix for the production costs.
   b. What is the expected immediate additional cost for each step of the process?
   c. What is the expected net cost per part, including accounting for rework?

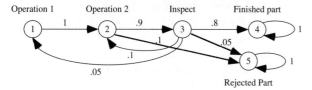

■ **FIGURE 4.3** A production process

10. Extending the previous problem, now turn attention to revenues (as opposed to the costs). Suppose that a scrapped part can be sent to a recycling facility, for which a salvage value of $3 will be received. Completed parts that have passed inspection are worth $100. No revenues are received while a part is in process.

   a. Construct an appropriate reward matrix for the revenues.
   b. What is the expected immediate additional revenue for each step of the process?
   c. What is the expected net revenue per part, including accounting for rework?

11. Referring to the accounting problem that you were asked to formulate in Chapter 2, problem 11, and calculate some things in Chapter 3, problem 11, let us now add monthly interest to outstanding accounts to track the value of payments. Part of the reason for charging interest is to compensate the lender for the risk that the debt will not be paid. The longer the debt goes unpaid, the greater the amount that will be collected if it is collected; however, the longer it goes unpaid, the lower the probability that it will be collected. Hence, it is not obvious what, for example, a three-month-old account is actually worth. We may use any convenient number as the starting value of the debt, so let us use $1. Assume that each month of delay adds 2 percent of the previous month's balance.

   a. Show how to express the compounding interest in the form of a reward matrix. (Hint: Only a few of the transitions have rewards associated with them.)
   b. What is the expected total reward for $1 of new account?
   c. What is the expected value of a three-month-old account?

12. Extending the previous problem, now assume that there exists an option to sell aging accounts to a collection agency at fifty cents on the dollar. That is, a collection agency is willing to take over the obligation at any point within six months of its beginning, paying half of the original size of the account. (It is counting on being able to collect using more aggressive tactics.) Is there any age at which it makes sense for the original holder of the debt to sell the account to the collection agency? If so, in what month should this happen?

13. In problem 13 of both Chapter 2 and Chapter 3, you were asked to model transfers of patients with hospital wards. Suppose now that the hospital administration wants to estimate the costs associated with hospital stays based upon the length of stay in each ward and the costs associated with transfers between wards (including admission and discharge). The previous model represented time using transfers as the epochs (not days), so we cannot use a costs per day in the various wards. Instead, we would need data on the average cost for entire length of stay in each ward. Assume that a stay in intensive care costs $6,000, in surgery costs $4,000, and in recovery costs $5,000. Also, there is extensive paperwork associated with each transfer and with discharge or death, and these costs are show in the following table:

Transfer	Cost
Intensive Care to Surgery	$400
Surgery to Intensive Care	$400
Intensive Care to Recovery	$300
Recovery to Intensive Care	$400
Death	$300
Discharge	$250

   a. Construct an appropriate reward matrix to reflect all of these costs.
   b. What is the expected total cost for the entire hospital stay of a patient who enters by way of the emergency room?
   c. What is the expected remaining cost for a patient in recovery?

CHAPTER 5

Continuous Time Markov Processes

CHAPTER CONTENTS

5.1. An Example 141
5.2. Interpreting Transition Rates 146
5.3. The Assumptions Reconsidered 149
5.4. Aging Does Not Affect the Transition Time 150
5.5. Competing Transitions 152
5.6. Sojourn Times 153
5.7. Embedded Markov Chains 154
5.8. Deriving the Differential Equations 155
5.9. Solving the Differential Equations 157
5.10. State Probabilities 159
5.11. First Passage Probability Functions 159
5.12. State Classification 160
5.13. Steady-State Probabilities 161
5.14. Other Computable Quantities 163
5.15. Case Study 165
5.16. Birth-Death Processes 167
5.17. The Poisson Process 169
5.18. Properties of Poisson Processes 171
5.19. Khintchine's Theorem 172
5.20. Phase-Type Distributions 173
5.21. Conclusions 175
5.22. Recommended Reading 175

This chapter extends the concepts and methods of Markov chains to the continuous time version. For some reason lost in history, we speak of Markov *processes*, rather than *chains*, in the continuous time case. This distinction in terminology is not very important, but the differences between the continuous time and the discrete time cases are.

Continuous time stochastic processes are similar in most respects to discrete time stochastic processes. Additional complexities occur, however, because each infinitesimal instant is available as a possible transition time. For example, it will not make sense to speak of a one-step transition matrix because time is not measured in steps. We will, however, be able to contain all of the data necessary to describe the transitions in a square matrix of dimension equal to the number of states. Also, the structural aspects that are apparent in a transition diagram will be the same. So, aside from the few additional complications in the mathematics, the continuous time case is really as easy to use in modeling.

The task of creating a continuous time model is very similar to that of creating a discrete time one: After deciding that you are going to treat time as continuous, you must identify the states; then you determine which transitions are possible. You might even draw a state transition diagram to visualize the states and transitions. After that, however, the differences appear. Instead of using one-step transition probabilities as the basic data, we will be forced to deal with transition *rates*. These are somewhat more difficult to obtain and to understand than the discrete time transition probabilities. Instead of matrix multiplication as a method for obtaining transition probabilities, it will be necessary to solve a set of differential equations. However, once past those obstacles, it will turn out that steady-state probabilities, mean first passage times, and all of the other quantities that were obtainable by solving linear equations are just as easy to obtain for the continuous time case as they were for the discrete time case.

5.1 An Example

EXAMPLE 5.1

To facilitate understanding, we will begin with a very small example that we can solve completely. The development of the equations will accentuate the comparison to the discrete time case, so that the concepts already learned can be translated easily. There are ther ways to reach the same results, but they might obscure the connections to the discrete case. We will skip several important details in order to get to results quickly. After the example is completely developed, we will present the theory in a way that includes full generality.

Consider the following situation that concerns the operation of an automatic loom used to weave cloth. Normally, the loom will operate without human intervention. Occasionally, however, a thread breaks and the shuttle may jam. There is an attendant standing by whose sole responsibility is to unjam the shuttle, tie threads, and put the loom back into operation. If you would rather think in terms of some other kind of machine that alternates between a "working" and "down" state, feel free to do so. We will only be keeping track of whether or not the machine is operating; what it does when it is operating is immaterial.

When time is continuous, we do not have as many options for selecting how to treat it. It will just match the real numbers, usually imagined as a continuous line. The units we use to measure time are a matter of choice: hours, minutes, seconds, and so forth. Pick one that makes the computations easy; in this example, the choice of hours makes sense. By the way, we will want to think of decimal units for fractions of an hour, not "clock" units based on sixty minutes. For example, an hour and a half would be 1.5, not 1:30.

Ignoring shift changes, lunch hours, and coffee breaks, the system will always be in one of two states. At state 1, the loom is shut off and the attendant is working to repair it; at state 2, the loom is operating and the attendant is idle. A typical realization of the process would resemble Figure 5.1. The upper lines, representing continuous intervals of time during which the loom is working, could be called "operating times." The lower lines could be called "repair times."

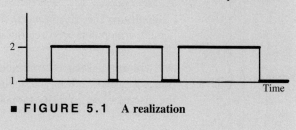

■ **FIGURE 5.1** A realization

The transitions can occur at literally any instant, so we cannot represent the stochastic process as a discrete sequence of random variables. Instead, we must define it as an uncountably infinite family of random variables. That is, at any specific time t on the continuum of values from 0 to infinity, there is a random variable X(t).

Definition A *continuous time stochastic process* is an uncountably infinite set of random variables $\{X(t), t \geq 0\}$ indexed by a continuous variable t. In the absence of any simplifying assumption, all of these random variables are related to one another.

One of the ways to simplify (there are some others, which are not quite so convenient), is to use Markov's assumption.

> **Definition** A *continuous time Markov process* is a continuous time stochastic process in which, for all state sequences $j_0, j_1, j_2, \ldots, j_n$ and all times $t_0 < t_1 < t_2 < \ldots, < t_n$, the following equation holds
>
> $$\mathbf{P}(X(t_n) = j_n | X(t_{n-1}) = j_{n-1}, X(t_{n-2}) = j_{n-2}, \ldots, X(t_0) = j_0)$$
> $$= \mathbf{P}(X(t_n) = j_n | X(t_{n-1}) = j_{n-1})$$

In words, the assumption is much the same as in the discrete case. It says that the probability of being in a particular state at a particular time depends only on the most recent information, not anything older. However, the Markov assumption is quite a bit stronger in the continuous case than it was in the discrete case. We must assume that at *every* point in time (not just transition epochs), the future beyond that point depends only on the state at that time. Nothing from the past—not even the length of time it has been in the present state—can influence the future. We will have more to say about this assumption shortly, but for now we will just accept it and move on.

We will also want to make use of a stationarity assumption.

> **Definition** A Markov process is *stationary* if for all states i and j and all positive time intervals t and s,
>
> $$\mathbf{P}(X(t+s) = j | X(s) = i) = \mathbf{P}(X(t) = j | X(0) = i)$$

This just says that the probabilities for the same transition in the same time interval are the same whenever in absolute time they occur. Again, we will return to a more thorough discussion of stationarity later. For now, just assume that the probability mechanism that is used to describe the process remains constant over time.

As we did in the discrete case version, we can draw a picture of the allowable transitions, using nodes for states and arrows for transitions.

> **Definition** The *transition diagram* for a continuous time Markov process is a directed graph with points representing states and arcs representing possible transitions.

We will use the same terminology for this picture in both the discrete and continuous time cases, but caution you against treating them exactly alike. The values on the arcs for the discrete time case are probabilities; in the continuous time case, they are something different.

Up to this point, the only distinctions between the continuous time version and the discrete time version have been adjustments in definitions and notation. Conceptually, they are still identical, so we can still think in the same way to achieve a formulation of a model. At this point, however, important differences emerge. In the discrete case, every discrete point in time was an epoch—a point in time at which a transition could occur. In continuous time, the epochs are only those instants at which transitions actually occur.

In a continuous time Markov process, there will be no transitions from any state to itself. Instead, we would simply say that there is no transition; or that the process stays in the state until it leaves. The continuous time transition diagrams will not have loops. (Recall that in discrete time, the means for keeping a process in a state for more than one epoch was to define a "virtual" transition from that state to itself.) Technically, it is possible and sometimes convenient to define virtual transitions in continuous time (that is, transitions that establish epochs in time but do not change the state), but that complicates both the formulation and the analysis. For that reason, we will ignore that variation and will assume that all transitions are to distinct states.

In a continuous time transition diagram, the values associated with the arrows cannot be transition probabilities. Instead, they will be transition *rates*. In any continuous time Markov process, any particular transition between distinct states, say from i to j, is characterized fully by one parameter, called a transition rate.

> **Definition** A *transition rate* for a direct transition from state i to a different state j, denoted λ_{ij}, is a constant ≥ 0 that expresses the speed of transition to state j given that the process is in state i.

That definition is a bit imprecise, but we will eventually provide a more complete explanation. The set of all of the transition rates (for all i and j, with $i \neq j$) describe the entire continuous time Markov process. If we construct a transition diagram, we can label each arrow with the corresponding λ_{ij} to express the data of the model. We will come back later to the question of how to estimate these transition rates from real-world data. For now, just accept the fact that the λ_{ij} provide the numerical data to describe the process, much as the p_{ij} did in the discrete case. They will always be positive values, possibly larger than 1 (they are *not* probabilities), or they will be 0 when there is no possibility of a direct transition from i to j. So all of the arrows that appear in the transition diagram should be labeled with positive values.

For the two-state example, the transition diagram would look like Figure 5.2. We will also want to create a matrix of these values, which will then provide all of the data necessary to calculate anything of interest, just as the transition matrix **P** did in the discrete case. First, however, we have to say something about the diagonal elements λ_{ii}. As already mentioned, there cannot be transitions from any state to itself in the continuous time case, so we will not have loops in the transition diagram or transition rates for the diagonal positions. Those positions are available for something else. For mathematical reasons that will emerge later, it turns out to be convenient to define the diagonal term in a particular way.

> **Definition** The diagonal term λ_{ii} of the transition rate matrix Λ is given by
> $$\lambda_{ii} = -\sum_{j \neq i} \lambda_{ij}$$

In words, the diagonal element is set equal to the negative value of the sum of the other transition rates in the same row. Or, in terms of the transition diagram, we add up the total of all transition rates out of state i, take the negative of that value, and use that for λ_{ii}. Another way to look at it is to set λ_{ii} to that value which forces the row sum to be 0. These diagonal terms will always be negative, unless the state is absorbing, in which case it will be 0.

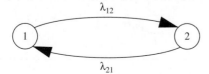

■ **FIGURE 5.2** A transition diagram

Definition The transition rate matrix, denoted Λ, is the matrix of all λ_{ij}, where $i \neq j$, and the diagonal terms are given by
$$\lambda_{ii} = -\sum_{j \neq i} \lambda_{ij}$$

EXAMPLE 5.2

Returning to the example, let the transition rates be three for the repair transition and two for the breakdown transition. That is, let

$$\lambda_{12} = 3 \qquad \lambda_{21} = 2$$

We are using whole numbers just for convenience; they are not likely to be so simple in real applications. Then the matrix Λ will be

$$\Lambda = \begin{bmatrix} -3 & 3 \\ 2 & -2 \end{bmatrix}$$

Once you have Λ, the formulation is complete. Before moving on to calculations, you should pause to consider whether the Markov and stationarity assumptions are warranted in the particular situation you are modeling. That consideration will require a full understanding of what those assumptions mean in continuous time, so we will return to a discussion of those issues. For now, we press on to compete the overview.

There will not be any one-step or n-step transition probabilities in continuous time, because there are no steps. In their place, we have a function of continuous time.

Definition A transition probability function, denoted $p_{ij}(t)$, is defined by
$$p_{ij}(t) = P(X(t) = j | X(0) = i)$$

A particular $p_{ij}(t)$ represents the conditional probability that the system is in state j at time t given that it was in state i at time 0. It does not say anything about what happened between time 0 and t, so the process might have visited numerous other states. These $p_{ij}(t)$ functions are the continuous time analogs of what would have been n-step transition probabilities in the discrete case—the $p_{ij}(n)$ viewed as functions of n. For every value of t, the value of each of these $p_{ij}(t)$ functions must be a probability, so none of them could either become negative or larger than 1.

We can also tell from the definition what the values of these functions must be at time $t = 0$.

Result For $i \neq j$, $p_{ij}(0) = 0$, and for $i = j$, $p_{ii}(0) = 1$

This result simply says the obvious: that the process cannot be in two different states at the same time. But it is useful to have those values as initial conditions when you are solving differential equations.

For every value of t, the sum of the probability functions over all second subscripts (and a fixed first subscript) must equal 1, because the process must be in some state at each t.

Result In any continuous time Markov process and for any i and any t,
$$\sum_j p_{ij}(t) = 1$$

Since there are only two states in our small example, we are speaking of four functions: $p_{11}(t)$, $p_{12}(t)$, $p_{21}(t)$, and $p_{22}(t)$. In larger problems with S states, there would be S^2 such functions. All of these functions are determined by the values contained in the matrix Λ, though it is not at all obvious how. Later in the chapter, we will go though the derivation of the equations. It will turn out that they are of the form

$$\frac{d}{dt}p_{ij}(t) = \sum_k p_{ik}(t)\lambda_{kj}$$

EXAMPLE 5.3

In order to finish out the example, let us accept the fact that the $p_{ij}(t)$ functions are given as the solutions to these differential equations, where the constant coefficients come from the Λ matrix.

$$\frac{d}{dt}p_{12}(t) = 3p_{11}(t) - 2p_{12}(t)$$
$$\frac{d}{dt}p_{11}(t) = -3p_{11}(t) + 2p_{12}(t)$$
$$\frac{d}{dt}p_{21}(t) = -3p_{21}(t) + 2p_{22}(t)$$
$$\frac{d}{dt}p_{22}(t) = 3p_{21}(t) - 2p_{22}(t)$$

Thus, we have a system of linear first-order differential equations with constant coefficients. In this particular example, they are simple enough to solve directly (though in larger problems they may pose a challenge to solve). Making use of whatever manipulations or solution technique we find most convenient, and using the initial conditions $p_{ij}(0) = 0$ for $i \neq j$ and $p_{ij}(0) = 1$ for $i = j$, we obtain the solutions (we are skipping a whole lot of math to display the final result):

$$p_{11}(t) = \frac{2}{5} + \frac{3}{5}e^{-5t} \qquad p_{12}(t) = \frac{3}{5} - \frac{3}{5}e^{-5t}$$
$$p_{21}(t) = \frac{2}{5} - \frac{2}{5}e^{-5t} \qquad p_{22}(t) = \frac{3}{5} + \frac{2}{5}e^{-5t}$$

These are conveniently expressed in the matrix form (a matrix of functions):

$$\mathbf{P}(t) = \begin{bmatrix} p_{11}(t) & p_{12}(t) \\ p_{21}(t) & p_{22}(t) \end{bmatrix} = \begin{bmatrix} \frac{2}{5} + \frac{3}{5}e^{-5t} & \frac{3}{5} - \frac{3}{5}e^{-5t} \\ \frac{2}{5} - \frac{2}{5}e^{-5t} & \frac{3}{5} + \frac{2}{5}e^{-5t} \end{bmatrix}$$

If you want to see what these functions look like, we can display them as a matrix of graphical functions, as in Figure 5.3.

Notice that the functions behave as they should in order for them to represent probabilities. They lie uniformly within the interval [0, 1] for all values of $t \geq 0$, the sum over the second subscript equals 1 for all values of $t \geq 0$, and the initial conditions are satisfied. Observe also that the limit of each function as t goes to infinity is immediately apparent, both in the function itself and in the graph. Convergence to this value is smooth and monotonic (as opposed to

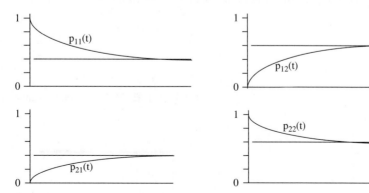

■ **FIGURE 5.3** The four functions graphed

discontinuous, oscillating, or both), and functions with the same second subscript but different first subscripts converge to the same value. These points, as illustrated by this particular example, are generally true in processes of this kind.

Having these four functions enables us to evaluate the probability, at any specific point in time t, of being in state j given that the process started in state i. That situation is parallel to having all powers of the **P** matrix in discrete time, so that you could look up the transition probability for any value of n.

5.2 Interpreting Transition Rates

Now that you have seen one small example carried out to solution, let us go back to fill in some of the details we skipped. One glaring omission was the meaning and source of the transition rates. We said that they provide the basic data to determine the process, but have not yet said where they come from or what they mean.

In order to help interpret the λ_{ij}, modify the process slightly to isolate just one transition. In the loom problem, instead of having an alternating pattern of failures and repairs, we will have just the one transition from state 1 to state 2, as shown in the transition diagram in Figure 5.4. If we were dealing with a larger process, we could get the same picture by isolating any single transition. The only difference is that states 1 and 2 could be any states i and j.

The realization of this process would look something like that shown in Figure 5.5.
The Λ matrix would have to be

$$\Lambda = \begin{bmatrix} -3 & 3 \\ 0 & 0 \end{bmatrix}$$

because there is now no transition from 1 back to 0. The differential equations (we have still not explained why we are using these equations, but will come back to that also) are now

$$\frac{d}{dt}p_{11}(t) = -3p_{11}(t)$$
$$\frac{d}{dt}p_{12}(t) = 3p_{11}(t)$$

There are only the two equations, because there are no transitions starting with 2 as the initial state.

We can solve the first differential equation easily because there is just one function and the differential equation is a familiar one. (You may recognize it more easily in the form $f'(t) = kf(t)$.) The solution is just the exponential function:

$$p_{11}(t) = e^{-3t}$$

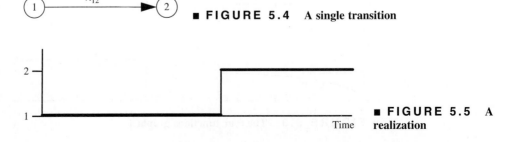

■ FIGURE 5.4 A single transition

■ FIGURE 5.5 A realization

The constant of integration is evaluated using the fact that $p_{11}(0) = 1$. From this we can also conclude, since there are only two states,

$$p_{12}(t) = 1 - p_{11}(t) = 1 - e^{-3t}$$

In these equations, the constant 3 is the transition rate from state 0 to 1.

In the general case, for any continuous time Markov process, if we isolated a particular transition from i to j and kept the symbol λ_{ij} for the transition rate, we would have

$$p_{ii}(t) = e^{-\lambda_{ij}t}$$

and

$$p_{ij}(t) = 1 - p_{ii}(t) = 1 - e^{-\lambda_{ij}t}$$

Now we can place a direct interpretation on the transition rate. Let T_{ij} be a random variable representing the time that the process takes to make the transition from i to j (with the understanding that that is the only transition possible). Then the probability distribution function of T_{ij} can be related to the solutions we just obtained by realizing that

$$P(T_{ij} > t) = p_{ii}(t)$$

In words, the time to make the transition exceeds t if, and only if, the process is still in the initial state i at time t.

So the complementary cumulative distribution function of T_{ij} is $e^{-\lambda_{ij}t}$ and the cumulative distribution function is 1 minus that quantity. Those functional forms happen to specify a recognizable distribution—namely, the negative exponential distribution. In fact, we now know a lot about T_{ij} and a lot more about continuous time Markov processes in general.

> **Result** The transition time T_{ij} for any specific transition in a Markov process is a negative exponentially distributed random variable with parameter λ_{ij}.

The mean of any negative exponentially distributed random variable is the reciprocal of the distribution parameter, so

$$E(T_{ij}) = \frac{1}{\lambda_{ij}}$$

This tells us immediately what λ_{ij} is and gives us an indication of how to estimate it from data. Namely,

> **Result** For $i \neq j$, and where a direct transition from i to j is possible,
>
> $$\lambda_{ij} = \frac{1}{E(T_{ij})}$$

In words, the transition rate is the reciprocal of the mean time to make the transition. This relationship makes sense when you think about the ordinary association between rates and times in any other context. For example, if you were able to walk a mile in fifteen minutes ($\frac{1}{4}$ hour), you would be traveling at the rate of four miles per hour. Here, our rate of transitions was three per hour, which translates to a mean time of $\frac{1}{3}$ hour to make the transition.

We could estimate the transition rate statistically by observing many instances of the transition, averaging all the times observed for that same transition (that is, estimating the mean time $E(T_{ij})$), and then take the reciprocal. The units of λ_{ij} would be "transitions per time unit." For example, if the average time turned out to be $\frac{1}{2}$ minute, λ_{ij} would be two transitions per minute.

EXAMPLE 5.4

Returning to the loom example, which could actually represent any two-state Markov process, let us imagine that we have just begun to formulate the model and need to estimate the parameters. Take a look at Figure 5.1 again to see a realization. The lower bars represent times spent in state 1 prior to making the transition to state 2. We can use those observations to estimate the expected value of T_{12}. We do this with an ordinary sample mean; that is, we add up all the values and divide by the sample size. We would do the same with the upper bar values to estimate the expected value of T_{21}. Once we have the expected transition times, we convert them to rates in the obvious way.

Just to be sure this estimation process is clear, suppose we have a realization of a process we are trying to model and that realization contains n_{ij} instances of transitions from i to j. Let $T_{ij}(k)$ be the kth observed value of a transition time from i to j in the realization. We get (as a statistical estimate)

$$E(T_{ij}) \approx \frac{\sum_{k=1}^{n_{ij}} T_{ij}(k)}{n_{ij}}$$

Then we use

$$\lambda_{ij} = \frac{1}{E(T_{ij})}$$

Another way to interpret the transition rates is not very helpful in estimating the values from data, but has its uses in the theoretical development that we will be using to show where the differential equations come from. The $P_{ij}(t)$ functions are continuous, well-behaved functions of t. They possess derivatives at every value of $t \geq 0$. See the result below.

Result For every i, j (including the case where i = j)

$$\lambda_{ij} = \frac{d}{dt} p_{ij}(t)|_{t=0}$$

In words, the transition rate for a particular transition (from i to j) is the derivative of the function $p_{ij}(t)$ evaluated at t = 0. Or, if you prefer to think in more graphical terms, it is the slope of the function at the point t = 0. It turns out that the same relation holds even when i = j, so even the diagonal terms of Λ can be interpreted in this way. You may want to look at the solution to the two-state problem on page 144 to verify that this relation holds for that case.

Although the interpretation is true for the most general cases, it is not particularly useful as a way of obtaining the transition rates, because we need the rates to determine the functions. Still, it is a surprising and interesting fact that the entire Markov process is determined just by the slopes of the functions at time 0. And, as mentioned, the derivative interpretation is useful in Section 5.8, "Deriving the Differential Equations."

5.3 The Assumptions Reconsidered

As was the case in discrete time, the derivation of the differential equations emphasizes the assumptions, but the result is in a form that permits use without understanding. That is, the λ_{ij}'s could be measured and the equations could be expressed and solved in the complete absence of awareness of the assumptions. Furthermore, to state the assumptions is not enough. You must be sufficiently aware of their implications to judge whether or not they are reasonable in particular contexts. The Markov and stationarity assumptions are both strong assumptions—especially so when time is treated as continuous. There are many situations where they are inappropriate. In fact, as you begin to appreciate how strong they are, you may begin to doubt whether any real-life process behaves as a stationary Markov process. It is an empirical fact, however, that many such real-life processes do exist. So the object now is to develop sufficient understanding to be able to distinguish between those processes that might properly be modeled as stationary Markov processes and those that should not.

Let us consider what the Markov and stationarity assumptions mean in the particular case of our loom example. When we assume that the loom failure process satisfies the Markov property, we are asserting that if at *any* time we know whether the loom is operating or shut off, we can determine all probabilities associated with the process from that time on. We do not need to know, nor does it even help us to know, the sequence of states leading up to the present one, how long the process spent in each of these states, or even how long it has been in the present state. It is common to personify this property and say that the process has no memory, or is forgetful at every point of time. Of course, in most cases, the process will be incapable of having memory in the literal sense (unless you are modeling human or animal behavior, in which case you should be especially careful). What we really mean when we say this is that the process operates *as if* it has no memory.

The process will be stationary if the breakdown and repair rates do not depend on the time of day or day of the week. It would not be stationary if, for example, the attendant worked slower at certain times.

We have seen that the Markov and stationarity assumptions imply that the times between events must be negative exponentially distributed. The parameters of these distributions, the λ_{ij}'s, may be dependent on the state occupied, i, and the next state, j, but all of the distributions *must* be of the negative exponential form. No other distribution family can even be considered for describing the times between events.

Recall that the negative exponential density function is of the form $f(t) = \lambda e^{-\lambda t}$. Figure 5.6 plots this function for three values of λ. Notice that the function intercepts the vertical axis at λ, that it diminishes monotonically to zero (asymptotically), and that the rate of convergence is proportional to λ. The total area under the curve is, of course, always equal to 1, as it must be for any density function. Varying λ, the only parameter of the distribution, cannot alter the basic form shown in Figure 5.6. The mean is $1/\lambda$ and the

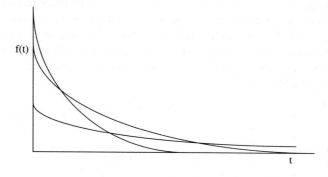

■ **FIGURE 5.6** The negative exponential density function

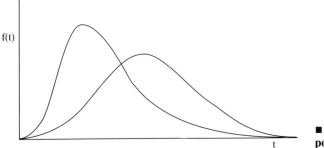

FIGURE 5.7 Other possible density functions

variance is $(1/\lambda)^2$. Thus the mean and variance are not separately adjustable, as one might prefer.

In many applications, by contrast, the times between events are most naturally conceived of as having a density function of the general form shown in Figure 5.7 (perhaps a gamma or Weibull). That is, one tends to think in terms of some nominal value, the mean, plus or minus some relatively minor variation. Or, put another way, the most likely values are considered to be clustered about the mean, and large deviations from the mean are viewed as increasingly unlikely. However, the form of the negative exponential density function implies that the most likely times are close to *zero,* and very *long* times are increasingly unlikely. If this characteristic of the negative exponential distribution seems incompatible with the application you have in mind, a simple Markov model may be inappropriate. However, there is a trick that can sometimes help you to model the kind of distributions shown in Figure 5.7 without losing the desirable Markov methodology. This trick will be explained in Section 5.20, "Phase-Type Distributions."

5.4 Aging Does Not Affect the Transition Time

The forgetfulness property of the negative exponential distribution has an interesting—some might even say strange—implication. If we say that we are in a state and some time has already passed since we entered it, and we ask about the *remaining* time until a transition occurs, the answer is, "The same as when we started." That is because the process is forgetful; it does not have any awareness of the elapsed time. If some more time passes, and we ask the question again, the answer is the same. No matter how much time has already passed, the distribution of remaining time remains exactly the same negative exponential with exactly the same parameter (and hence mean) value. This is not the sort of answer you expect when you are waiting for something to happen. Normal human intuition would suggest that if you begin an activity that has a finite expected time to completion, then the remaining time to completion should become shorter as time passes. It may be difficult to imagine any other possibility.

We can prove that a negative exponential random variable has this property, and the proof is instructive. So let T be such a random variable, and let λ be its parameter. (We could add subscripts to make it refer to any particular transition, but will instead keep it simple.) Now if a certain time s has already passed, the complementary cumulative distribution of remaining time T − s would be

$$P(T - s > t | T > s)$$

or

$$P(T > (t + s) | T > s)$$

But this is an ordinary conditional probability which we can evaluate as

$$P(T > (t+s) | T > s) = \frac{P((T > t+s) \cap (T > s))}{P(T > s)}$$

The numerator is the intersection of two events, one of which is contained within the other; in other words, if T is greater than $t + s$, it is automatically greater than s. (None of these values can be negative.) So the intersection becomes just $P(T > t + s)$. Now we just substitute from the complementary cumulative distribution function of the negative exponential, which is

$$P(T > t) = e^{-\lambda t}$$

That substitution gives

$$P(T > (t+s) | T > s) = \frac{e^{-\lambda(t+s)}}{e^{-\lambda s}}$$

Now the exponents cancel to eliminate everything involving s, leaving

$$P(T > (t+s) | T > s) = e^{-\lambda t}$$

That is the result we were seeking. The complementary cumulative distribution function of the remaining time, given that an interval of length s has already passed, is *exactly* what it was at the beginning. You can see from this argument that the exponential function had just the right form; nothing else would have worked. That is why the negative exponential distribution is the only one that is consistent with the Markov assumption.

The forgetfulness characteristic of the negative exponential distribution may help you decide whether it is appropriate to use a Markov model. To use a specific example, suppose that you want to represent some process in which the time between a change of state corresponds to the lifetime of an automobile tire. The state changes when the tire fails. Now there are at least two distinct kinds of tire failure. If what you have in mind is the kind of failure that is due to road hazards—that is, blowouts—then the distribution can legitimately be assumed to be negative exponential. There is no logical connection between road hazards and the age of the tire, so the fact that a blowout has not yet occurred would not suggest that the remaining time to failure is any more or less than when the tire was new. In other words, the lifetime distribution is forgetful. Since the only forgetful distribution is the negative exponential, the Markov assumption would be justified in this case. On the other hand, if you are interested in wear-out failures, the process would not be forgetful. How long the tire has been in use, as well as its age, would affect our estimate of its remaining life. In this instance, a simple Markov model would be inappropriate (though there are ways to achieve the same effect in a Markov model, using tricks explained in Section 5.20, "Phase-Type Distributions".)

Here is another way to think about the age-independence of transition times. In actuarial tables and in reliability data, it is common to express the time to death or failure in the form a *hazard rate function*, defined as the probability density function of the time, divided by the complementary cumulative distribution function. Formally,

$$h(t) = \frac{f(t)}{G(t)}$$

You may think of that (loosely) as the conditional probability that the event of interest will occur in the next time increment, given that it has not happened yet. For example, life insurance tables might chart the death rate for male individuals who have reached a given age, which would be a discrete version of the hazard rate function.

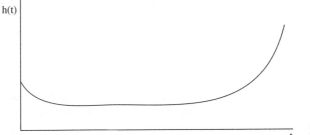

FIGURE 5.8 A typical hazard function

In graphical form, a typical hazard function might look something like Figure 5.8. This characteristic shape has been called a bathtub curve because of its resemblance to the cross-section of a bathtub. The declining values at the start of the curve (for small t) reflect a reduction in the risk of death or failure as the people or items survive factors affecting infant mortality or "burn-in" defects. At the other end of the curve, the risks increase because of age or wear. In the middle, the causes of death or failure are essentially random accidents that have nothing to do with age, so the curve is relatively flat.

The Markov assumption corresponds to a constant hazard function. In fact, for the negative exponential distribution,

$$h(t) = \frac{f(t)}{G(t)} = \frac{\lambda e^{-\lambda t}}{e^{-\lambda t}} = \lambda$$

On the other hand, any inter-event time that falls or rises (as in the ends of the bathtub curve) or in any way changes with time could not satisfy the Markov property. Knowing that can help you to decide whether the Markov assumption is appropriate.

5.5 Competing Transitions

We used a derivation of the equations that emphasized the similarity of the continuous time Markov process to the discrete time Markov chain. We then gradually discovered that the transition times had to be negative exponential in order to be consistent with the assumptions. There is a completely different way to do it that starts from the assumption that all transition times are independent negative exponential random variables. We will not go through that development here, but there are some lessons to convey from that perspective.

Imagine that the process is in some particular state i at any time t and you are anticipating what will happen next. For each of the possible transitions out of state i, there is an arrow in the transition diagram. Focus just on those arrows for now, as indicated by Figure 5.9. We can think of the next transition that occurs as being the winner in a race among

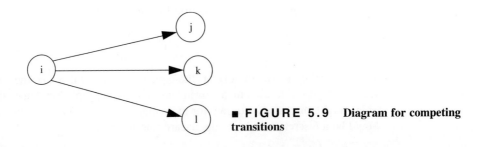

FIGURE 5.9 Diagram for competing transitions

the possible transition times, where each time is a negative exponentially distributed random variable having a parameter given by the arrow value. If we were programming a simulation, we would generate three random variates from the appropriate distributions, find the minimum sample value, and that would determine which state to go to next. The other sample values would be discarded.

Quite commonly, Markov models are formulated from knowledge that competing processes drive the transitions. For example, if the state were the number of items in stock—an inventory—then a demand process would be driving the state lower and a replenishment process would be driving it higher. Whichever of these two competing processes happens first would determine the next transition. In Chapter 6 we will see queueing models in which a service process is competing with an arrival process. If the service completes first, the number of customers in the system will decrease; if the next arrival occurs first, the number will go up. Failure and repair processes would be similar.

In all of these competing process models, the occurrence of an event in one process will cause a state transition, leaving the other process (or processes) with either an interruption (requiring a restart) or a remaining time to the next event. For example, if an arrival in a queueing process occurs, the customer that was in service at that moment will have some remaining service time. If the model were anything but Markovian, that would present a serious problem, because the remaining time would not be the same as if the same state were entered in a different way or at a different time. But, because of the forgetfulness property of the negative exponential distribution, remaining times are distributed exactly as if no time had passed, or as if all of the competing processes begin anew at the instant a new state is entered. Markov models handle this problem nicely; anything else would be much more difficult.

5.6 Sojourn Times

The sojourn time in a state is defined as the total time spent in the state before making the next transition, regardless of to which state it makes the transition. (The word "sojourn" means a temporary stay.) We will use τ_i, the Greek letter tau, to represent the random variable for sojourn time in state i and h_i for the mean of that random variable. Formally,

$$\tau_i = \min\{T_{ik}\}$$

where the T_{ik} are random variables for all of the possible transitions out of state i and the minimum is taken over all states k that can be reached directly from i.

The easiest way to find the distribution of τ_i is to use the complementary cumulative distributions of the T_{ik}. We know that each of these is given by

$$P(T_{ik} > t) = e^{-\lambda_{ik} t}$$

and they are all independent. Now think about the event represented by the relation $\tau_i > t$. If the sojourn time is greater than t, that means that all of the T_{ik} are greater than t (because if any were less than t the transition would have occurred). So the sojourn time is greater than t if, and only if, the first T_{ik} is greater than t, *and* the second one is also, *and* the third, and so on. The "ands" correspond to the intersection of events. If the events are independent, as they are here, we can multiply the probabilities. Hence,

$$P(\tau_i > t) = \prod_k P(T_{ik} > t)$$

$$P(\tau_1 > t) = \prod_k e^{-\lambda_{ik} t}$$

But since multiplying in this way becomes addition of the exponents,

$$P(\tau_i > t) = e^{\sum_k (-\lambda_{ik} t)}$$

$$P(\tau_i > t) = e^{-\left(\sum_k \lambda_{ik}\right) t}$$

But this means that the complementary cumulative distribution of the sojourn time τ_i has the form of a negative exponential random variable—one with the parameter equal to the sum $\sum_k \lambda_{ik}$. This is the result we were after.

> **Result** The sojourn time for any state i in a continuous time Markov process must be negative exponentially distributed with parameter
>
> $$\sum_{k \neq i} \lambda_{ik}$$

You may recall from Section 3.6 on page 82 that the sojourn time distribution in a discrete time Markov chain is geometric; this negative exponential distribution is the analog for the continuous time case. Once we know the distribution, we can get the mean, the variance, or any probability. For example, the mean sojourn time h_i is the reciprocal of the sum of the transition rates out of the state.

> **Result** The mean sojourn time in state i is given by
>
> $$h_i = 1 / \left(\sum_{k \neq i} \lambda_{ik} \right)$$

Notice that the sum that appears in the sojourn time results is the same sum we used when we determined the diagonal elements of Λ. The term λ_{ii} was the negative of that sum, so we can also express the mean sojourn time as shown below.

> **Result** The mean sojourn time in state i is given by
>
> $$h_i = -1/\lambda_{ii}$$
>
> This, of course, will be a positive number; the negative sign here is just to reverse the negative in the diagonal term.

5.7 Embedded Markov Chains

Every continuous time Markov process contains a discrete time Markov chain, called the *embedded chain*. The idea is to interpret the instants at which the transitions occur as the epochs of the discrete time chain. More precisely, the value of X_n is the state just after the *n*th transition. You lose information—the time between transitions—but the Markov and stationarity properties are preserved. Why would you want to transform to a less accurate model? The only reason would be to answer a different kind of question. For example, if you wanted to answer a question about the number of times a state is entered, as opposed to the total time spent in the state, it would be appropriate to construct the embedded chain.

To construct the one-step transition matrix, we have to calculate the probabilities p_{ij} for every pair ij for which transitions can occur. Hence, the transition diagram for the discrete time chain will have exactly the same structure as that for the Markov process, unless you have to add loops for absorbing states. The matrix $\mathbf{P}$ will have zeros everywhere that $\mathbf{\Lambda}$ does and also on the diagonal (unless the state is absorbing, in which case it will have a 1 on the diagonal). For any fixed i, corresponding to a row of $\mathbf{P}$, and any particular j, p_{ij} is the probability that j is the next state as opposed to any of the other possibilities. Thinking of the transition as a race among the various negative exponentially distributed transition time variables, the winner is determined by the minimum. Using arguments along the lines used to determine sojourn times, the probability p_{ij} is given by

$$p_{ij} = \frac{\lambda_{ij}}{\sum_k \lambda_{ik}}$$

unless the state is absorbing, in which case p_{ii} is automatically set to 1. This equation makes good sense if you just think about it. It is also easy to carry out. Just scale all of the λ_{ij} so that each row sums to 1, which you do by dividing each term by the sum of the entire row.

EXAMPLE 5.5

Here is a small example showing the original $\mathbf{\Lambda}$ matrix and the derived $\mathbf{P}$ matrix for the embedded chain:

$$\mathbf{\Lambda} = \begin{bmatrix} -5 & 2 & 3 \\ 1 & -3 & 2 \\ 0 & 1 & -1 \end{bmatrix}, \quad \mathbf{P} = \begin{bmatrix} 0 & \frac{2}{5} & \frac{3}{5} \\ \frac{1}{3} & 0 & \frac{2}{3} \\ 0 & 1 & 0 \end{bmatrix}$$

Please take special note that the two models give different results. For example, in this particular case, the steady-state probabilities for the continuous time version are $\pi_1 = 1/19$, $\pi_2 = 5/19$, $\pi_3 = 13/19$, while for the embedded discrete model they are $\pi_1 = 5/33$, $\pi_2 = 5/11$, $\pi_3 = 13/33$. The fundamental reason that they come out differently is that *they have different meanings*, based upon the different treatment of time. You will have to be careful in interpreting the results from the embedded Markov chain.

5.8 Deriving the Differential Equations

We now return to the issue of proving that the transition probability functions are determined by a particular set of differential equations. We are starting with only the Markov and stationarity assumptions.

We start from the continuous time expression of the Chapman-Kolmogorov equations, as shown below.

> **Result** In a continuous time Markov process, for any states i and j and any positive time intervals t and s, the following equation,
>
> $$p_{ij}(t+s) = \sum_k p_{ik}(t)p_{kj}(s)$$
>
> holds where the sum runs over all states.

This equation follows directly from the Markov and stationarity assumptions, in a theorem that parallels the proof for the discrete time version. That is, you condition on the state at time t and work out the conditional probabilities, simplifying expressions using the Markov and stationarity properties.

We will want to use this equation with a small time interval Δt in place of s. So we start with

$$p_{ij}(t + \Delta t) = \sum_k p_{ik}(t) p_{kj}(\Delta t)$$

Now, in order to distinguish the treatment of a special case, pull one term out of the sum, namely, the term where $k = j$:

$$p_{ij}(t + \Delta t) = p_{ij}(t) p_{jj}(\Delta t) + \sum_{k \neq j} p_{ik}(t) p_{kj}(\Delta t)$$

Next, we want to develop expressions to substitute for the transition functions for small Δt. Recall from calculus that the Taylor's series expansion of a function $f(x + \Delta x)$ is

$$f(x + \Delta x) = f(x) + f'(x) \Delta x + f'(x) \frac{(\Delta x)^2}{2!} + \cdots$$

Actually, we want the MacLaurin's series version, which is Taylor's series at the point $x = 0$. Apply this to $p_{kj}(\Delta t)$ to get

$$p_{kj}(\Delta t) = p_{kj}(0) + p'_{kj}(0) \Delta t + p''_{kj}(0) \frac{(\Delta t)^2}{2!} + \cdots$$

We can simplify the first two terms on the right-hand side. The higher-order terms involve quantities that will disappear when we take a limit later, so summarize them by the shorthand $o(\Delta t)^2$. This notation means "something that will vanish as fast as $(\Delta t)^2$ when Δt goes to zero."

We know that for $k \neq j$,

$$p_{kj}(0) = 0$$

We know this from the meaning of the notation; that is, the probability of being in state j at time zero given that you are in a different state k at time zero has to be zero. For similar reasons,

$$p_{jj}(0) = 1$$

We also know that the first derivatives of the $p_{kj}(t)$ functions evaluated at $t = 0$ are the transition rates. That is,

$$\frac{d}{dt} p_{kj}(t)|_{t=0} = \lambda_{kj}$$

and this is true even for $k = j$. Now substituting these simplifications back into the expression for $p_{ij}(t + \Delta t)$,

$$p_{ij}(t + \Delta t) = p_{ij}(t)[1 + \lambda_{jj} \Delta t + o(\Delta t^2)] + \sum_{k \neq i} p_{ik}(t) [0 + \lambda_{kj} \Delta t + o(\Delta t^2)]$$

which can be rearranged to give

$$p_{ij}(t + \Delta t) - p_{ij}(t) = p_{ij}(t) \lambda_{jj} \Delta t + \sum_{k \neq j} p_{ik}(t) \lambda_{kj} \Delta t + o(\Delta t^2)$$

(Here we just collapsed all of the terms involving powers of Δt higher than 1 into the last term.) Now we divide on both sides by Δt and put the sum back together to get

$$\frac{p_{ij}(t + \Delta t) - p_{ij}(t)}{\Delta t} = \sum_k p_{ik}(t)\lambda_{kj} + \frac{o(\Delta t^2)}{\Delta t}$$

Now if we take the limit of both sides as Δt goes to zero, we get the derivative on the left-hand side (from the definition of derivatives) and a sum that is independent of Δt on the right. The higher-order terms on the right vanish in the limit. This leaves us with the result below.

> **Result** For any continuous time Markov process, the transition probability functions satisfy
>
> $$\frac{d}{dt}p_{ij}(t) = \sum_k p_{ik}(t)\lambda_{kj}$$

This is the differential equation for one of the transition probability functions. Applying the same argument for every transition pair gives us the full set of differential equations. These are always linear, first-order differential equations with constant coefficients—the λ_{ij}'s.

Recognizing the above sum as matrix multiplication, we may express all of the differential equations at once in the matrix form.

> **Result** For any continuous time Markov process, the transition probability functions satisfy
>
> $$\frac{d}{dt}\mathbf{P}(t) = \mathbf{P}(t)\mathbf{\Lambda}$$
>
> where $\frac{d}{dt}\mathbf{P}(t)$ is the matrix whose (i, j)-th element is $\frac{d}{dt}p_{ij}(t)$, $\mathbf{P}(t)$ is the matrix whose (i, j)-th element is $p_{ij}(t)$, and $\mathbf{\Lambda}$ is the matrix whose (i, j)-th element is λ_{ij}.

EXAMPLE 5.6

For the two-state example studied earlier, λ_{01} was three and λ_{10} was two. In this case, the matrix $\mathbf{\Lambda}$ would be given by

$$\mathbf{\Lambda} = \begin{bmatrix} -3 & 3 \\ 2 & -2 \end{bmatrix}$$

and the four differential equations could be obtained from

$$\frac{d}{dt}\mathbf{P}(t) = \mathbf{P}(t)\mathbf{\Lambda}$$

You should check yourself on being able to do this matrix manipulation in symbolic form to get the set of differential equations shown on page 145 of this chapter. A calculator will not help here; you need to work with the symbols.

5.9 Solving the Differential Equations

In most elementary treatments until recently, students were advised to forget about trying to solve the differential equations. Even small problems posed difficulties and large ones were beyond the reach of all but the most sophisticated special-purpose computer programs. That

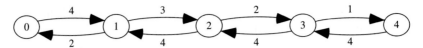

■ **FIGURE 5.10** A transition diagram

situation has changed in the last few years. It is now feasible to obtain numerical solutions to the differential equations for continuous time Markov processes using readily available personal computer software, such as MATLAB[1] or Mathematica[2].

Without getting into the details of any particular package, we want to indicate the possibilities that are opened by having such a resource at your command. There is a software package marketed under the name Stella[3] in its educational form and under the name iThink for business applications. It models flows using just a few primitive elements and a graphical format that simplifies model construction. Implicitly, in way that is well hidden from the casual user, the software numerically solves a set of differential equations. There are also very similar programs called Powersim[4] and Vensim[5] that offer free trial versions via the Internet.

Expressing the model in the conventions of this package results in a diagram which is strikingly similar to the transition diagram. For example, a continuous time Markov process having the transition diagram shown in Figure 5.10 could be modeled as shown in Figure 5.11. (It is interesting to note in passing that we are now engaged in the curious task of modeling a model.)

The boxes in Figure 5.11, which are "stocks" in the language of Stella, represent the states of the transition diagram. They hold "units of probability," which start out according to the initial distribution and flow through the "pipes," which are controlled by "valves." Each valve contains a simple formula that says that the flow though the pipe is equal to the transition rate times the stock in the box the flow is coming from. It takes only a few minutes to set up such a diagram. When you tell the program to run, it will automatically generate a table and graph of the results, as shown in Figure 5.12. For this run, we set up the initial conditions to reflect starting the process in state 0. We can read the $p_{0j}(t)$ functions directly from the plot. For example, we can see that the probability of being in state 1 rises sharply, then settles back to a slightly lower value where it stays. The probability of being in state 3 follows an S-curve to its asymptotic value. All of the functions reach steady-state values within about 3.5 time units. The program makes it easy to change parameter values and make multiple runs to gain insight into the behavior of the system.

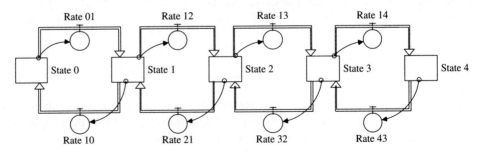

■ **FIGURE 5.11** A Stella diagram

---

[1] Copyright The Mathworks, Inc. (mathworks.com).
[2] Copyright Wolfram Research (wolfram.com).
[3] Copyright isee systems (iseesystems.com).
[4] Copyright Powersim Software (powersim.com).
[5] Copyright Ventana Systems, Inc. (vensim.com).

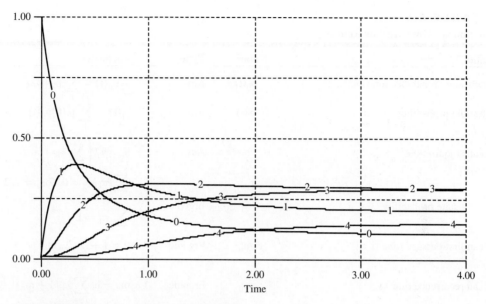

FIGURE 5.12 The solutions plotted

5.10 State Probabilities

There is a small and easy extension that should be covered here. If the initial state is unknown, but is given by a probability distribution, then the absolute probabilities are given by the following result.

> **Result** For any Markov process, the unconditional probability of being in state j at time t is
> $$p_j(t) = \sum_i p_i(0) p_{ij}(t)$$
> This relation follows from the law of total probability.

A matrix version of the same expression would give

> **Result** For any Markov process, the unconditional probability vector at time t is
> $$\mathbf{p}(t) = \mathbf{p}(0)\mathbf{P}(t)$$
> where the lowercase bold **p**(t) and **p**(0) symbols are *row* vectors defined in the obvious way. Note the order of the matrix multiplication on the right-hand side; the reverse order does not work.

5.11 First Passage Probability Functions

It makes perfectly good conceptual sense to define first passage and first return probability functions. However, to do anything with them, you have to be able to work with integral equations. That is not too difficult if one is familiar with LaPlace transform mathematics,

TABLE 5.1
Continuous Time Markov Results

Name	Notation	Type	Formula	Matrix Form
Transition probability function	$p_{ij}(t)$	Any	$\frac{d}{dt}p_{ij}(t) = \sum_k p_{ik}(t)\lambda_{kj}$	$\frac{d}{dt}\mathbf{P}(t) = \mathbf{P}(t)\mathbf{\Lambda}$
Absolute probability	$p_j(t)$	Any	$p_j(t) = \sum_i p_i(0)p_{ij}(t)$	$\mathbf{p}(t) = \mathbf{p}(0)\mathbf{P}(t)$
Mean sojourn time	h_i	Any	$h_i = 1/\left(\sum_{k \neq i}\lambda_{ik}\right)$	none
Steady-state probability	π_j	Ergodic	$\sum_k \pi_k \lambda_{kj} = 0,\text{ for } j = 1, 2, \ldots$ $\sum_j \pi_j = 1$	$0 = \boldsymbol{\pi}\mathbf{\Lambda}$ $\boldsymbol{\pi}\mathbf{1} = 1$
Mean first passage time ($i \neq j$)	m_{ij}	Ergodic	$0 = 1 + \sum_{k \neq j}\lambda_{ik}m_{kj}$	$\mathbf{m}_j = (-\mathbf{P}_{jj}^*)^{-1}\mathbf{1}$
Mean recurrence time	m_{ii}	Ergodic	$m_{ii} = h_i \sum_{j \neq i} \lambda_{ij}(1 + m_{ji})$	none
Mean time accumulated in a transient state j	e_{ij}	Terminating	$0 = \sum_{\text{Trans } k}\lambda_{ik}e_{kj} \text{ for } i \neq j$ $0 = 1 + \sum_{\text{Trans } k}\lambda_{ik}e_{ki}$	$\mathbf{E} = (-\mathbf{Q})^{-1}$
Expected duration	d_i	Terminating	$d_i = \sum_{\text{Trans } j} e_{ij}$	$\mathbf{d} = \mathbf{E}\mathbf{1}$
Absorption probability	a_{ij}	Terminating	$0 = \lambda_{ij} + \sum_{\text{Trans } k}\lambda_{ik}a_{kj}$	$\mathbf{A} = (-\mathbf{Q})^{-1}\mathbf{R}$
Hit probability	f_{ij}	Terminating	$0 = \lambda_{ij} + \sum_{\text{Trans } k \neq j}\lambda_{ik}f_{kj}$ $0 = \sum_{\text{Trans } k \neq j}\lambda_{ik}f_{ki}$	none
Conditional mean first passage time	$m_{ij}^{(c)}$	Terminating	$0 = a_{ij} + \sum_{\text{Trans } k}\lambda_{ik}a_{kj}m_{kj}^{(c)}$	none

but is a bit beyond what we are using here. For the record, the first passage probability function $f_{ij}(t)$ is the (possibly defective) density function for the random variable representing the time required to reach state j for the first time given that the process is in state i at time zero. If all we require are *mean* first passage times, we can solve linear equations to get them, as we did in the discrete case. The equations are similar in form, and are given in Table 5.1.

5.12 State Classification

We have now covered the general theory for arbitrary Markov processes; that is, the calculations that will be valid regardless of the structure of the process. Now we want to explore the calculations for ergodic and terminating processes. Both of these classes are useful in practical applications, but they have to be treated separately for the same reasons that they did in the discrete time case.

All of the state classification terms developed for the discrete time case apply to the continuous time case as well, with the exception of periodic states, which do not exist in the continuous case. Reachability and communication are determined by structure, which can be observed from the "connectedness" of states in the transition diagram. A process is irreducible if all states communicate. Recurrent and transient states are conceptually the same as they are in discrete time. That is, a state is recurrent if, once it occurs, it is certain to occur again; otherwise, it is transient. A recurrent state could be null (that is, be certain of occurring again once it has occurred, but have an infinite mean recurrence time). An ergodic process is an irreducible process in which all of the states are recurrent and nonnull. We do not have to worry about periodicity because that is something that could occur only when we treated time discretely.

5.13 Steady-State Probabilities

In most situations involving more than a few states, the differential equations expressing the $p_{ij}(t)$ cannot be solved without expending impractical efforts. However, their limiting values—the steady-state probabilities—can be found without solving the differential equations. In many practical situations, these will be sufficient. As in the discrete time case, the process must be ergodic for the procedure to work.

Assuming, then, that the process at hand is irreducible and all states are ergodic, we may derive a set of linear equations determining the steady-state probabilities. These are defined, as you should expect, as shown on next page.

Definition The steady-state probabilities are defined by

$$\pi_j = \lim_{t \to \infty} p_{ij}(t)$$

The argument for the steady-state equations goes as follows: From the Chapman-Kolmogorov equation

$$\frac{d}{dt} p_{ij}(t) = \sum_k p_{ik}(t) \lambda_{kj}$$

Taking limits on both sides,

$$\lim_{t \to \infty} \frac{d}{dt} p_{ij}(t) = \lim_{t \to \infty} \sum_k p_{ik}(t) \lambda_{kj}$$

Then by simple manipulation,

$$\frac{d}{dt} \lim_{t \to \infty} p_{ij}(t) = \sum_k \lim_{t \to \infty} p_{ik}(t) \lambda_{kj}$$

Actually this last step contains a flaw. It is not generally true that the limit of a derivative is equal to the derivative of the limit. However, these $p_{ij}(t)$ functions are suitably well behaved to ensure that the equality will hold in this case. Then, using the definition of the π_j's,

$$\frac{d\pi_j}{dt} = \sum_k \pi_k \lambda_{kj}$$

Finally, because the derivative of a constant is zero, we obtain the following equation.

162 Chapter 5 ■ Continuous Time Markov Processes

> **Result** For any ergodic continuous time Markov process, the steady-state equations are
> $$0 = \sum_k \pi_k \lambda_{kj}$$
> for each state j.

There will be an equal number of equations and unknowns in the system of linear equations, but, as in the discrete case, the equations will be dependent. Still assuming that the process is irreducible and the states ergodic, there will be exactly one dependency. The additional equation necessary to determine the unique solution is again, as it was in the discrete time case,

$$\sum_j \pi_j = 1$$

and this is still called the normalizing equation.

> **Result** For any ergodic continuous time Markov process, the steady-state equations are
> $$\mathbf{0} = \boldsymbol{\pi}\boldsymbol{\Lambda}$$
> $$\boldsymbol{\pi}\mathbf{1} = 1$$
> where $\mathbf{0}$ is a row vector of 0's, $\boldsymbol{\pi}$ is the row vector of steady-state probabilities, $\boldsymbol{\Lambda}$ is the matrix of transition rates, and $\mathbf{1}$ is a column vector of 1's.

Observe that the discrete time version of the steady-state equations has a different form than that of the continuous case. The vector $\boldsymbol{\pi}$ appears on both sides of the equations in the former case, and on only one in the latter. So it is not just a matter of replacing $\mathbf{P}$ by $\boldsymbol{\Lambda}$. Perhaps surprisingly, the very same "flow equilibrium" concept that enables one to write out the steady-state equations from the transition diagram works in both cases. This time, the total flow rate out of a point j to other points k is

$$\sum_{k \neq j} \pi_j \lambda_{jk} = \pi_j \sum_{k \neq j} \lambda_{kj} = \pi_j(-\lambda_{jj})$$

and the total flow rate into j from other points k is

$$\sum_{k \neq j} \pi_k \lambda_{kj}$$

If you set these two terms equal to one another, you get

$$\pi_j(-\lambda_{jj}) = \sum_{k \neq j} \pi_k \lambda_{kj}$$

which is just another way of expressing

$$0 = \sum_k \pi_k \lambda_{kj}$$

Once you are used to the idea, it is easy to write the steady-state equations just by looking at the transition diagram.

EXAMPLE 5.7

To illustrate the solution procedure, consider once again the loom example. In that case, the matrix of transition rates was

$$\Lambda = \begin{bmatrix} -3 & 3 \\ 2 & -2 \end{bmatrix}$$

so the steady-state equations would be

$$[0 \quad 0] = [\pi_1 \quad \pi_2] \begin{bmatrix} -3 & 3 \\ 2 & -2 \end{bmatrix}$$

or

$$0 = -3\pi_1 + 2\pi_2$$
$$0 = 3\pi_1 - 2\pi_2$$

It is immediately apparent that these two equations are dependent, since one is just the negative of the other. Using either one, we can conclude that

$$\pi_2 = \frac{3}{2}\pi_1$$

Now by the normalizing equation,

$$\pi_1 + \pi_2 = 1$$
$$\pi_1 + \frac{3}{2}\pi_1 = 1$$
$$\pi_1 \left(1 + \frac{3}{2}\right) = 1$$
$$\pi_1 = \frac{2}{5}$$

And, finally,

$$\pi_2 = \frac{3}{2}\pi_1 = \frac{3}{5}$$

You can go back now to Figure 5.3 on page 145 to verify that the functions $p_{11}(t)$ and $p_{21}(t)$ both converge to 2/5, the value of π_1, and the functions $p_{21}(t)$ and $p_{22}(t)$ both converge to 3/5, the value of π_2. In practical applications, you will almost always want to get the steady-state results by this linear equation method, rather than by solving the differential equations.

5.14 Other Computable Quantities

Most applications of continuous time Markov processes appearing in the literature involve only the steady-state probabilities. For example, one rarely encounters examples of continuous time terminating processes, despite the fact that discrete time terminating models are quite common. There is really no reason that this should be so, because the standard quantities such as mean first passage times and absorption probabilities are as easy to compute in the continuous case as they are in the discrete time case.

For ergodic processes, the mean first passage time m_{ij} is defined as the expected total time to reach state j for the first time, starting from the initial state i. That sounds just the same as it did in the discrete time case, but notice that we are no longer counting epochs—we are now recording the passage of real (clock) time. Hence the units of m_{ij} would be minutes or hours or whatever time units we are using, but *not* transitions. The equation for mean first passage times is below.

> **Result** In any ergodic continuous time Markov process, the mean first passage times m_{ij} (for $i \neq j$) satisfy
>
> $$0 = 1 + \sum_{k \neq j} \lambda_{ik} m_{kj}$$

This equation works *only* for $i \neq j$; it will *not* provide mean recurrence times. Keeping j fixed and varying i over all of the states except j will give $S - 1$ independent equations whose solution gives all the mean first passage times to state j.

If we want to use a matrix form, we have to replace both the *j*th row and *j*th column in Λ by zeros, and put a 1 in the (j, j) position, leaving a modified matrix we will call $\Lambda_j^\dagger$. (The 1 on the diagonal does not actually do anything computationally; it merely prevents the matrix from becoming singular so that it can be inverted.) Then the matrix form of the equations is

$$\mathbf{0} = \mathbf{1} + \Lambda_j^\dagger \mathbf{m}_j$$

which can be solved in matrix form to obtain the result below.

> **Result** The vector of mean first passage times to state j is given by
> $$\mathbf{m}_j = (-\Lambda_j^\dagger)^{-1} \mathbf{1}$$

If that notation looks a bit awkward, here in words is how you can implement it in a spreadsheet. Starting from $-\Lambda$, replace the elements in row j and column j by zeros. Then place a 1 in the cell where those two intersect—the diagonal. Invert that matrix, and finally sum across each row. The result will be the column vector whose elements are m_{ij} for all i except j. (You should just ignore whatever appears for m_{jj}. It will have the value 1, but that is not the mean recurrence time; it is just a placeholder.)

A problem with mean recurrence times that requires a different approach from the one that worked in discrete time stems from the fact that the fundamental definition of recurrence is more complicated in continuous time. Before the process can return to a state, it first has to leave. So one way to obtain the mean recurrence time is to condition on the next state that is entered:

$$m_{ii} = \Sigma\, E\,(\text{time to reach } j + \text{time to return to } i)\, P\,(j \text{ is the next state})$$

where the sum runs over all states that are reachable in one transition from i. (We do not want the sum to include other states because the transition time to them would formally be infinite.) After substituting the mean transition time and transition probabilities for the embedded Markov chain, we have

$$m_{ii} = \sum \left(\frac{1}{\lambda_{ij}} + m_{ji} \right) \left(\frac{\lambda_{ij}}{-\lambda_{ii}} \right)$$

This expression can be simplified a little by substituting the expression for mean sojourn time, h_i,

$$m_{ii} = h_i \sum (1 + \lambda_{ij} m_{ji})$$

Then, since the sum runs over all the states that are reachable in one step from i, the number of 1's in the sum is $|B_i|$, or the number of elements in the bundle of i.

> **Result** The mean recurrence time to any state i in an ergodic Markov process is
> $$m_{ii} = h_i \left(|B_i| + \sum \lambda_{ij} m_{ji} \right)$$

In this version of the formula, the sum can be allowed to run over all states because the irrelevant ones will cause no harm.

We have seen that the results for ergodic Markov processes are similar (with small adjustments) to those of ergodic Markov chains. The same similarity holds for terminating

Markov processes. Recall that in discrete time, we represented the end of a terminating process somewhat artificially by putting loops on the absorbing states, as if "staying" in a state meant repeating it as epochs continued. We did that in order to force the transition matrix **P** to have every row sum to 1. In the continuous time case, we do not have to do that. Once an absorbing state is reached, the process just stays there. In other words, an absorbing state can be recognized immediately in the transition diagram as a state that has no outgoing arrows, or in the transition rate matrix as a state whose row contains only zeros. When we arrange the transition rate matrix Λ to have the transient states first, followed by the absorbing states, we will get the form

$$\Lambda = \begin{bmatrix} \mathbf{Q} & \mathbf{R} \\ \mathbf{0} & \mathbf{0} \end{bmatrix}$$

where each of the submatrices are of appropriate dimensions.

Although absorption probabilities a_{ij} in continuous time mean exactly what they do in discrete time (just the probability of ending up in a particular absorbing state), the quantities that relate to time must be interpreted carefully. The mean time spent in a transient state j, denoted e_{ij}, means the expected total accumulation of clock time in the state, not just the expected *number* of times (a count) that it is visited. Similarly, d_i adds up the total time until absorption occurs, not the number of transitions. Conditional mean first passage times also refer to clock time.

Hit probabilities for $i \neq j$ work out fine, but when $i = j$, we have the same problem that we did with mean recurrence times. A separate equation is needed to handle the f_{ii} case. A matrix expression for these could be shown, but there are so many modifications to the vectors and matrices that they would only be confusing.

We will not take time to derive any of the formulas that enable you to calculate all these quantities, but they are all easily set up and solved. In fact, with the exception of the differential equations for the transition probabilities, they are just as easily obtained as the corresponding quantities in the discrete time case. Notice, however, that there are some subtle differences in the formulas for the two cases.

5.15 Case Study

EXAMPLE 5.8

Day trading is a form of short-term speculation that involves buying and selling stocks, options, commodity futures, or other financial instruments over very brief intervals—usually only minutes or hours. In earlier times, only large financial institutions had the means to participate in such trades, but the rise of the Internet opened them up to individuals. The general idea is to exploit the volatility of ordinary price changes in real time, rather than trying to predict long-term trends.

Imagine a computerized trading system that is based upon the following ideas. The price of a stock at any moment is determined by the balance between two competing forces—one pushing the price higher and the other pushing it lower. One force might prevail slightly over the other to produce a short-term trend either upward or downward. However, random perturbations also affect the price, so countertrend movements are also possible. Suppose that we had a way to detect these microtrends, at least on some occasions. For example, we might have a computerized price-tracking program that logs the times between upward steps and between downward steps. When the (moving) average time between downward movements is at least two minutes greater than average time between upward movements, our program would instantly buy a predetermined number of shares at the current price. At the same time, it would place orders to sell if the price drops by fifty cents per share or if it rises by seventy-five cents per share. Within a matter of minutes, the entire buy-and-sell transaction will be completed. Sometimes, you will lose fifty cents per share; other times you will gain seventy-five cents per share. If you win as often as you lose, you will come out ahead. Even though the expected gain per share is small,

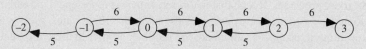

FIGURE 5.13 A transition diagram for day trading

you could make a lot of money if the scheme worked and you could make many plays per day with many shares per play.

In order to evaluate this plan, let us assume that the times between upward and downward movements of twenty-five cents per share are negative exponentially distributed, the estimated mean time between upward moves is ten minutes and the estimated mean time between downward moves is twelve minutes. Those mean times correspond to transition rates of six and five movement events per hour, respectively. We will define six states: -2, -1, 0, 1, 2, and 3. The state numbers refer to the number of "ticks" in the price below or above the purchase price, with state 0 as the initial state. The continuous time Markov process would be a terminating process, with the transition diagram shown in Figure 5.13.

Now there are a few questions we might want to answer about this plan. The primary one is whether the chance of winning is at least as great as the chance of losing. However, we might also be interested in the expected duration of the time when we are holding the stock, partly because our money will be tied up in this transaction and unavailable for other trades, and partly because we should be concerned about the dynamics changing if we hold it too long. The first question is answered by the absorption probability a_{03} and its complement. One of the problems has you set up the equations and solve for this value, but that answer is good: The probability of gaining is greater than the probability of losing. The expected duration of the transaction is equivalent to the expected time until absorption (regardless of which absorbing state is reached), so it is given by d_0. This value is a little more than half an hour. At this point, it appears that the plan might work.

However, if you play with the parameter values (guided by the suggestions offered in one of the problems), you will find that small changes can turn success into failure. That reversal of fortune should serve as a warning. The outcomes depend critically upon data values as well as the assumptions. In the real world of financial trading, information is so volatile that you should be skeptical of any model that depends upon stationarity assumptions. For example, take a look at Figure 5.14. What trends or patterns do you think you can detect?

It so happens that the apparent patterns in this chart are pure illusion. The graph was obtained by plotting a random sequence of independent increments (I know because I generated it). There is no trend or any other kind of bias; it is equally likely that the next observation will be higher or lower than the current one. Despite what patterns you may think you see in the graph, there is literally no information whatsoever in that record that can enable you to predict whether it will next go up or down. Of course, sometimes such records *do* contain useful predictive information. However, it can be extremely difficult to distinguish real patterns from illusory ones. The human brain is wired to detect patterns, which is why we are so capable at recognizing faces and at understanding language and at reading handwriting despite the noise in those signals. Unfortunately, that ability tends to lead us astray when we detect *apparent* patterns in *random* noise.

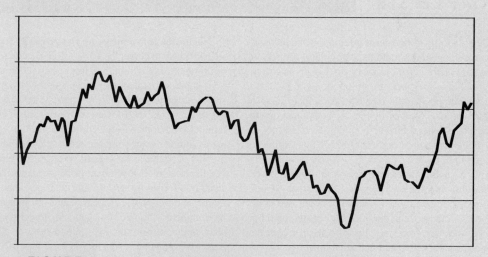

FIGURE 5.14 Stock price over time

> That is not to say that stochastic modeling is useless for such applications; in fact, it is one of the primary tools for sophisticated traders. The point is that you will need much more advanced knowledge to handle anything that complex, and you should acquire that knowledge if you have any interest in engaging in speculative investing. Day trading in particular is an extremely risky venture that should be undertaken only by professionals. Fortunes have been made by some traders who employed advanced modeling techniques; unfortunately, their success has spawned a lot of fraudulent get-rich-quick schemes that seem impressive because they use a lot of technical jargon. If you are ever tempted, ask yourself this question, "If it really worked, why would they sell it to you?"

5.16 Birth-Death Processes

At this point, we have covered the basic theory of continuous time Markov processes, at least to the degree that we did so for discrete time Markov chains. From this point on, we are going to develop results for special cases that prove to be particularly useful.

Nearly all simple queueing models (as well as a number of other common models) are special cases of the birth-death process, which is itself a special case of the general continuous time Markov process. A birth-death process is characterized as a Markov process in which all transitions are to the next state immediately above (a "birth") or immediately below (a "death") in the natural integer ordering of states. That is, a birth-death process does not jump states.

The transition diagram of a birth-death process would look like Figure 5.15. There may be either a finite or an infinite number of states, and the transition rates associated with the lines shown are arbitrary.

The transition rate matrix is also of recognizable form. Because jumps are forbidden, $\lambda_{ij} = 0$ except when $|i - j| \leq 1$. This implies that Λ is tridiagonal:

$$\Lambda = \begin{bmatrix} \lambda_{00} & \lambda_{01} & 0 & 0 & \cdots & 0 \\ \lambda_{10} & \lambda_{11} & \lambda_{12} & 0 & \cdots & 0 \\ & \lambda_{21} & \lambda_{22} & \lambda_{23} & \cdots & 0 \\ & & & & \cdots & \\ & & & & \cdots & \\ & & & & \cdots & \\ 0 & & \cdots & & \cdots & \end{bmatrix}$$

Furthermore, since the diagonal terms must equal "minus-the-sum-of-the-other-elements-in-the-same-row," we must have

$$\lambda_{ii} = -(\lambda_{i,i-1} + \lambda_{i,i+1})$$

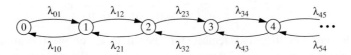

■ **FIGURE 5.15** Transition diagram of a birth-death process

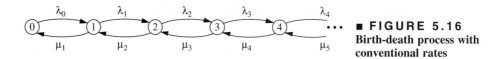

FIGURE 5.16 Birth-death process with conventional rates

That is, the matrix must be of the form:

$$\Lambda = \begin{bmatrix} -\lambda_{01} & \lambda_{01} & 0 & 0 & \cdots & 0 \\ \lambda_{10} & -(\lambda_{10}+\lambda_{12}) & \lambda_{12} & 0 & \cdots & 0 \\ & \lambda_{21} & -(\lambda_{21}+\lambda_{23}) & \lambda_{23} & \cdots & 0 \\ & & & & \cdots & \\ & & & & \cdots & \\ & & & & \cdots & \\ 0 & & & \cdots & & \cdots \end{bmatrix}$$

Again, there may be either a finite or an infinite number of states, and the specific values for the transition rates are arbitrary.

It is common to simplify the notation, using λ_j for the birth rate $\lambda_{j,j+1}$ and μ_j for the death rate $\lambda_{j,j-1}$. No generality is lost, but the special notation does tend to conceal the fact that the birth-death process is just a special case of the general Markov process. The transition diagram would look like Figure 5.16, and the transition matrix would have corresponding changes.

The case study of the previous section was a finite birth-death process that had absorbing states at both ends. More commonly, birth-death processes will be ergodic, and the steady-state probabilities will be of most interest. The steady-state equations for the birth-death process have the characteristic form:

$$0 = -\lambda_0 \pi_0 + \mu_1 \pi_1$$
$$0 = \lambda_0 \pi_0 - (\lambda_1 + \mu_1)\pi_1 + \mu_2 \pi_2$$
$$0 = \lambda_1 \pi_1 - (\lambda_2 + \mu_2)\pi_2 + \mu_3 \pi_3$$
$$0 = \lambda_2 \pi_2 - (\lambda_3 + \mu_3)\pi_3 + \mu_4 \pi_4$$

and so forth. It is always possible to solve the steady-state equations for the birth-death process easily, because the structure allows one to solve for π_1 in terms of π_0, from the first equation, which yields

$$\pi_1 = \frac{\lambda_0}{\mu_1} \pi_0$$

This can then be substituted into the second equation to allow a solution for π_2 in terms of π_0, and so forth. We can find the general recursion relation

$$\pi_j = \frac{\lambda_{j-1}}{\mu_j} \pi_{j-1}$$

which is convenient in spreadsheet calculations, or we can express the solution in the "solved" form

$$\pi_j = \frac{\lambda_{j-1}\lambda_{j-2}\lambda_{j-3}\cdots\lambda_0}{\mu_j \mu_{j-1} \mu_{j-2} \cdots \mu_1} \pi_0$$

which may simplify considerably for special cases of the transition rates. Finally, the normalizing equation will determine the value of π_0. We will see several examples of this procedure in Chapter 7, dealing with queueing models.

5.17 The Poisson Process

The Poisson process is a special case of the birth-death process. It might be more descriptive to call it a pure birth process, since the death rates, μ_j, are all zero. The birth rates, the λ_j, are all equal to a constant value λ. Thus the behavior of the Poisson process is governed by the single parameter λ, and the structure is such that the state index can only increase (see Figure 5.17).

The Poisson process is often used to model the kind of situation in which a count is made on the number of events occurring in a given time. For example, it is used in the simplest queueing models to represent the arrivals of customers to a service facility. The state at time t would correspond to the number of arrivals by time t.

The matrix of transition rates Λ would be

$$\Lambda = \begin{bmatrix} -\lambda & \lambda & 0 & 0 & \cdots & 0 \\ 0 & -\lambda & \lambda & 0 & \cdots & 0 \\ 0 & 0 & -\lambda & \lambda & \cdots & 0 \\ & & & & \cdots & \\ & & & & \cdots & \\ & & & & \cdots & \\ 0 & & & \cdots & & \end{bmatrix}$$

Since the applications usually involve counting events, and the count would ordinarily begin at zero, we will assume that the initial state is zero. The differential equations determining the $p_{0j}(t)$ would be

$$\frac{d}{dt} p_{00}(t) = -\lambda p_{00}(t)$$
$$\frac{d}{dt} p_{01}(t) = \lambda p_{00}(t) - \lambda p_{01}(t)$$
$$\frac{d}{dt} p_{02}(t) = \lambda p_{01}(t) - \lambda p_{02}(t)$$

In general,

$$\frac{d}{dt} p_{0j}(t) = \lambda p_{0j-1}(t) - \lambda p_{0j}(t)$$

but note that the first equation, for $j = 0$, is a special case.

The first equation involves only $p_{00}(t)$ and is of a particularly simple form. It can be integrated directly to yield the solution

$$p_{00}(t) = e^{-\lambda t}$$

(The constant of integration is evaluated from $p_{00}(0) = 1$.) This solution can be substituted into the next equation, giving

$$\frac{d}{dt} p_{01}(t) = \lambda e^{-\lambda t} - \lambda p_{01}(t)$$

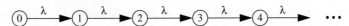

■ **FIGURE 5.17** Transition diagram for the Poisson process

This now involves only $p_{01}(t)$. Tables of standard differential equations or integration by parts will yield the solution

$$p_{01}(t) = \lambda t e^{-\lambda t}$$

This can then be substituted into the next equation to give

$$\frac{d}{dt} p_{02}(t) = \lambda^2 t e^{-\lambda t} - \lambda p_{02}(t)$$

which can be solved to give

$$p_{02}(t) = \frac{(\lambda t)^2}{2!} e^{-\lambda t}$$

Continuing in this manner, we find that the general solution is

$$p_{0j}(t) = \frac{(\lambda t)^j}{j!} e^{-\lambda t} \quad \text{for all } j \geq 0$$

(To verify that it is, indeed, the general solution, it can be substituted into the general differential equation.)

The form of the expression is that of a Poisson distribution, which accounts for the name given to the model. One way to interpret the result is to fix t, which has the effect of making λt a fixed parameter. Then the set $\{p_{00}(t), p_{01}(t), p_{02}(t), \ldots\}$ would give the probability distribution of the state at the fixed time t. In terms of a count of events, we would say that the number of events occurring in a fixed time interval t is Poisson distributed with parameter λt. Furthermore, since the mean of a Poisson distribution is equal to the parameter, λt can be interpreted as the expected number of events occurring in time t.

There are other ways to characterize the same process. Suppose that T_n represents the random variable corresponding to the time between event $(n-1)$ and event n. In particular T_1 would be the time until the first event. Then

$$p_{00}(t) = \boldsymbol{P}\{\text{no events in time } t\}$$
$$p_{00}(t) = \boldsymbol{P}\{T_1 > t\}$$

So

$$\boldsymbol{P}\{T_1 \leq t\} = 1 - p_{00}(t) = 1 - e^{-\lambda t}$$

This distribution can be recognized as the negative exponential distribution with parameter λ. By a similar argument, it can be shown that

$$\boldsymbol{P}\{T_n \leq t\} = 1 - p_{n-1,n-1}(t)$$

and, by solution of the appropriate differential equation, it can further be shown that

$$p_{n-1,n-1}(t) = e^{-\lambda t}$$

Consequently,

$$\boldsymbol{P}\{T_n \leq t\} = 1 - e^{-\lambda t}$$

In words, the times between events in a Poisson process are all negative exponentially distributed with the same parameter λ. Since the mean of a negative exponential random variable is the reciprocal of the parameter, λ can be interpreted as the reciprocal of the expected time between events.

5.18 Properties of Poisson Processes

The Poisson process has a number of mathematically convenient properties that contribute to its usefulness in modeling. Sometimes a situation calls for separating or filtering events according to some criterion. For example, all customers of a certain type might be directed one way and others a different way. Or, in a production process, an inspection might direct defective products one way and good ones another. The question is, "Can a Poisson stream of events be 'separated' like this without losing the desirable characteristics of Poisson processes?"

Although we will make no attempt to prove it, the answer to this question is yes, provided that the separation is done in a particular way. Whenever an event occurs in the parent, or input, process, it will be assigned to one or the other of the output processes. The rule that determines whether a particular event is assigned to process B or to process C must be probabilistic and independent of everything else. (See Figure 5.18.) In other words, it is as if the assignment were determined by the toss of a (possibly biased) coin. Other conceivable rules, such as alternating the assignments, will not produce Poisson processes as outputs. If λ_A is the rate for the input stream, and p_B denotes the probability that an event is assigned to stream B, then the rate for process B is $p_B \lambda_A$, just as you would probably suspect. It is also true, as you might imagine, that the result can be extended to the separation of a Poisson process into more than two processes.

There are several useful ways to interpret the separability property of Poisson processes. If the input process represents arrivals of customers to a bank, then the output processes might be the arrivals to individual tellers. Of course, to preserve the Poisson process, the customers must be separated on a completely random basis. In particular, we could not allow the customer to choose their tellers on the basis of how long the lines are. Another possible interpretation of the separating device is that of a filter. Perhaps the input process represents parts arriving at an inspection station that separates them into acceptable parts and rejects. Provided that the probability of acceptance is independent of everything, both the stream of accepted parts and the stream of rejects would retain the Poisson character.

Next, imagine that we have two independent processes operating simultaneously, and consider the effect of merging these processes as shown in Figure 5.19. More precisely, what we mean by merging the processes is technically called *superposition*. The superposition of two processes, A and B, produces a third process, C, in which an event occurs whenever an event occurs in either A or B. Figure 5.20 shows how events in both A and B are "copied" in C.

It is a very convenient fact that the superposition of two independent Poisson processes produces a Poisson process. This fact can be quite tricky to prove if you think in terms of the

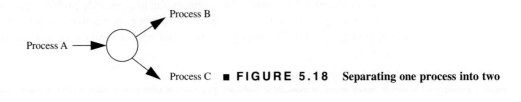

■ **FIGURE 5.18**   Separating one process into two

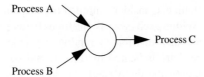

■ **FIGURE 5.19**   Merging two processes

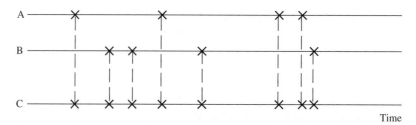

■ FIGURE 5.20 The superposition of two processes

times between events, but becomes almost trivial if you think in terms of counting the number of events occurring in time t. Let $N_A(t)$, $N_B(t)$, and $N_C(t)$ be the random variables counting the number of events occurring in processes A, B, and C, respectively, over the same time interval t. Then since process C counts events occurring in either A or B,

$$N_C(t) = N_A(t) + N_B(t)$$

Of course, both $N_A(t)$ and $N_B(t)$ are Poisson distributed random variables. But we already know that the sum of independent Poisson random variables is a Poisson random variable. Since this is true for all t, C is a Poisson process.

Furthermore we know that the expectation of a sum is the sum of the expectations. Consequently, if λ_A and λ_B are the parameters or rates of occurrence of the A and B processes,

$$E[N_A(t)] = \lambda_A t$$
$$E[N_B(t)] = \lambda_B t$$

and, therefore,

$$\begin{aligned} E[N_C(t)] &= E[N_A(t) + N_B(t)] \\ &= E[N_A(t)] + E[N_B(t)] \\ &= \lambda_A t + \lambda_B t \\ &= (\lambda_A + \lambda_B)t \end{aligned}$$

In other words, the rate of occurrence in process C is equal to the sum of the rates of occurrence in processes A and B. One could not ask for a nicer result. It should also be apparent from the argument used that the result extends in the obvious way to the superposition of more than two independent Poisson processes.

This superposition property of the Poisson process is useful in a variety of contexts. The processes A and B might represent, for example, the arrivals of orders for some good to each of two different stores that are supplied from the same inventory. The superposition process would represent the total demands from inventory. Or, in a queueing problem, the component processes may represent arrivals from two different sources, or perhaps of two different types (for example, male and female). The superposition would represent all of the arrivals to the system.

5.19   Khintchine's Theorem

Perhaps the most remarkable property of the Poisson process is indicated by a rather deep limit theorem formulated by A. Y. Khintchine. Consider a number of independent processes, each of which generates events over time. Within each process, the times between successive events are assumed to be independent, identically distributed random variables. In Figure 5.21, the interevent times for the first process are indicated by T_1, T_2, and so on. It is these random variables that are assumed to be identically distributed. We do not assume that the interevent

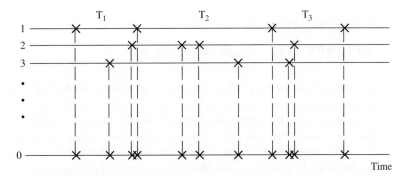

■ **FIGURE 5.21** Khintchine's theorem

times for different processes have the same distribution. Of course, if the interevent times are negative exponentially distributed for any process, then that process would be a Poisson process, but we specifically refrain from making any such assumption. The forms of the interevent time distributions are arbitrary. The technical terminology for a process of the kind described—that is, one with independent, identically distributed interevent times—is a renewal process.

Khintchine's theorem states that, provided certain reasonable conditions are satisfied, the superposition of a large number of independent renewal processes is approximately a Poisson process. Because we already know that the superposition of Poisson processes is a Poisson process, the real point of the theorem is that the component processes need not be Poisson. Regardless of the distributions of the times between events within processes, if there are enough of them, the superposition is approximately Poisson.

Khintchine's theorem is somewhat analogous to the central limit theorem. It will be recalled that the latter ensures that the sum of a sufficiently large number of independent random variables will be approximately normally distributed regardless of what the distributions of the constituent random variables may be. Khintchine's theorem states that the superposition (a kind of sum) of a sufficiently large number of independent renewal processes will produce a Poisson process regardless of what the distributions of the constituent renewal processes may be. Like the central limit theorem, Khintchine's theorem explains why such a remarkably simple process with such convenient properties should happen to occur so often in real life. It also serves to justify the Poisson assumption in cases where the process of interest can be thought of as consisting of the superposition of a large number of independent renewal processes.

To illustrate this last point, consider the arrivals of customers to a supermarket. If you think in terms of an individual customer, the process describing his or her arrival is almost certainly not Poisson. That is, it seems very unlikely that a negative exponential distribution would adequately serve to describe the times between one customer's visits to the supermarket. On the other hand, it does not seem unreasonable to assume that successive intervals between visits are independent and identically distributed, and that the arrival patterns for different customers are independent. Since the arrivals to the store are given by the superposition of the arrival processes of all of the customers, Khintchine's theorem provides reason to expect that the arrivals do indeed form a Poisson process.

5.20 Phase-Type Distributions

In Section 5.3, "The Assumptions Reconsidered," we discussed the desirable characteristics and also the limitations of the negative exponential distribution. We also mentioned that there was a trick to get beyond the limitations. In this section, we open up the range of modeling

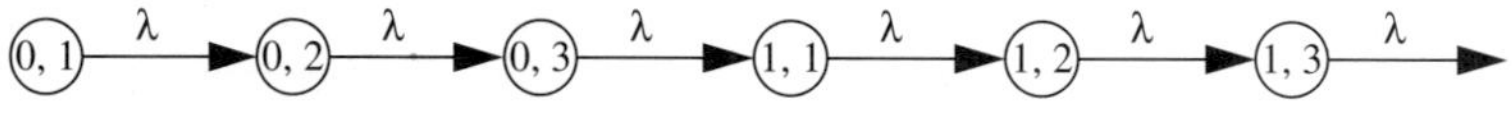

```
(0,1) →λ→ (0,2) →λ→ (0,3) →λ→ (1,1) →λ→ (1,2) →λ→ (1,3) →λ→
```

FIGURE 5.22   Erlang-3 counting process

possibilities enormously. There is a way to represent a richer class of distributions by putting negative exponentials together. In fact, any distribution can (at least in principle) be represented to an arbitrarily close degree by a member of this extended family. Of course, this extension does not come without a price. The state space will be larger—perhaps much, much larger.

We start with a simple extension to explain the basic idea. Assume that we want to model a counting process—one that starts at zero and increases states one step at a time, like a Poisson process—but one that has Erlang-3 distributed times between events. One way in which this situation could occur is that we want to count every third event in a Poisson process. But it would not have to be a situation that involves a Poisson process at all. It might be that we simply want to model a process in which the times between events have been found to fit an Erlang-3 distribution. In either case, the modeling problem is the same. At first glance, we might think that we cannot use a Markov model at all, because the transition times must be negative exponential to satisfy the Markov property. But, remembering that an Erlang-3 distribution is equivalent to the sum of three negative exponential random variables, we can "trick" the model to make it Markovian.

The trick is to think of the Erlang-3 distribution as consisting of three phases, each of which is a time interval of negative exponential duration. Then, augment the state definition to include a "phase counter." That is, the state definition will tell us not only how many events have been counted, but also in which phase the event generator is currently. Figure 5.22 conveys the idea if you know how to read the states.

If the process is now in state (i, j), it means that you have so far counted i events in the counting process you are modeling, and the next count is currently in phase j, which means that j − 1 negative exponential intervals have occurred since the last count.

Now if you examine the transition diagram and ignore the state labels, it looks exactly like the diagram for a Poisson process. Hence the solution to the differential equations will be exactly the same. The only change is in how we interpret the results. The probability that j counted events have occurred by epoch n is the marginal probability

$$p_{0(j,1)}^{(n)} + p_{0(j,2)}^{(n)} + p_{0(j,3)}^{(n)}$$

That is, we have to add over the three phases. We would have to make similar adjustments for mean first passage times and other measures.

It is obvious how to extend this same idea for Erlang distributions of any number of phases. The trick also suggests some immediate easy generalizations. For example, we would not have to use the same λ for each phase. By using different values, we can model distributions that are the sums of negative exponentials with distinct parameters. That family of distributions—called the generalized Erlang—gives you more separately adjustable parameters, and therefore more flexibility in modeling. One of the remaining limitations, however, is that all members of the generalized Erlang family have coefficients of variation (the ratio of standard deviation to mean) of value less than 1. Nothing you can do to the parameters will change that basic fact. So they are all less variable than the negative exponential, which has a coefficient of variation equal to 1.

Is there any other trick we can use to model distributions that are more variable than the negative exponential? Yes, indeed, there certainly is. All that is required is to realize that the embedded sequence of phases is just a very special case of a continuous time terminating Markov process. In fact, you can form networks of negative exponentials to get essentially any distribution shape you want. All you have to do is maintain, as part of the state definition, an indicator of where you are in the network of phases. In fact, any continuous time terminating Markov process could be embedded within the one of interest. You complete a transition in the real process when the embedded process "finishes," which is to say reaches an absorbing state.

Distributions that can be represented in this way—that is, as the time to absorption in a continuous time Markov process—are called *phase-type distributions*. It is a rich class, with many desirable properties. For example, any linear combination of phase-type random variables gives another phase-type random variable. *Any* distribution can be approximated to an arbitrarily close degree by a phase-type distribution.

We will not take the space here to explain fully how to use phase-type distributions. The crucial point is that the Markov assumption is not nearly as limiting as it first appears to be. With the use of a few formulation tricks, you can build a Markov model to represent almost any kind of stochastic process. Of course, the cost is a larger state space and increased complexity in the transitions. In practice, there will be a trade-off between model fidelity and computational complexity.

5.21 Conclusions

We have seen that the behavior over time of a stationary Markov process is completely characterized by, in the discrete case, the one-step transition matrix, **P**, and in the continuous case, the matrix of instantaneous transition rates, Λ. Given the elements of the appropriate matrix (plus, in some cases, information about the initial conditions), it is possible to calculate virtually everything of interest about the process. For the most part, the mathematics required is no worse than linear algebra. Even in the continuous case, in which differential equations may appear, they are first-order, linear differential equations with constant coefficients. In short, there are no theoretical barriers to solving practically any kind of problem that can be phrased in terms of stationary Markov processes.

Although there is some danger of overextending the techniques to model situations where the stationarity or Markov assumption, or both, are not warranted, the methods are sufficiently robust to permit an enormous variety of real-world processes to be represented and studied. Chapter 6 develops just one category of application.

5.22 Recommended Reading

Many, but not all, of the same books that are useful in learning Markov chains also cover continuous time Markov processes. However, the applications seem to take on a very different tone. Whereas discrete time applications range over an enormous range of topics, the continuous time ones focus on a few: queueing, reliability, computer system performance analysis, and a few others. The next chapter will develop the queueing category.

1. Berger, Marc A., *An Introduction to Probability and Stochastic Processes*. Springer-Verlag, New York, 1993.
2. Bhat, U. N., *Elements of Applied Stochastic Processes*. Wiley, New York, 1972.
3. Cinlar, E. *Introduction to Stochastic Processes*. Prentice-Hall, Englewood Cliffs, NJ, 1975.
4. Clarke, A. B., and R. Disney, *Probability and Random Processes for Engineers and Scientists*, 2nd ed. Wiley, New York, 1985.
5. Cox, D. R., and H. D. Miller, *Theory of Stochastic Processes*. Wiley, New York, 1965.
6. Feller, W., *An Introduction to Probability Theory and Its Applications*, vol. II, 2nd ed. Wiley, New York, 1957.

7. Grassmann, Winfried K., *Stochastic Systems for Management.* North-Holland, New York, 1981.
8. Heyman, D. P., and M. J. Sobel, *Stochastic Models.* North-Holland, New York, 1990.
9. Khintchine, A.Y., *Mathematical Methods in the Theory of Queueing.* Hafner, New York, 1960.
10. Neuts, Marcel F., *Matrix-Geometric Solutions in Stochastic Models.* Dover, New York, 1981.
11. Parzen, E., *Stochastic Processes.* Holden-Day, New York, 1962.
12. Ross, S., *Introduction to Probability Models.* Academic, New York, 1972.
13. Tijms, Henk C., *Stochastic Modeling and Analysis: A Computational Approach.* Wiley, New York, 1986.
14. Wolff, Ronald W., *Stochastic Modeling and the Theory of Queues.* Prentice-Hall, Englewood Cliffs, NJ, 1989.

Chapter 5 Problems

1. During the course of a day, the vehicle traffic at a certain location varies randomly, changing among light, medium, and heavy, where these terms have definite, measurable meanings. Imagine a three-state continuous time Markov process where changes in traffic density are represented by transitions. Assume that the only way to get from light to heavy or vice versa is by way of the medium state. Of course, to be Markovian, the transition times must be negative exponentially distributed.

   Explain how to estimate the needed parameter values for the model, referring specifically to how the traffic data would be collected and manipulated.

   Assuming the values are $\lambda_{LM} = 4$, $\lambda_{ML} = 3$, $\lambda_{MH} = 2$, $\lambda_{HM} = 1$ in units of transitions per hour show how to answer these questions:

   a. If the traffic is light now, what is the expected length of time it will stay light?
   b. If the traffic is medium now, what is the expected length of time it will stay medium?
   c. If the traffic has been medium for the last ten minutes, what is the expected length of time until it changes?
   d. If the traffic has been medium for the last ten minutes, what is the expected length of time until it becomes light?
   e. What percentage of the time is the traffic heavy?

2. During the winter months in a particular city, snowfall occurs at the average rate of two inches per week. Melting and evaporation will reduce existing accumulation at the rate of three inches per week. In both cases, the time in days to increase or decrease an inch is negative exponentially distributed. That is, if no melting were to occur, the time until one more inch would fall would be negative exponentially distributed. Similarly, if there were no new snow, the time to decrease an inch would be exponential. Of course, both natural processes occur together.

   a. Develop a continuous time Markov model of the changing depth of snow on the ground in inches. Provide a transition diagram or matrix and identify all parameters.
   b. Is the Markov assumption reasonable? Discuss factors that would tend to affirm or refute the assumption. (Don't worry about stationarity.)
   c. Show how to compute the fraction of time there is more than three inches of snow on the ground.

3. Consider four rats in a cage. Each rat alternates between periods of eating and periods of not eating. The eating period lasts an exponentially distributed amount of time with mean half an hour, and the noneating period lasts an exponentially distributed amount of time with mean three hours. Assuming the rats act independently of one another, what proportion of time are none of the rats eating? (Hint: this is more difficult than it appears. Think carefully about the transition rates.)

4. An office supply store keeps a small inventory of printer cartridges. The ink in them can dry out, so they have an expiration date, which in turn means that the store manager does not want to keep too many on hand. On the other hand, the manager does not want to be out of stock when a customer wants to buy a cartridge. Assume that customer demands occur one at a time (that is, customers never want to buy more than one at a time) according to a Poisson process with a mean rate of two per week. Assume that the store manager can order replacement stock one carton at a time, where a carton contains six cartridges. After placing an order for replacement, the store has to wait for a period of time (order processing plus delivery) that is negative exponentially distributed with a mean of one week. Such an order will be placed by the manager as soon as the stock falls to one cartridge.

   a. Set up an ergodic Markov process model to track the level of inventory. The states should correspond to the number of cartridges in stock. Each customer demands decreases the stock by one unit. The replacement event will increase the stock by six.
   b. What is the probability that a customer wanting one of these cartridges finds the store out of stock?

5. The batteries in a smoke detector are good for a length of time that is a random variable. This random variable is not likely to be negative exponentially distributed, but suppose for the sake of convenience that it is. Also assume that the mean lifetime is six months. Suppose that someone comes around to test the batteries at random intervals that are also negative exponentially distributed, with a mean of four months. If the batteries are good, they are not replaced; if they are bad, they are replaced. Fires (or smoke-producing events) occur according to a Poisson process with a rate of 0.1 per year (corresponding to an average of ten years between events). What is the probability that a fire occurs at a time when the smoke detector's batteries are bad?

6. A computer lab has a dozen machines, but at any given time some of them may be out of commission. Usually the problem is something as simple as the operating system locking up, which requires that a staff person reboot the machine and make sure that all the standard settings are correct. Sometimes the problem may be more serious, taking more time to correct. Overall, the repair time for any individual computer is negative exponentially distributed with a mean of thirty minutes. Once a computer is operating, the time until failure is negative exponentially distributed with a mean of ten hours.

  a. If there is only one staff person to do the repairs, what is the average number of operating computers in the lab over the long run?

  b. If there were two staff people to do the repairs, how much would that same average be?

7. A truck trailer has four wheels on one axle, two on each side. Each tire is independently subject to failure (blowouts) at the rate of one failure per 20,000 miles, on the average. Assume that the failure process is Markovian, and let R(m) denote the probability that none of the tires has blown after m miles. (Hint: Think of mileage as time, and think very carefully about your state definition.)

  a. Explain how to obtain R(m).

  b. What is the expected number of miles until the first blowout?

At least one tire on each side must *not* have failed in order to keep going. That is, one blowout on either or both sides is acceptable, but two on the same side forces a stop. Assume that flat tires are not repaired, and let Q(m) denote the probability that the trailer is still operable after m miles.

  c. Explain how to obtain Q(m).

  d. What is the expected number of miles until the first forced stop?

  e. How reasonable is the Markov assumption?

8. A very simplified description of the activity of a single nerve cell (neuron) would be as follows: It collects input signals until a certain threshold is reached, at which time it fires, producing an output signal that is passed along to other cells. After firing, there is a brief recovery period, during which it does not recognize input signals. To simplify even further, assume that the input signals occur according to a Poisson process with rate λ, that exactly N input signals must be received and recognized before the neuron fires, and that the recovery time is negative exponentially distributed with mean $1/\mu$.

  a. Develop a continuous time Markov process model, showing either a transition diagram or an instantaneous transition matrix. Be sure to define your states.

  b. Show how to compute the fraction of input signals that are recognized over the long run.

  c. Show how to compute the (steady-state mean) firing rate of the neuron, that is, the number of times it will fire per unit time.

9. Customers arrive at a bank machine according to a Poisson process at a rate of four per hour. Half of the customers are men and half women. Suppose that during an eight-hour period, twenty male customers arrived. What is the expected number of women customers that arrived during that same period?

10. A company has positions for three employees. Applicants for these jobs appear at a Poisson rate of two per year. If all jobs are filled, they look elsewhere. Those who hold a job do so for an exponentially distributed time with mean one-half year.

  a. Determine the proportion of applicants that are turned away.

  b. What is the average number of employees?

11. Imagine a pure death process (that is, a birth-death process with no birth transitions), starting in state 5, and having a constant death rate $\mu = 1$ per hour. Show how to find the probability that the process has reached state 0 within eight hours.

12. Birds arrive at a birdfeeder according to a Poisson process at a rate of six per hour.

  a. What is the expected time you would have to wait to see ten birds arrive?

  b. What is the probability that the elapsed time between the second and third bird exceeds fifteen minutes?

  c. If you have already waited five minutes for the first bird to arrive, what is the probability that the bird will arrive within the next five minutes?

13. This problem concerns the inventory stocking policies for type O negative blood at a particular hospital. Although some blood types are compatible with others, a person having type O negative cannot receive any other type. Thus we can isolate this type and treat it separately from others. The supply of this type occurs essentially at random from public donations at a mean rate of, say, four pints per day. The demands are also essentially random in nature, at a rate of three pints per day. On the average, more is received than is used, and some will be discarded. It is expensive to keep (on the order of $5 per pint per day) because special procedures needed to protect the medical quality. Therefore, it has been decided to establish a maximum inventory level, N, such that whenever the available supply is N, additional donations will be refused or used to replace "old" blood whose shelf life has made it less desirable than new blood. Because of the random nature of both supply and demand, there will be occasions when there is a need for blood and none is on hand. Whenever this happens, blood can be obtained on very short notice from another hospital at a cost of $30 per pint. There is no medical problem associated with doing this; it is purely a matter of cost. Using a continuous time Markov model, show how to determine the optimal N.

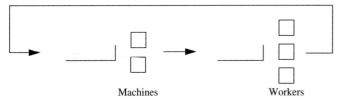

■ **FIGURE 5.23** Recycling molds

14. Consider an inventory demand process for some item that can be purchased or rented. That is, the item might or might not be returned after some period. Suppose that the demands follow a Poisson process with rate μ and the returns form another Poisson process with rate λ. Of course, the demand rate is greater than the return rate. Suppose that the replenishment policy is to order a quantity of Q (for example, 100) units as soon as the stock level reaches r (for example, 10). Backorders are not taken, so that demands that occur when the stock is empty are simply rejected. The lead time for the replenishment order to arrive is L time units. Using a continuous time Markov model with the state defined as the number of units on hand,

   a. Show how to compute the expected time until the stock would run out (starting from the time the order is placed) if the replenishment did not occur.

   b. Show how to compute the probability that the stockout condition will occur before the replenishment occurs.

15. A certain industrial process involves the use of molds that must be cleaned between uses. Hence, the molds go through a closed cycle as shown in Figure 5.23. There are two machines that use the molds, and there are three workers assigned to clean them. There are a total of eight molds in the system. Although it is not likely to be realistic, assume that the service times for both the forming and cleaning are negative exponential. The basic objective is to keep the forming process supplied with clean molds as often as possible.

   a. Show a Markov transition diagram for the process. Be sure to define your state.

   b. Define an appropriate performance measure for the system, and explain how to calculate it.

   c. Assuming that service times cannot be improved, there are two ways to get better performance: put more molds into the system or assign another worker to the cleaning job. Explain how to compare these two alternatives.

16. Three copies of a popular library book, "How to Overcome Test Anxiety," are held on reserve. This means that anyone who wants to read the book has to request it at the reserve desk and must return it there the same day. (No one is allowed to remove such books from the library.) Demands for the book occur throughout the day according to a Poisson process, with a mean rate of two per hour. On the average, a user keeps the book out for one and a half hours. The random variable representing the time a book is out is negative exponentially distributed, and the times for different users are independent.

   If the morning starts out will all three books on the shelf, how long will it be, on the average, until the first time all three are borrowed? (That is, all three are in use at the same time, and therefore none is available for another user.)

17. Three diesel generators supply electricity to the small settlement at the South Pole. Usually, all three run continuously, but if one fails the other two can take over the power load. However, if two fail, the one remaining generator can only handle half of the power load. At that point, the camp will have to cut back activities to only essential (life-support) functions. If all three go out, the camp is in deep trouble. Suppose that the causes of generator failure are purely chance events, so that the length of time between failures can be modeled as a negative exponential time with mean 200 days or 4,800 hours. Also assume that only one member of the team is qualified to perform repairs, taking an exponential time with mean one day.

   a. Model this process as a continuous time Markov Process. That is, define your time units and states, and show the transition diagram for a continuous time Markov process to model the availability of the generators, complete with all transitions and numerical rates.

   b. What is the expected number of days that the camp can operate before reaching the state of having no power (and therefore faces a crisis), assuming that the process starts will all three generators running?

CHAPTER 6

Queueing Models

CHAPTER CONTENTS

6.1. An Example 180
6.2. General Characteristics 182
6.3. Performance Measures 186
6.4. Relations Among Performance Measures 188
6.5. Little's Formula 190
6.6. Markovian Queueing Models 191
6.7. The M/M/1 Model 193
6.8. The Significance of Traffic Intensity 198
6.9. Unnormalized Solutions 200
6.10. Limited Queue Capacity 202
6.11. Multiple Servers 204
6.12. Is It Better to Merge or Separate Servers? 207
6.13. Which is Better: More Servers or Faster Servers 208

6.14. Case Study: A Grain Elevator 209
6.15. The M/M/c/c and M/M/∞ Models 210
6.16. Finite Sources 212
6.17. The Machine Repairmen Model 214
6.18. Numerical Calculations Using a Spreadsheet 214
6.19. Queue Discipline Variations 217
6.20. Non-Markovian Queues 218
6.21. The M/G/1 Model 219
6.22. Approximate Solutions for Other Models 220
6.23. Conclusion 221
6.24. Recommended Reading 221

In Great Britain, *queue* is a common word that means either a waiting line (noun) or the act of joining a line (verb). In the United States, the word is rarely used in ordinary conversation. Consequently, the subject of queueing theory may seem rather esoteric to some readers.[1] Actually it deals with phenomena that we all encounter as part of our everyday lives. In fact, one is almost certain to encounter some form of queue, or waiting line, in every waking hour. Aside from obvious examples, such as standing in line at a ticket window or grocery checkout counter, we also experience many

---

[1] You may be surprised to see the word spelled "queueing" rather than "queuing," as the rules of English spelling would suggest. Although both are correct, the majority of mathematicians who work in the area keep the final "e" before the "ing," at least partially because they like the fact that it is one of very few words in the English language that have five consecutive vowels.

more subtle instances of queueing behavior. For example, whenever we make a telephone call, there is a brief delay as a free communication channel is located. As you drive a car, the traffic congestion you encounter at intersections or on the open highway is a form of queueing behavior. The flow of material through a manufacturing operation or of paper orders through a sequence of processing stages represent other forms of queues. In short, the study of queueing theory has to do with congestion and delays that inevitably accompany conflicting demands for resources of all kinds.

To have any hope of understanding what creates and influences queues, you must deal with variability. The phenomenon does not even exist if you eliminate or choose to ignore variation in the times between events. Hence, we use probability distributions to represent that variation. In this case, the probabilities do not have to reflect any uncertainty; the same queueing behavior could result even if the times could be known in advance.

In this chapter, you will learn how to construct and solve equations that describe queueing behavior for a wide variety of situations. We cannot hope to exhaust even the elementary models, because the number of possible variations is enormous, as you will soon discover. But we will accomplish enough to permit you to apply a significant number of standard formulas and to derive your own in many other cases. After establishing some basic results, we will examine a case study with several illuminating aspects. You will be impressed with the ability of the theory to help in a situation where common sense fails. If you want to peek ahead to page 209–10, just for motivation, feel free to do so.

The models can contribute both computational tools and a deeper intuitive understanding of queueing phenomena. Although the formulas can be impressively powerful, it is probably more valuable in the long run to develop your intuition. With a good conceptual understanding of how the variations effect performance, you will be able to apply the results of queueing theory even in the absence of numerical data. For this reason, once we derive results we will scrutinize them for the design or performance implications they carry. The questions we will address include, "Is it better to have more service lines or faster servers?" and "Is it better to merge or separate the waiting lines?" These questions have definite answers that are not obvious until you know how to analyze queues. Once you know what the answers are, you can find many situations in real life to apply that knowledge. We will also expose a common misunderstanding about how to manage the capacity of a service system. Most people assume that you should attempt to balance the arrival rate to the full capacity of the system, in order to achieve full utilization of the resources. Queueing theory reveals that this is a dangerous mistake when random variations are present. The ability to model the stochastic behavior that leads to congestion and delays provides insight that is unavailable to those who think only of deterministic behavior or averages.

6.1 An Example

EXAMPLE 6.1

Let us begin with a straightforward example that possesses a full range of possibilities for modeling, but offers no major difficulties in interpretation. Assume that we are studying a small retail bakery shop that sells bread, rolls, cookies, and cakes to individual customers. The customers arrive at random moments throughout the day and are given service by a single clerk. The service consists of taking the order, selecting and bagging the goods, accepting payment, making change, and all of the other normal things you would expect. What is important from the current modeling perspective is not the actual activities of providing service, but rather the length of time consumed by the transaction. Just so there is no ambiguity, we will say that the service begins at the moment the customer arrives, provided that the clerk is not already occupied. If the clerk is occupied with a customer, the service begins as soon as the clerk becomes free. A service is completed when the clerk is free to attend to the next customer.

If a customer happens to arrive at a moment when the clerk is already occupied with serving a prior arrival, then the new arrival must wait. Of course, it is possible that several customers may back up. That is, a series of random arrivals may just by chance occur very close together, or a particular service time might be unusually long, or both. To keep things simple, we will assume that customers are served in the order of their arrival. To ensure this, the bakery might employ a take-a-number system. Even if the shop is occasionally crowded with customers awaiting service, we will assume that there is unlimited waiting room; no arriving customer is denied access. We will later relax these assumptions, but they are both convenient and reasonable for a start.

The whole idea of developing a mathematical model to represent this situation is to permit the calculation of quantities such as the average delay time, the fraction of time the clerk is busy, or the average length of the line. These performance measures are the result of rather complex interactions among the elements of the system, and are therefore hard to estimate without the aid of a model. Of course, to be able to use any of the formulas that are developed, it will be necessary either to collect statistical data from the system or to assume values for certain input variables. It is worth mentioning in advance (since many students lose this perspective as they learn about the modeling possibilities), that there is no need to model what we can measure directly. So, in the case of an existing bakery shop, if one wanted to know the average customer delay time, the easiest and most reliable method would be to go to the shop and time the delays. The real value of a model would become apparent, however, if one wanted to evaluate the effect of a proposed change in the system. It is generally easier, cheaper, faster, and safer to use a mathematical model to predict the consequences than to make the changes and observe the results.

There is a commonly used pictorial representation of a queueing system, which for the bakery shop would take the form of Figure 6.1. In the figure, the box represents the server, the X's are the customers, and the angled trough is the queue. The picture shows a situation in which there is one customer in service and two waiting. The open-ended left-hand side of the trough suggests unlimited waiting room; if there is only a limited capacity, the queue would be shown with a bar at the left end. The ordering of the customers in the queue suggests that they are served in order of arrival.

If the bakery shop had, say, three clerks all serving at the same line, then the picture of the system would look like Figure 6.2. In this version of the model, we assume that the servers are identical; any of them could handle any of the customers. As soon as a server is freed up, the customer at the front of the queue would move (either physically or logically) into the vacant service position. The take-a-number system

■ **FIGURE 6.1** A single-server queueing system

■ **FIGURE 6.2** A three-server system

■ **FIGURE 6.3** A parallel queueing system

that is common in bakery shops and certain other retail businesses maintains, in effect, a single waiting line.

In many other circumstances, such as grocery stores, toll booths, and fast-food restaurants, each server has a separate line, as pictured in Figure 6.3. If the system is like this, an arriving customer joins one of the queues (perhaps the shortest one) and is ultimately served at the front of that particular queue. Although it may seem at first like a subtle distinction, such differences can have a profound effect on system behavior. Perhaps you have noticed that many banks and airports have replaced separate waiting lines with a single one. Once we have developed results, we will be able to compare these alternative system designs.

Returning to the bakery shop example, imagine that we observed and recorded the times at which the successive customers arrive. These times might be recorded in a tabular format, as shown in Table 6.1. Imagine also that we recorded

■ **TABLE 6.1**
Bakery Shop Example

Customer	Arrival Time	Service Time Interval (in minutes)
1	9:03	4
2	9:05	1
3	9:07	2
4	9:09	3
5	9:10	1
6	9:12	2
7	9:13	1
8	9:15	2
9	9:20	2
10	9:22	2
11	9:23	3
12	9:26	1
13	9:27	1
14	9:30	2
15	9:33	4
16	9:36	4
17	9:40	2

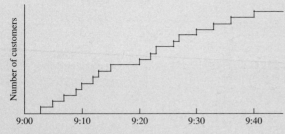

■ **FIGURE 6.4** Cumulative arrivals

the service time intervals for each customer, as shown in the third column. We have used integer (minute) values in this table, to simplify graphing, although in practice you would probably measure decimal values.

Figure 6.4 shows one way to plot the information about arrival times. Formally, we could say that a graph like this plots an instance of the stochastic process $\{A(t)\}$, where $A(t)$ represents the cumulative number of arrivals in time t. We have thus far made no particular assumptions about the nature of this process, except that arrivals occur one at a time. That is, the steps rise just one unit each time there is an arrival event.

It is convenient to think of the service time intervals as short blocks of time that can then be added to the graph of Figure 6.4 to show when each customer's service began and ended. In constructing this new graph, it is important to treat the customers in order and to remember that a particular service interval cannot begin until the prior customer's is finished. The resulting graph would look like Figure 6.5.

Now the graph looks like a series of stacked blocks. If you look horizontally across one level, you can see what an individual customer experienced. You can see when he or she arrived, how long a delay was encountered, when service began, how long it lasted, and when it was completed. If you look vertically at a line cutting through the blocks at a particular point in time, you can see the conditions that existed at that time. You can identify which customer, if any, was being served, and which ones, if any, were waiting. The vertical distance between the top and bottom stair-stepped lines can be interpreted as the number of customers present in the system at any time. Of course, the completions line can never rise above the arrivals line, for that would mean either (depending on how you look at it) that there was a negative number of customers in the system, or that some customer completed service before arriving.

Another way to graph the same information is to track the current state of the system, that is, the number of customers present at any point in time. That number would include one in service, if there is one, and whatever number might be waiting. When there is a service completion and no arrival, the number would decrease. The record would look something like Figure 6.6.

Remember, these charting conventions are for recording *past* performance of an observed system and what we have shown is only *one* possible realization out of an infinite number of possibilities. Our real task is to develop tools to predict the performance of systems that may not even exist. Using just a little information about the arrival pattern and service requirements, and taking into account the random nature of both, we want to be able to calculate various performance measures that depend upon the overall working of the system. But before getting too involved in the details, it is worth your time to improve your insight into queueing behavior by exploring some of the possibilities using this graphical tool. There are some brief problems at the end of the chapter to get you started.

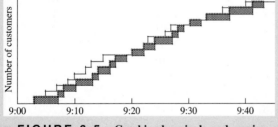

■ **FIGURE 6.5** Combined arrivals and service times

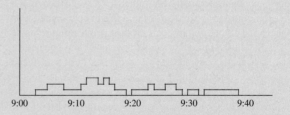

■ **FIGURE 6.6** The number present over time

6.2 General Characteristics

Before delving into the mathematics of queueing models, we should clarify some of the basic concepts, assumptions, and notation. The broadest concepts will be discussed first, with gradually increasing restrictions added as we work toward specific results.

6.2 General Characteristics

> **Definition** A *customer* is anything that arrives, occupies a server for some period of time, and departs.

It need not be human, or for that matter a physical object. It could be a job, an order, a transaction, a message, or some similar abstraction. One must be careful not to place too narrow an interpretation on the word customer, even in cases where the meaning may seem obvious. For example, a family of four buying tickets to a movie would count as only one customer, because they arrive, are served, and depart as a single unit.

> **Definition** A *server* is whatever resource is kept busy by a customer. A *service time* is a nonnegative time interval for whatever takes place that occupies both a customer and a server. The *number of servers* (sometimes called channel capacity) is the maximum number of customers that can be served simultaneously.

A service cannot take place unless both a customer and a server are present. If there are no customers, the server will be idle. A server is an abstract concept that should not automatically be assumed to correspond to anything physical. It could be a person, or a machine, or some tangible object, but it need not be. One server takes care of one customer. In most queueing models, it is assumed that one server will not help another, even when there are fewer customers in the system than there are servers.

An example of when you might be misled about the number of servers could occur when a crew of workers takes care of a single customer. In such a case, the entire crew would count as only one server. Occasionally, you may be hard-pressed to decide which of several possibilities determines the number of servers. For example, a loading dock for truck trailers may have room for ten trailers, but only three crews to do the loading. In such cases, you will usually want to use the smaller number (three, in this instance), but you should always refer to the original definition (the number of customers that can be served simultaneously) to make a precise determination.

In some of the more obscure applications of queueing theory, you may even have to decide which of two possibilities are the customers and which are the servers. A classic example is the so-called machine repairman problem that will be discussed in Section 6.17. You may have a choice that is governed more by your point of view than by the physical characteristics of the problem.

By convention, the classical notation for the number of servers in a queueing system is a lowercase c. In remembering this, it may be helpful to know that the notation originally referred to channels in the telephone systems that were the principal application of early queueing theory.

> **Definition** The *arrival process* is the stochastic process that describes how the customers appear for service.

Generally speaking, the arrival process is specified by naming some kind of a stochastic process, along with whatever parameters and simplifying assumptions are appropriate. Almost always, but not necessarily, customers are assumed to arrive one at a time. Usually, the arrival process is taken to be at least a renewal process (which means that the times between arrivals are independent random variables), and furthermore the process is assumed stationary (which means that the parameters do not vary over time). Nonstationary processes would be useful to model, for example, rush-hour traffic or slow periods, but the additional complications in the analysis and parameter estimation are sufficiently troublesome to cause one to avoid them whenever possible.

Usually, arrival processes are assumed to be state-independent, which means that they operate without any influence from the queueing system. Customers would just appear one after another, having no knowledge or expectation of the state of the system. Sometimes, however, we want to model state-dependent arrivals, such as when the sight of a long waiting line might discourage customers.

The simplest arrival process to deal with is a Poisson process, in which the times between arrivals are independent negative exponentially distributed random variables. The same process, viewed another way, gives a Poisson distributed random variable for the number of customers arriving in any time interval (hence the name for the process). Aside from several mathematical properties that make the Poisson arrival process convenient to use in queueing models, the data requirements are absolutely minimal. A single parameter, the arrival rate, is enough to describe the entire process.

Especially when the arrival process is Poisson, but usually even in other cases, the *arrival rate* is conventionally represented by the Greek lowercase lambda (λ). If one were estimating this parameter from statistical data, it would generally be preferable to deal with the reciprocal of lambda, which would be the mean time between arrivals. That is, you would collect a number of interarrival times and average them to get the mean time between arrivals; the mean arrival rate would be 1 divided by this value. For example, a five-minute average interarrival time would correspond to an arrival rate of 1/5 arrivals per minute. Of course, you may use any convenient time unit (seconds, minutes, hours, and so forth), as long as you are careful to maintain consistency throughout your model.

> **Definition** The *service process* is described by a service time distribution, which could be any probability distribution that is non-negative. Almost always, the service times for different servers are assumed to be identically distributed independent random variables, and the service times for different customers are assumed to be independent. Also, they are independent of the arrival process.

The service process is often regarded as being similar to the arrival process, but there are important differences. Strictly speaking, it is not a stochastic process, because it is interrupted by the idle periods when there are no customers present.

The service time distribution could depend on the customer or server or both. Sometimes one thinks of the service time as being determined by the customer, such as when varying products arriving at a machine have different processing times. In other cases, the variability is associated with how the server performs on a consistent stream of customers, such as when a human operator has to solve problems. These differences do not affect the queueing model, because we simply represent the service times as random variables. What actually happens during this time, or why the distribution is what it is, are not pertinent.

Sometimes the parameters of the service time distribution may reflect a dependence on the number of customers already present. For example, the servers may work faster when the queue grows long.

The easiest service time distribution to work with is the negative exponential distribution, which is fully characterized by a single parameter, the *mean service rate* (or its reciprocal, the mean service time). The usual notation for a service rate is a Greek lowercase mu (μ).

> **Definition** A *queueing system* consists of one or more servers, an arrival process, and a service process, along with some additional assumptions about how the system works.

The word "queue" is sometimes used to describe the whole system, but we will be careful to reserve the use of this word for just that part of the system that holds the excess customers who cannot gain immediate access to a server. Thus, the number of customers in the system at any given time will equal the number of customers in the queue plus the number of customers in service. Of course, these numbers will vary over time as customers come and go, so they are formally stochastic processes.

If there is a limited amount of waiting room, or for any other reason customers are prohibited from joining a queue once it reaches a certain length, we say that the queue or system has *finite capacity*. Although you may be tempted to regard every real-life system as having finite capacity (simply because nothing in real life is infinite), you should realize that specifying a capacity adds a complication to the analysis. We do so, therefore, only when we have reason to believe that the limited capacity is a significant factor in determining the behavior of the system. An uppercase N is the usual notation for the capacity of the system; the capacity of the queue would be $N - c$..

> **Definition** The phrase *queue discipline* refers to the rule by which the next customer to be served is selected.

The most common queue discipline is first come-first served, often abbreviated FCFS. In some inventory applications, the same rule is called first in-first out, or FIFO. In most cases where the customers are human, a policy of fairness dictates that customers should be served in the order of their arrival, which is equivalent to a FCFS discipline. However, there are exceptions to the general rule, and there can even be cases where other queue disciplines make more sense. Last come-first served, abbreviated LCFS, and random service selection, abbreviated RSS, are other simple rules. More complicated ones involve various kinds of priorities or classifications.

> **Definition** The *source population* of customers is the set from which the arrivals come. The phrase *finite source population* refers to the case of a limited number of potential customers.

The population is usually assumed to be infinite, so that the arrival process is unaffected by the number of customers already in the system. Sometimes, though, it is useful to be able to limit the number of potential arrivals to some finite value. For example, a maintenance facility may service only twenty trucks in a fleet; if they are all in the shop, no additional arrivals are possible. This is somewhat different from a limited queue capacity, as we shall see.

By now, you must have realized that many variations in queueing models are possible. The assumptions can combine in so many ways that it can be a challenge just to keep straight which particular combination is under discussion. Fortunately, this problem was recognized long ago, and a standard notation was developed. It is called Kendall's notation, after the mathematician D. G. Kendall who proposed an early version of the classification scheme. Briefly, the idea is to specify a particular queueing model by a code consisting of six terms separated by slashes, as shown in Figure 6.7. The first term is an alphabetic code for the type of arrival process assumed. The second term is a similar code for the service process. The third, fourth, and fifth terms are either numbers or variables for the number of servers, capacity of the system, and size of the source population, respectively. The final term is an alphabetic code for the queue discipline, such as FCFS or LCFS. The last two or three terms may be omitted if one intends them to assume their usual values of infinite capacity, infinite source population, and first come-first served discipline. The first three terms are always given.

The most commonly used letter codes are shown below.

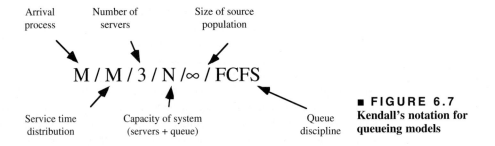

FIGURE 6.7 Kendall's notation for queueing models

> **Notation**
>
> M for negative exponential interarrival or service times (the M stands for Markov).
> D for deterministic interarrival or service times.
> E_k for Erlang-k interarrival or service times.
> GI for general independent interarrival times.
> G for general service times.
> FCFS for first come, first served.
> LCFS for last come, first served.
> RSS for random service selection.

To give a few examples, the designation M/D/2/N would indicate a Poisson arrival process, deterministic service times, two servers, and a system capacity of N; GI/E_3/c/10/10/LCFS would indicate general independent interarrival times, Erlang-3 service times, c servers, a system capacity of ten, a finite source of size ten, and a last come-first served queue discipline. The simplest models, which will be explored in some depth, are the M/M/1, the M/M/c, the M/M/c/c, and the M/M/∞. We will also be able to say a little about the M/G/1 model.

6.3 Performance Measures

Whatever the assumptions of a particular model may be, there are several commonly desired performance measures. Depending on the circumstances that motivated the construction of the model in the first place, one or several of these may stand out as being the key values and others may be of little concern. In different circumstances, the relative importance attached to the various measures may reverse. For example, in one instance, the primary concern may be for the mean delay encountered by customers; in another, it may be for the utilization of the servers.

Although there are times when one would like more detailed information, the results that appear in this chapter are confined to steady-state performance. This term is often misunderstood. It does not mean that the queueing system has somehow settled down to a regular pattern of behavior. If we were observing the system in real time and gathering statistics, we would see perpetual fluctuations in the individual observations. For example, if we were to record the number of customers in the queue every ten minutes, these values would continue to vary. However, if one were to calculate a running average of those numbers, the average would settle down to a consistent value after a sufficiently large number of observations. That is, further observations would have no significant effect on the statistic. In our analytical work, we are not calculating statistics from observations, but the concept is similar. We are determining values for performance measures that refer to expected behavior of the system over the long term.

Perhaps the most basic information we can have about the steady-state performance of a queueing system is the probability distribution of the number of customers present in the system. From this distribution, almost all of the other common measures can be computed. Later we will demonstrate methods for computing the distribution, but for now let us assume that it has already been determined, in order to show how it can be used. Regardless of which model we are dealing with (number of servers, and so forth), let π_j represent the steady-state probability that j customers are present in the system. The index j will run from zero to whatever value corresponds to the capacity of the system, N or ∞.

A direct interpretation of π_j would answer the question, "What is the probability of finding j customers in the system at some arbitrary time long after the last observation?" However, an indirect interpretation is often more useful. It can be proved that π_j is equal to the fraction of time, over the long run, that the system has exactly j customers present. For example, π_0 is the fraction of time the system is empty. Or, for another example, in a finite capacity system, π_N would be the fraction of time the system is full.

Recalling that c represents the number of servers, the sum $(\pi_c + \pi_{c+1} + \pi_{c+2} + \cdots)$ can be interpreted as the probability, or fraction of time, that all servers are busy. The same value could also be described as the probability of delay or as the fraction of arriving customers who are forced to join the queue. Several other useful quantities can be constructed from subsets of the π_j distribution.

Two other distributions can be easily derived from π_j. The number of busy, or occupied, servers is a random variable that is closely related to the number of customers in the system. Let b_j represent the probability that j servers are busy. That same distribution can be used for the number of customers in service, since that number exactly matches the number of busy servers.

The size of the queue (the customers waiting for service, not counting any that might be receiving service) is a random variable whose steady-state distribution is denoted by q_j. At any point in time, the number of customers in the system will be equal to the number in the queue plus the number in service. This basic fact links the three probability distributions, their means, and numerous other measures.

Several mean values are so commonly computed that they have standard notation. The mean number of customers in the system (that is, the mean of the π_j distribution) is denoted by L. Of course, we can relate L to the π_j's by the basic definition of a mean,

$$L = \sum_j j\pi_j$$

but this is useful only after we know the π_j's, which will change form as the model assumptions vary. Hence, the formula for L in terms of the model parameters will change, depending upon the particular model in use, but the same letter is always used.

Similarly, the mean number of busy servers, denoted by B, is obtainable from the b_j distribution in the usual way. In the next section, we will be able to derive a very simple expression for B which applies very generally and avoids the use of the distributions entirely. We will use L_q to represent the mean queue length, from the q_j distribution. We can often use a simple relation to obtain the value from L.

The performance measures defined so far all relate to the number of customers somewhere in the system. Usually a customer is more interested in the time it takes to get through the queue or system. We use W to represent the steady-state mean time that a customer spends in the system, and W_q for the mean time in the queue.

The throughput, or output rate, or production rate of the queueing system, formally defined as the mean number of service completions per unit time, will be denoted by R. Since

■ TABLE 6.2
Summary of Performance Measures

Symbol	Meaning
π_j	Probability j customers present
b_j	Probability j servers busy
q_j	Probability j customers in queue
L	Mean length of system
B	Mean number of busy servers
L_q	Mean length of queue
U	Mean utilization
R	Mean throughput rate
W	Mean waiting time in system
W_q	Mean waiting time in queue

we are speaking of steady-state results, the throughput can also be thought of as the rate at which customers enter the system. For infinite capacity systems, every customer is accepted and eventually served, so in that case R would equal λ, the raw arrival rate. However, finite capacity systems will reject customers who happen to arrive when the system is full. For those systems, R will be something less than λ.

This is quite a bit of notation to absorb all at once, but if you refer to Table 6.2, you will notice that there are mnemonic cues for each variable to help you remember. All of the capital letters refer to steady-state mean values; the lowercase letters indicate probabilities.

6.4 Relations Among Performance Measures

As you might expect, there are many relationships among the performance measures. Some of these are quite obvious; others are not. It is important to know these relationships, so that you can easily derive unknown values from known ones. Usually only one lengthy algebraic computation is enough to produce, for example, L, after which everything else is very easily produced using very minor transformations. The same relations would hold for statistically determined measurements. So, for example, if one were observing a queueing system over a period of time to estimate a value for L, other performance measures could be derived without recording additional statistics. None of these relations require the Markov assumption.

One of the simplest relations, which will hold for any queueing model possessing a steady-state solution, is shown below.

> **Result** In any queueing system,
> $$L = L_q + B$$

In words, the mean number in the system equals the mean number in the queue plus the mean number in service. A similar relation in terms of times is shown below.

> **Result** In any queueing system,
> $$W = W_q + \frac{1}{\mu}$$
> where $1/\mu$ is the mean service time, whether or not the distribution is negative exponential.

In words, the expected time to pass through the system is the expected time spent in the queue plus the expected time spent in service. Both of these relations follow from the general property that the mean of a sum is the sum of the means (which is always true).

The three steady-state distributions (for the number in the system, the number in the queue, and the number of busy servers) are all similarly related. If you have the π_j's, you can easily get the q_j's and b_j's just by thinking about what they mean. In particular, there will be zero customers in the queue whenever there are c or fewer customers in the system, and there will be j customers in the queue whenever there are $c + j$ customers in the system.

Result In any queueing system with c servers,

$$q_0 = \pi_0 + \pi_1 + \cdots + \pi_c$$
$$q_j = \pi_{c+j} \text{ for } j > 0$$

Similarly, there will be j busy servers whenever there are j customers in the system (and when j is less than c) and all c servers will be busy whenever there are c or more customers in the system.

Result In any queueing system with c servers,

$$b_j = \pi_j \quad \text{for } j < c$$
$$b_c = \pi_c + \pi_{c+1} + \cdots$$

Both of these results come from understanding equivalent events, and do not depend in any way upon distribution assumptions.

If we know the mean number of busy servers, B, we can easily compute the average utilization, U, from

$$U = \frac{B}{c}$$

which says that the average utilization equals the mean number of busy servers divided by the number of servers. If there is only one server, we can write

$$U = B = 1 - \pi_0$$

The throughput rate is slightly more complicated. In steady-state, the mean number of service completions, or departures, per unit time must equal the mean number of arrivals per unit time (for, if not, the mean number in the system could not be a stable value). Hence, the result below.

Result In any infinite capacity queueing system with state-independent arrivals

$$R = \lambda$$

However, finite capacity systems will not accept every arrival. Only those customers who arrive when the system is less than full are allowed into the system. In such cases, one must adjust the pure arrival rate (the attempts) downward to obtain the *effective* arrival rate. If arrivals "see" random states (a characteristic of all of the models developed in this chapter, but not all queueing models), the fraction of attempts that are successful is $1 - \pi_N$. See the result below for all of these finite capacity models.

> **Result** In any finite capacity queueing system with state-independent arrivals
> $$R = \lambda(1 - \pi_N)$$

6.5 Little's Formula

It might be expected that mean waiting times are related to mean numbers of customers, in the sense that long lines would tend to imply long waiting times. Few people would guess, however, that there is a very simple direct relationship linking L, R, and W. The equation, known as Little's formula, is shown below.

> **Result** In any queueing system where steady-state results exist and arrivals are state-independent
> $$L = RW$$

Once any two of these quantities are fixed, the third is determined. If any one is fixed, the other two must vary in direct proportion. In particular, see the result below.

> **Result** In an infinite capacity system, where $R = \lambda$,
> $$L = \lambda W$$

Incidentally, when you see Little's formula referenced in other books, it will most likely be stated in this latter form. However, experience has shown that students have difficulty remembering to modify λ in finite capacity cases, so in this book we have elected to sacrifice tradition for clarity.

Little's formula applies on a very broad scale to all kinds of problems. In fact, any system that transforms input to output over time and possesses steady-state performance measures corresponding to L, R, and W will obey that law. At one extreme, the system could involve an arbitrarily complicated network of interacting processes; at the other, it might involve no queueing at all. The rule can be applied to any portion of a system, or to collections of systems. Figure 6.8 is helpful in relating the terms. L represents the steady-state mean number of "somethings" in a system; R is the mean rate of entry or exit; and W is the mean time spent in the system.

As an immediate example, we could restrict attention to just the queue portion of a queueing system to obtain the useful relation shown below.

■ **FIGURE 6.8** Little's formula

> **Result** In any queueing system where steady-state results exist and arrivals are state-independent
> $$L_q = RW_q$$

Another very convenient application of Little's formula will yield a simple expression for B, the mean number of busy servers. If we think of the box as containing just the servers (excluding the queue), and recognize that the mean time in this box is just the mean service time, we get the result below.

> **Result** In any queueing system where steady-state results exist and arrivals are state-independent
> $$B = \frac{R}{\mu}$$
> or, for an infinite capacity system in which $R = \lambda$ (with any number of servers)
> $$B = \frac{\lambda}{\mu}$$

To cite a more remote application, we could be dealing with an entire factory. Little's formula would state that the in-process inventory for the factory as a whole (L) equals the production rate of the factory (R) times the average flowtime of jobs through the factory (W). Alert readers will realize that Little's formula is a powerful tool.

6.6 Markovian Queueing Models

There is a class of queueing models for which complete detailed solutions are quite easy to obtain and furthermore allow many variations. The class consists of those models that can be treated as special cases of the continuous time Markov process discussed in Chapter 5. To maintain independence of the chapters, our treatment here will not assume that you have already studied Markov processes in detail. However, you will find that both your understanding and ability to develop new models will be greater if you take the time to learn that material.

To be a Markovian queueing system, both interarrival times and service times must be negative exponentially distributed. As already mentioned, this distribution family has the advantage of being fully characterized through only one parameter, which is conveniently related to the mean. Of course, that also implies that one has very limited control over the distribution; for example, one cannot separately adjust the mean and the variance.

One might well ask how reasonable or realistic are those assumptions. The answer is, usually, not very. Hardly ever will you encounter in real life a situation where the negative exponential assumptions can be rigorously defended. Actually, however, the proper question to ask is not how good the assumptions are, but how good the results are. In other words, we should be content to use the assumptions, even knowing that they are not exactly correct, if they enable us to obtain acceptable results. From this point of view, the Markovian assumptions seem better than you might expect. To be sure, there are instances when the use of the negative exponential distribution could produce very misleading results, and one should always be cautious in drawing conclusions in critical situations. Still, the simple Markovian models work surprisingly well in many practical situations.

When the queueing model is Markovian, you can set up and solve a set of linear equations whose solution provides the steady-state distribution of the number of customers in the system, that is, the π_j distribution. All of the other performance measures we have mentioned can then be computed from these. To be able to write down the proper equations, you must understand the material presented in the last chapter or learn the simplified method presented here.

To start, you first must identify the possible states of the queueing system. Normally these will be integers corresponding to the possible numbers of customers present in the system (that is, 0, 1, 2, . . . , N), although more complicated state descriptions may sometimes be necessary. Then, for each state, you must identify the possible transitions out of that state. For example, an arrival to the system would normally change the state from j to $j + 1$ (unless j is already the maximum value in a finite capacity system), and a service completion would normally change it to $j - 1$. Just think of all of the events that could occur when you are in a particular state, and trace the change of state that each would cause. Do not bother considering all of the ways to enter each state; these will take care of themselves.

The transition diagram provides a convenient visual portrayal of the structure of the states and transitions. Each state is represented by a small labeled circle, and each transition by an arrow, in the obvious way. Figure 6.9 shows a portion of a typical diagram.

The next step, which is the only one posing even slight difficulties, is to figure out the appropriate transition rate to associate with each transition. We will be going through several cases individually. Once you get the idea, you will be able to handle all kinds of variations easily. In the very simplest case, the M/M/1 model, each arrival transition (j to $j + 1$) is assigned the ordinary arrival rate, λ, and each service-completion transition (j to $j - 1$) is assigned the service rate μ. So, for this particular model, the transition diagram looks like Figure 6.10.

From the transition diagram, you can set up the steady-state equations for the π_j using the principle of flow balance. The logic of this principle is intuitively appealing, easy to remember, and simple enough to use confidently. Most importantly, it yields the correct equations in every case. To state the principle, we must introduce the concept of "flow" along a transition arc. If an arc is from point i to point j, with an associated transition rate of λ_{ij}, then the flow along the arc is $\lambda_{ij}\pi_i$. When we first write this, of course, we do not know the value of π_i, so that variable will appear as an unknown in the equation. The principle of flow balance declares that, in steady-state, the total flow into every state must equal the total flow out. When the principle is applied to the states of the transition diagram, one by one, each state produces one equation, which is linear in the unknown π_j's. Collectively, the full set of equations determines the full π_j distribution, with one minor qualification.

Each equation will be linear, and there will be exactly as many equations as there are unknowns. The equations will be consistent, so there is sure to be a solution. The problem is that the equations will contain one dependency. That is, any one of the equations could be eliminated without loss of information. (This property follows from the basic properties of transition rates, and is more fully explained in Chapter 5.) However, there is one more equation that must be satisfied by the π_j values, and that is the requirement that the

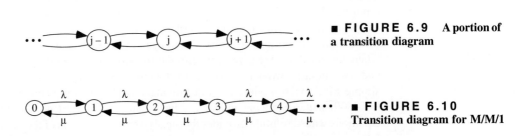

■ FIGURE 6.9 A portion of a transition diagram

■ FIGURE 6.10 Transition diagram for M/M/1

probabilities sum to 1. This equation, called the normalizing equation, will be independent of all the others. Consequently, the set of flow balance equations (less any one) together with the normalizing equation jointly determine a unique solution. Moreover, you may be certain in advance that the solution will be of proper form for a probability distribution. In short, nothing can go wrong.

Frequently (virtually always in simple queueing models), the transition diagram is of the birth-death form. This means that the states correspond to integers and that all transitions are to the state just above (a birth) or to the state just below (a death). Whenever the number of customers in the system is an adequate state description and customers arrive and are served individually, the transition diagram will be of the birth-death form. A model in which, for example, customers are serviced in pairs would not have the birth-death form, because a single transition could change the state from i to i − 2. All of the relations and methods discussed thus far would still apply; in particular, the flow balance equations would be valid. However, there is a definite advantage to dealing with the birth-death form when it comes to solving the equations.

The equations for a birth-death model have a characteristic form that makes them particularly easy to solve by hand. The first equation, which is constructed by examining the transitions in and out of state 0, will have only two unknowns, π_0 and π_1. This happens because state 0 is connected only to state 1. The coefficients of the unknowns will vary depending upon model assumptions, but it will always be possible to solve for π_1 in terms of π_0. That is, π_1 can be written as some multiple times π_0. The second equation, which comes from observing the flow into and out of state 1, will involve three unknowns, π_0, π_1, and π_2. By subtracting our first equation, we can eliminate π_0, and then solve for π_2 in terms of π_1. Then substituting our solution for π_1 in terms of π_0, we obtain π_2 in terms of π_0. Continuing in this way, each successive equation introduces a new unknown that can be reduced to some multiple of π_0. Eventually, we have all of the unknowns expressed in terms of the one undetermined value of π_0. The final step uses the normalizing equation to find the value of π_0. Since all of the probabilities must sum to 1, and each probability is now expressed as a multiple of π_0, this one equation can be manipulated to yield an expression for π_0. If the model has an infinite number of states, there will be an infinite series to sum.

The description of the method of solution is perhaps more difficult to follow than the computations are to carry out. It really amounts to no more than simple algebraic substitution. Once you have gone through an example or two (referring, if necessary, to the paragraphs above), it will be obvious how to proceed. After we examine a few cases, we will introduce a general solution that covers all birth-death models in a form that is well suited to computer implementation.

If the equations are not of the birth-death form, you may have difficulty solving them by hand. In such cases, you may be forced to use a computer program to get numerical solutions. It is possible that difficulties of another kind may appear in any such attempt, because queueing equations are quite often ill-conditioned. Nothing in this book will demand sophisticated knowledge of numerical methods, but you should be aware of the possibility in more general cases.

6.7 The M/M/1 Model

The easiest Markovian model to set up and solve is the single-server, infinite capacity version. The bakery shop example could serve as a concrete application of this model. Of course, we must assume that the arrivals occur according to a Poisson process (or, equivalently, that the interarrival times are independent and negative exponentially distributed). This assumption is quite reasonable for the bakery shop example. We must also assume that service times are negative exponentially distributed, which is not very likely to be satisfied by the type of service performed in a bakery shop. However, our objective at this time is more to develop the models than to select the best one, so we will proceed with the awareness that the assumptions are imperfect. Eventually, we will develop a more suitable model for the bakery shop.

In accordance with the usual notational conventions, we use λ for the mean arrival rate and μ for the mean service rate. These will be carried through the analysis in symbolic form, to permit the derivation of general formulas. However, if it helps to have specific numbers the first time through, you could imagine that customers arrive at the mean rate of fifteen per hour ($\lambda = 15$) and the mean service rate is twenty per hour ($\mu = 20$). Such rates would be obtained from stopwatch type data if you observed an average time between arrivals of four minutes and an average service time of three minutes. Realistically, of course, one could not expect to deal with such nice round numbers.

The state of the queueing system will be the number of customers present, which will be indicated by an integer number from 0 to infinity. The assumption that customers arrive and are served individually gives the birth-death structure to the transition diagram. The arrival rate and service rate are associated with the appropriate transition arcs in the obvious way, yielding the transition diagram already shown in Figure 6.10.

Applying the principle of flow balance to each state, starting with state 0, we get the steady-state equations:

$$\lambda \pi_0 = \mu \pi_1$$
$$(\lambda + \mu)\pi_1 = \lambda \pi_0 + \mu \pi_2$$
$$(\lambda + \mu)\pi_2 = \lambda \pi_1 + \mu \pi_3$$
$$(\lambda + \mu)\pi_3 = \lambda \pi_2 + \mu \pi_4$$

and so on. If you add the first equation to the second and cancel common terms on both sides of the equation, you get:

$$\lambda \pi_1 = \mu \pi_2$$

which can substitute for the second equation (since we have added an equal quantity to both sides). If we then add that new second equation to the original third equation, we get:

$$\lambda \pi_2 = \mu \pi_3$$

which becomes our new third equation. Continuing in this manner, we see that all of the equations can be put into the form:

$$\lambda \pi_j = \mu \pi_{j+1}$$

Using the solution strategy outlined above for birth-death type equations, we solve the first equation for π_1 in terms of π_0, giving

$$\pi_1 = \frac{\lambda}{\mu} \pi_0$$

Substituting this in the second equation and solving for π_2 yields

$$\pi_2 = \left(\frac{\lambda}{\mu}\right)^2 \pi_0$$

Continuing in this manner, we find that, in general,

$$\pi_j = \left(\frac{\lambda}{\mu}\right)^i \pi_0$$

To be certain that this form is, indeed, the solution for the entire infinite set of equations, it should be verified by substituting it into the general equation.

At this point, we can see that the parameters λ and μ appear only together as a ratio. It is conventional and convenient to replace the ratio by the single parameter ρ (a lowercase Greek rho), where $\rho = \lambda/\mu$. (Warning: ρ is a parameter whose definition may vary from one model to another. Here it is defined as λ/μ, but in other queueing models it may be defined differently.) This parameter ρ is called the traffic intensity, and we will have a good deal to say about it later. For now, just treat it as a symbol that was introduced to simplify the expressions.

This substitution simplifies the expression for π_j to:

$$\pi_j = \rho^j \pi_0$$

The remaining unknown, π_0, is evaluated by imposing the normalizing equation:

$$1 = \pi_0 + \pi_1 + \pi_2 + \cdots$$
$$1 = \pi_0 + \rho\pi_0 + \rho^2\pi_0 + \cdots$$
$$1 = \pi_0[1 + \rho + \rho^2 + \cdots]$$
$$1 = \pi_0\left[\frac{1}{1-\rho}\right]$$

The infinite series that appears within the brackets is a simple ratio series that will converge to the finite sum provided that the quantity ρ is less than 1. (The implications of this restriction will be explored more fully later.) Consequently, the solution for π_0 is

$$\pi_0 = 1 - \rho$$

Putting this expression back into the solution for the general state probability produces the final solution shown below.

Result For the M/M/1 queueing system, the steady-state distribution is

$$\pi_j = \rho^j(1-\rho)$$

The same formula holds for all j from 0 to infinity. This probability distribution happens to be of a common family—the geometric distribution. The summation of the infinite series produced the requirement that $\rho < 1$.

Since we have the complete probability distribution of the number of customers in the system, we can calculate the mean, L.

$$L = \sum_j j\pi_j$$

$$L = \sum_{j=0}^{\infty} j\rho^j(1-\rho)$$

$$L = (1-\rho)\sum_{j=0}^{\infty} j\rho^j$$

The infinite series is a familiar one. It converges to a finite sum only if $\rho < 1$, but this condition has already been assumed.

$$L = (1-\rho)\left[\frac{\rho}{(1-\rho)^2}\right]$$

After canceling the common term, we get a simple formula as shown below.

Result For the M/M/1 queueing system, the steady-state mean number in the system is
$$L = \frac{\rho}{1-\rho}$$

We can also obtain the queue length distribution, q_j, from the π_j, as shown below.

Result For the M/M/1 queueing system, the steady-state distribution for the number of customers in the queue is
$$q_0 = \pi_0 + \pi_1 = (1-\rho) + \rho(1-\rho) = 1 - \rho^2$$
$$q_j = \pi_{j+1} = \rho^{j+1}(1-\rho) \text{ for } j \geq 1$$

From this distribution, we can get the mean number of customers in the queue, L_q.
$$L_q = 0q_0 + 1q_1 + 2q_2 + \cdots$$
$$L_q = \rho^2(1-\rho) + 2\rho^3(1-\rho) + \cdots$$

Result For the M/M/1 queueing system, the steady-state mean queue length is
$$L_q = \frac{\rho^2}{1-\rho}$$

Another way to get the same result is to recognize that the difference between L and L_q—that is, $(L - L_q)$—is equal to B, the mean number of busy servers, which is equal to ρ. That trick will avoid the infinite series summation that always appears in the calculation whenever you go back to the distribution.

W and W_q, the mean waiting time and mean delay time, respectively, are obtainable from Little's formula. They are shown below.

Result For the M/M/1 queueing system,
$$W = \frac{1}{\lambda}L = \frac{1}{\lambda}\left(\frac{\rho}{1-\rho}\right) = \left(\frac{1}{1-\rho}\right)\left(\frac{1}{\mu}\right)$$
$$W_q = \frac{1}{\lambda}L_q = \frac{1}{\lambda}\left(\frac{\rho^2}{1-\rho}\right) = \left(\frac{\rho}{1-\rho}\right)\left(\frac{1}{\mu}\right)$$

The form of these two expressions allows you to see them as multiples of the mean service time $1/\mu$.

We can also get, for this model, the complete probability distribution of waiting time (the time to pass through the entire system), so that one could compute, for example, the probability of having to spend more than a particular time waiting for service. Unfortunately, only this M/M/1 model has a simple result; you will have to go to more advanced books to get waiting time distribution results for any of the other models. In fact, the following argument leaves out some of the technical details that are crucial to a real proof, but it does give an outline of the logic.

6.7 The M/M/1 Model

Let Z be the random variable representing the total time to pass through the system (in steady-state). We know, of course, $E(Z) = W$, but now we want the entire distribution in the form of a density function, a cumulative distribution function, or a complementary cumulative distribution function (ccdf). It turns out that the ccdf is the most convenient form to work with. Consider the probability that Z exceeds t and condition on the number of customers present at the time a randomly chosen customer arrives. That is, we are using the law of total probability in the form

$$P(Z > t) = \sum_{k=0}^{\infty} P(Z > t | k \text{ customers present}) P(k \text{ customers present})$$

The steady-state probabilities of finding k customers present in the system are already known; they are the π_j. The conditional probabilities can be constructed by a verbal argument. If the newly arrived customer finds k customers ahead of him, and assuming the first come-first served discipline is adhered to, he will not begin service until k negative exponential service times are completed. He will not *complete* his service until $k + 1$ service times are completed. (Any partial service time that might have been completed at the time of arrival can be ignored, because the negative exponential distribution is forgetful.) Viewed another way, the new arrival will still be there at time t if less than $k + 1$ service times are completed in time t. But since the service times are negative exponential, the number completed in time t (assuming no interruptions, of course) would be Poisson distributed, so

$$P(Z > t | k \text{ customers}) = \sum_{i=0}^{k} \frac{(\mu t)^i}{i!} e^{-\mu t}$$

Substituting these, along with the π_j into the conditional statement above,

$$P(Z > t) = \sum_{k=0}^{\infty} \left(\sum_{i=0}^{k} \frac{(\mu t)^i}{i!} e^{-\mu t} \right) \rho^k (1 - \rho)$$

$$= e^{-\mu t}(1 - \rho) \sum_{i=0}^{\infty} \frac{(\mu t)^i}{i!} \sum_{k=i}^{\infty} \rho^k$$

$$= e^{-\mu t}(1 - \rho) \sum_{i=0}^{\infty} \frac{(\mu t)^i}{i!} \left(\frac{\rho^i}{1 - \rho} \right)$$

$$= e^{-\mu t} \sum_{i=0}^{\infty} \frac{(\rho \mu t)^i}{i!}$$

$$= e^{-\mu t} e^{\rho \mu t}$$

Finally, we combine the exponents to get a single exponential function shown below.

> **Result** For the M/M/1 queueing system,
> $$P(Z > t) = e^{-(1-\rho)\mu t}$$

The complementary cumulative distribution of Z has the form of a negative exponential distribution with the parameter $(1 - \rho)\mu$. So, in words, we have found that the waiting time distribution for the M/M/1 model is another negative exponential distribution. This is a very nice result that enables us to calculate the probability of having to wait more than any particular time. Unfortunately, as already mentioned, we cannot get similarly tidy results for any other model.

EXAMPLE 6.2

To illustrate the use of the M/M/1 results, and at the same time to gain some insight into an important aspect of queueing behavior, let us return to the bakery shop example. We assumed there that the arrival rate λ was fifteen customers per hour and the mean service rate μ was twenty customers per hour, yielding a traffic intensity ρ of 0.75. For this set of conditions, then, the steady-state distribution of the number of customers in the shop is

$$\pi_j = (0.75)^j (0.25)$$

We can also say immediately that the server will be idle 25 percent of the time (since $\pi_0 = 0.25$), and therefore busy 75 percent of the time. The long-term average number of customers in the shop will be L = 3, and the long-term average number of customers waiting will be $L_q = 2.25$. The average time spent in the shop, W, can be found from Little's formula to be twelve minutes, which means that the average time spent waiting for service, W_q, is nine minutes. If you wanted the know the probability of having to spend more than fifteen minutes in the shop, you would evaluate

$$P(Z > 0.25) = e^{-(1-0.75)20(0.25)} = e^{-1.25} = 0.2865$$

6.8 The Significance of Traffic Intensity

The parameter ρ is called the *traffic intensity*. It is defined (for the M/M/1 case) as the ratio of the arrival rate λ to the service rate μ, but there are also other ways to interpret it. Expressing it in the form

$$\rho = \frac{\left(\frac{1}{\mu}\right)}{\left(\frac{1}{\lambda}\right)}$$

shows that it is the ratio of the mean service time to the mean time between arrivals. So when we say that ρ must be less than 1, we are really saying that the mean service time must be less than the mean time between arrivals. It could also be thought of as the arrival rate, λ, times the mean service time, $1/\mu$, so one could interpret ρ as the mean number of arrivals during a period equal to the mean service time. All of these interpretations give some logical explanation for the requirement that appeared in the solution procedure that ρ be less than 1. In simple words, the model will not give an answer if the customers are arriving faster, on the average, than they can be served.

When we have multiple servers, ρ is defined differently. When there are c servers, we want the traffic intensity to have the same intuitive interpretation, so we define it as $\rho = \lambda/(c\mu)$ (except when there are infinite servers, in which case this would not make sense, and we return to the original definition). This could be viewed as just an extension of the definition, because c = 1 is just a special case. But in any case except for the infinite server case, the traffic intensity is defined as the ratio of the arrival rate to the *total* service rate, that is, the rate at which all of the servers can process customers. By adopting that convention, a traffic intensity of 1 will always be the (unreachable) limit for stable queueing behavior.

Additional interpretations of ρ can be obtained from the steady-state solution. The term π_0 is the steady-state probability that the system is empty, which can also be thought of as the fraction of time (over the long run) that the server is idle. Hence, the fraction of time the server is busy, or the utilization, U, is $1 - \pi_0$. But since, for this model, $\pi_0 = 1 - \rho$, we have U = ρ. Similarly, we can argue that ρ is also equal to B, the mean number of busy servers, since there is only one. Once again, these interpretations suggest that, in order to maximize the use of the

6.8 The Significance of Traffic Intensity

server, the traffic intensity should approach, but not exceed, the value 1. However, when we have probed a little deeper into the behavior of the queue as a function of the traffic intensity, we will expose some dangers in drawing such a conclusion.

It appears that the bakery shop as originally described provides a reasonable quality of service, although the waiting times might seem a bit long, particularly considering that the server is idle fully one-fourth of the time. Suppose that the arrival rate started to increase, perhaps because the bakery gained in popularity or because a sale was advertised. What would happen to the average queue length, waiting time, and so forth? In particular, suppose that the arrival rate increased from fifteen to eighteen customers per hour, which is a 20 percent increase. The traffic intensity, which is also equal to the server utilization, will rise to 0.9, which seems to be an improvement in the use of the server. The clerk will still be idle 10 percent of the time. However, the average number of customers in the shop will increase to $L = 9$, tripling what it was formerly. Again using Little's formula, the average time in the shop, W, is found to be thirty minutes, of which twenty-seven minutes is time in the queue. In other words, an apparently modest 20 percent increase in the arrival rate, which still left a fair amount of idle time for the server, tripled the average delay time experienced by the customers.

Figure 6.11 graphs both L and L_q as a function of the traffic intensity (or utilization rate), ρ, over the full range of permissible values, $0 < \rho < 1$. This figure reveals what happens as we attempt to obtain full utilization of the server—the expected number in the queue and in the system grows to infinity. It would be bad enough if the queue might occasionally grow to infinity, or if there were some probability that it might grow to infinity, but this result is even worse. It says that the *mean* queue length grows to infinity.

You can see the same outcome in the mean waiting time, W, the mean time in queue, W_q, or even in the distribution of waiting time. As the traffic intensity increases to 1, the queue explodes. Although we got this result for the M/M/1 model only, it is typical of all queueing models.

If this result seems so extraordinary that it casts doubt on the model, you must realize that there is a good reason why we do not witness infinite queues in everyday life. Whenever they start to grow too long, people intercede and alter the system. In other words, even if we manage

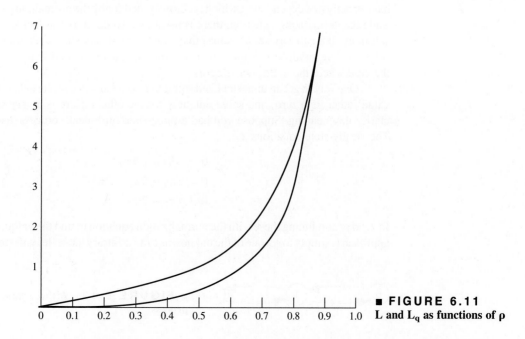

■ FIGURE 6.11
L and L_q as functions of ρ

to achieve $\rho = 1$, we will never see that system reach steady-state. The model does correctly reflect what would happen if the system were allowed to operate.

Closer examination of the curves reveals even further cause for concern. In order to limit the mean queue length to a reasonably small size, say five, the utilization must be a disappointingly low 0.85. That is, we must be prepared to tolerate 15 percent idleness of the server in order to avoid a queue which averages more than five. Of course, even with an average queue length of five, it may, occasionally, be considerably longer. Furthermore, the mean queue length is very sensitive to small changes in ρ in this range.

Thus the stochastic model has revealed that there is an unavoidable conflict between the desire to obtain full utilization of a server and the desire to keep the mean queue length short. The intuitive notion that the system is "in balance" when the arrival rate is equal to the service rate is clearly and dangerously wrong. The ideal ratio of arrival rate to service rate is something less than 1; its specific value depends on the relative costs of idleness versus congestion.

A misunderstanding of the queueing phenomenon has undoubtedly been the source of a good many disputes between labor and management, between customers and agents, and so on. Most people encountering a queue interpret the situation too narrowly. For example, a manager who sees that a worker is idle on one occasion may jump to the conclusion that the feed rate should be increased. He may very well discover later that a very long queue has developed, and be forced to reduce it again. Over a period of time, as he persists in trying to achieve a balance between the feed rate and the worker's capability, and is consistently frustrated, he may reach the conclusion that the worker is deliberately thwarting his efforts. The worker, over the same period of time, may reach the conclusion that the manager is unreasonably demanding. In fact, both the occasional idleness and the occasional backlogs may be attributable solely to the variability in interarrival and/or service times.

6.9 Unnormalized Solutions

The solution method for the M/M/1 queue provides a template for numerous variations, some of which will look much more complicated. It will be helpful in what follows to use the concept of an *unnormalized solution*. The steady-state equations for any well formed Markov process have exactly one dependency, which is compensated for by the normalizing equation. If you set aside the normalizing equation, there is no unique solution, but in fact an infinite number of solutions. We call *any* set of values that satisfy the steady-state equations an unnormalized solution. Any multiple of an unnormalized solution is another unnormalized solution (except in the case where the multiplier is zero).

One way to get an unnormalized solution is to set any one of the unknowns to an arbitrary value (other than zero) and solve uniquely for the others. Here is a very simple example to clarify the meaning. Suppose you had a three-state birth-death process, as in Figure 6.12. The steady-state equations are

$$0 = -2\pi_0 + 1\pi_1$$
$$0 = 2\pi_0 - 5\pi_1 + 2\pi_2$$
$$0 = 4\pi_1 - 2\pi_2$$

Instead of combining these with the normalization equation to find the unique solution, our new approach is simply to set one of the unknowns to an arbitrary value, leaving only two unknowns.

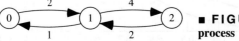

■ **FIGURE 6.12** A three-state birth-death process

Those variables are not the same as the original, so let us distinguish them by adding a caret mark. A good choice might be to set $\hat{\pi}_0$ to the value 1. We would then have the equations

$$0 = -2 + \hat{\pi}_1$$
$$0 = 2 - 5\hat{\pi}_1 + 2\hat{\pi}_2$$
$$0 = 4\hat{\pi}_1 - 2\hat{\pi}_2$$

There are only two independent equations, but we have only two remaining unknowns, so there is now a unique solution, namely:

$$\hat{\pi}_0 = 1$$
$$\hat{\pi}_1 = 2$$
$$\hat{\pi}_2 = 4$$

Obviously, these are not probabilities because they do not sum to 1. However, they do satisfy all of the equations except for the normalization equation. If we had set $\hat{\pi}_0$ to some other value, say 3, we would get a different unnormalized solution, namely,

$$\hat{\pi}_0 = 3$$
$$\hat{\pi}_1 = 6$$
$$\hat{\pi}_2 = 12$$

which, as you can see, is just a multiple of the first solution. In fact, any unnormalized solution will be just a multiple of any other. They vary only in how they are scaled. In order to find the one out of this entire set of solutions that satisfied the requirement that the terms sum to 1, all we have to do is find the correct multiplier. Call that still unknown multiplier 1/G. That is,

$$\pi_j = \frac{1}{G}\hat{\pi}_j$$

where the π_j are the actual probabilities we were seeking, and the $\hat{\pi}_j$ are the terms of any unnormalized solution. Of course, G will depend upon *which* unnormalized solution we decide to use. You will see in a moment why we define our multiplier as 1/G rather than G.

In order for the normalization equation to be satisfied,

$$\sum_{\text{all } j} \pi_j = 1$$

which implies

$$\sum_{\text{all } j} \frac{1}{G}\hat{\pi}_j = 1$$

$$\frac{1}{G} \sum_{\text{all } j} \hat{\pi}_j = 1$$

$$G = \sum_{\text{all } j} \hat{\pi}_j$$

In words, we find the normalization constant G by summing all of the terms of any unnormalized solution. In the case of the numerical example above, $G = 1 + 2 + 4 = 7$, so we get for our normalized solution,

$$\pi_0 = \frac{1}{7}$$
$$\pi_1 = \frac{2}{7}$$
$$\pi_2 = \frac{4}{7}$$

If you perform the same transformation on any other unnormalized solution, you will get the same final result. For example, if we used the other one above, we would get

$$\pi_0 = \frac{3}{21}$$
$$\pi_1 = \frac{6}{21}$$
$$\pi_2 = \frac{12}{21}$$

which of course reduces to the same final values.

We will call G the normalization constant. It will always be equal to the sum of the terms in any unnormalized solution, although its specific value will depend upon which unnormalized solution you choose. You may think of this trick as using the steady-state equations to find the numerators of your probabilities, and then just sum to find the denominator.

6.10 Limited Queue Capacity

When waiting space is limited, a number of minor modifications must be made to the previously developed model. In terms of the queueing notation introduced earlier, we are now concerned with the M/M/1/N system, in which N represents the maximum number of units or customers allowed in the system. Because the arrival process would ordinarily continue to generate arrivals even when the system was full, we must assume that these blocked arrivals are lost to the system.

A schematic version of the finite queue is shown in Figure 6.13. Notice that the limited capacity of the queue is indicated by a bar at the left end of the trough (although the precise value of the limit is not shown).

The associated transition diagram is given in Figure 6.14, which is identical to Figure 6.10, except that it is finite.

The steady-state equations are identical to those used previously, except that the last equation is now a special case:

$$\lambda \pi_0 = \mu \pi_1$$
$$(\lambda + \mu)\pi_1 = \lambda \pi_0 + \mu \pi_2$$
$$\dots$$
$$\lambda \pi_{N-1} = \mu \pi_N$$

The fact that the last equation is of an altered form does not influence the solution for the π_i in terms of π_0, because the equations contain a dependency. That is, we do not require the last

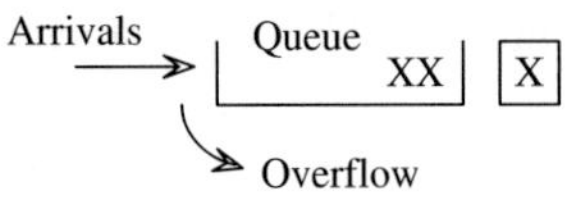

■ **FIGURE 6.13** A finite queue

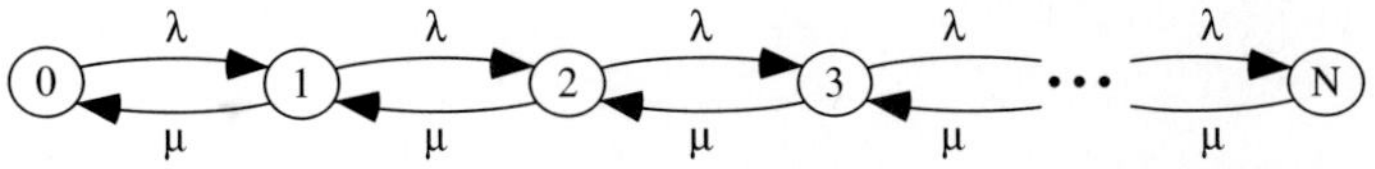

■ **FIGURE 6.14** The transition diagram for a finite queue

equation. Using the concept of an unnormalized solution, we can set π_0 to 1. We obtain the unnormalized solution,

$$\hat{\pi}_j = \rho^j$$

where, again, ρ is defined as λ/μ. (It is easy to verify that this solution is consistent with the neglected last equation.) So far, then, the limitation on system capacity has not made any difference.

At this point, however, the normalization constant G is a finite sum rather than the infinite sum that appeared before.

$$G = \hat{\pi}_0 + \hat{\pi}_1 + \hat{\pi}_2 + \cdots + \hat{\pi}_N$$

$$G = 1 + \rho + \rho^2 + \cdots + \rho^N$$

$$G = \left[\frac{1 - \rho^{N+1}}{1 - \rho}\right]$$

$$\frac{1}{G} = \frac{1 - \rho}{1 - \rho^{N+1}}$$

This time the series is finite, so no convergence requirement appears to restrict the value of ρ. The formula given for the finite sum is valid for any value of ρ, with the exception of $\rho = 1$, when G is trivially $N + 1$, so

$$\frac{1}{G} = \frac{1}{N + 1}$$

for this special case. In summary, see the result below.

Result For the M/M/1/N queueing system

$$\pi_j = \rho^j \left[\frac{1 - \rho}{1 - \rho^{N+1}}\right] \quad \text{for } \rho \neq 1$$

$$\pi_j = \frac{1}{N + 1} \quad \text{for } \rho = 1$$

The mean number in the system, L, can be obtained from the distribution, using well-known finite-sum formulas shown below.

Result For the M/M/1/N queueing system

$$L = \frac{\rho}{1 - \rho}\left[\frac{1 - \rho^N}{1 - \rho^{N+1}}\right] - \frac{N\rho^{N+1}}{1 - \rho^{N+1}} \quad \text{for } \rho \neq 1$$

$$L = \frac{N}{2} \quad \text{for } \rho = 1$$

If the traffic intensity ρ is low and N is large, L will be close to

$$\frac{\rho}{1 - \rho}$$

as you would expect. If ρ is much greater than one, L approaches N. This, too, makes sense. A heavily overloaded system will tend to remain nearly full. This is one way to get good utilization of the server. Of course, when the arrival rate is much greater than the service rate, most arrivals will be blocked and therefore never be served by the system. Depending on the application, this loss may be anything from perfectly acceptable to intolerable.

Incidentally, the fraction of customers or items lost due to blocking is easily quantified. π_N gives the probability that the system is full.

Result For the M/M/1/N queueing system

$$\pi_N = \rho^N \left[\frac{1-\rho}{1-\rho^{N+1}} \right] \quad \text{when } \rho \neq 1$$

$$\pi_N = \frac{1}{N+1} \quad \text{when } \rho = 1$$

The same quantity can be interpreted as the fraction of time, over the long run, that an arriving customer will find the system full; hence, it is the fraction of arriving customers who are blocked. The number blocked per unit time would be this fraction times the number of arrivals per unit time, or $\pi_N \lambda$. This number may be useful in determining how many more customers would be served if the waiting space, N, were increased. One of the problems at the end of the chapter explores the idea further.

The case $\rho = 1$ is interesting, in that it can provide additional insight into why a balance between the arrival rate and service rate is generally not desirable. When $\rho = 1$, the distribution of the number in the system is uniform; that is, all states are equally likely. The mean would be N/2, or "half full." For any finite N, some items are being blocked. In an industrial setting where the items are workpieces that must be processed, the only recourse (short of unbalancing the arrival and service rates) is to increase the waiting room. But as N is increased, the states remain equally likely and the mean is still half full. In the limit, as N goes to infinity, the probability of being in any particular state goes to zero and the mean goes to infinity (or half of infinity, which is the same thing).

6.11 Multiple Servers

Another important and easily handled variation involves multiple servers drawing from a single queue, as depicted in Figure 6.2. Such a system would be designated by M/M/c/N, where c is the number of servers. The system capacity N would certainly have to be at least c and could go up to infinity. One important special case occurs when the maximum number permitted in the system equals the number of servers, which is called the M/M/c/c mode or the "no waiting" case. Another special case occurs when there are an unlimited number of servers and an infinite capacity, called the M/M/∞ model. Those two special cases will be treated in a later section. Here we will focus on the more general multiple-server model, where the system capacity N is larger than the finite number of servers c (so that a queue can occur).

Assuming that all servers operate at the same mean rate, μ, there is no need to keep track of which servers are busy. It will be sufficient, in describing the state of the system, to indicate the number of customers in the system. Despite the fact that service may take place with a number of customers simultaneously, departures will occur one at a time. (The probability of two or more services being completed at exactly the same instant is zero.) Consequently it will still be appropriate to model this system as a birth-death process. What is different now is that the transition rates will be state dependent. If there is only one customer in the system, he will receive service from one of the servers at the rate, μ. If there are two customers present, they will both receive service, each at the rate μ. The transition from state 2 to state 1 will occur as soon as either completes service, so the rate at which this occurs is twice as great as the rate would be if only one server were working, or 2μ. Similarly, the transition rate from three to two is 3μ, and so on. (If you find it difficult to think in terms of rates, you may want to convince yourself that these rates are

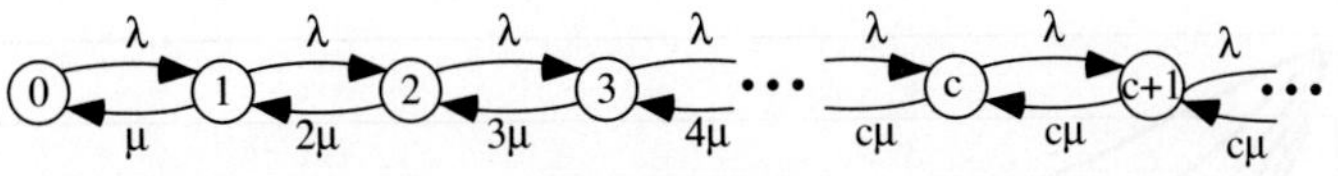

6.11 Multiple Servers

FIGURE 6.15 The transition diagram for multiple servers

correct by translating them to mean times between transitions.) The transition rates corresponding to service completions continue to increase in this way until all c servers are occupied, after which they remain at the value $c\mu$. The transition diagram is shown in Figure 6.15.

The steady-state equations would be

$$\lambda \pi_0 = \mu \pi_1$$

$$(\lambda + j\mu)\pi_j = \lambda \pi_{j-1} + (j+1)\mu \pi_{j+1} \quad \text{for } j < c$$

$$(\lambda + c\mu)\pi_j = \lambda \pi_{j-1} + c\mu \pi_{j+1} \quad \text{for } j \geq c \quad (\text{and } j < N)$$

If the total system capacity, N, is finite, there would be one more equation for the last state,

$$\lambda \pi_{N-1} = c\mu \pi_N$$

The solution of these equations offers no difficulties in principle, but the solutions do require some skill with summing of finite or infinite series. If you enjoy that sort of thing, there are many problems to try your hand at, just by varying c and N. If you would prefer to bypass the algebra, Table 6.3 is provided to cover the cases where $N = \infty$ and $c = 1, 2, \ldots, 6$. The formulas are all in terms of a traffic intensity parameter ρ, which is defined by $\rho = \lambda/c\mu$ (notice the denominator includes the number of servers as well as the service rate), and in each case the convergence requirement $\rho < 1$ is assumed. Formulas for π_0 and L_q are given, from which you can easily obtain all of the steady-state probabilities and other performance measures using the relations in Sections 6.4 and 6.5. As you can see, the expressions grow considerably more complicated as the number of servers increases; however, the functions are smooth and regular as the plotted values show.

TABLE 6.3
Results for M/M/c ($\rho = \lambda/c\mu < 1$)

Number of servers	π_0	L_q
1	$1 - \rho$	$\dfrac{\rho^2}{1 - \rho}$
2	$\dfrac{1 - \rho}{1 + \rho}$	$\dfrac{2\rho^3}{1 - \rho^2}$
3	$\dfrac{2(1 - \rho)}{2 + 4\rho + 3\rho^2}$	$\dfrac{9\rho^4}{2 + 2\rho - \rho^2 - 3\rho^3}$
4	$\dfrac{3(1 - \rho)}{3 + 9\rho + 12\rho^2 + 8\rho^3}$	$\dfrac{32\rho^5}{3 + 6\rho + 3\rho^2 - 4\rho^3 - 8\rho^4}$
5	$\dfrac{24(1 - \rho)}{24 + 96\rho + 180\rho^2 + 200\rho^3 + 124\rho^4}$	$\dfrac{625\rho^6}{24 + 72\rho + 84\rho^2 + 20\rho^3 - 75\rho^4 - 125\rho^5}$
6	$\dfrac{5(1 - \rho)}{5 + 25\rho + 60\rho^2 + 90\rho^3 + 90\rho^4 + 54\rho^5}$	$\dfrac{324\rho^7}{5 + 20\rho + 35\rho^2 + 30\rho^3 - 36\rho^5 - 54\rho^6}$

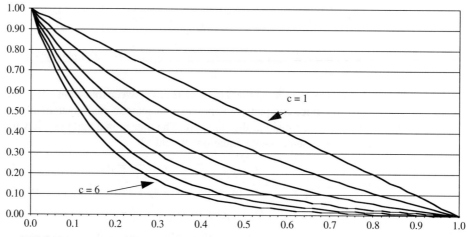

■ **FIGURE 6.16** π_0 as a function of ρ (where $\rho = \lambda/c\mu$)

By examining the graphs for π_0 and L_q, shown in Figures 6.16 and 6.17, you can get an approximate idea of the value of π_0 or L_q for any value of ρ and any number of servers.

You can see from Table 6.3 that the expressions increase in complexity quite a bit as the number of servers increases. They are even worse when the capacity is limited to some finite N (larger than c). However, we can always compute numerical results for the finite cases easily, using a method explained later in Section 6.18, "Numerical Calculations Using a Spreadsheet."

Furthermore, there is an approximate solution that works quite well for the infinite capacity cases.

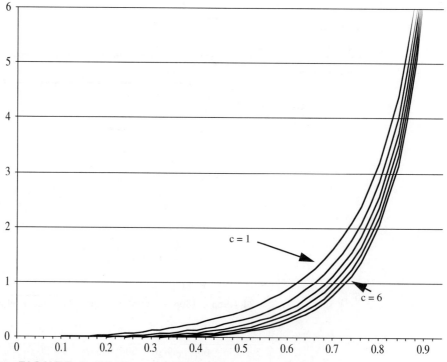

■ **FIGURE 6.17** L_q as a function of ρ (where $\rho = \lambda/c\mu$)

> **Result** For a M/M/c queueing system, an approximate formula for the steady-state number in the queue is
> $$L_q = \left(\frac{\rho^{\sqrt{2(c+1)}}}{(1-\rho)}\right)$$
> where $\rho = \lambda/(c\mu)$.

When $c = 1$, this expression reduces to the correct formula. For higher values of c, the formula is inexact, but tracks the correct curves quite closely. For example, when $c = 6$ and $\rho = 0.9$, the formula gives 6.74 and the correct result is 6.66. (The approximate curve always lies on or slightly above the correct one.) Of course, the approximation is a lot easier to evaluate than solving the exact equations, deriving the correct formula, and then evaluating that.

6.12 Is It Better to Merge or Separate Servers?

EXAMPLE 6.3

The results of the last section give us the opportunity to compare two different ways of organizing resources. The question might be posed as follows: Given equivalent traffic conditions, is it better to have one line feeding a group of servers, or a separate line for each server? For example, if we were given the choice of designing a system for three servers in either of the configurations shown in Figure 6.18, which is the better choice?

Let us understand first what equivalent traffic conditions mean. The overall arrival rate λ should be the same for both alternatives, but in the latter case it will be divided among the c servers, so each of the independent queues would have an arrival rate of λ/c. The service rates should be the same for both alternatives, so each server will operate at the rate μ. When we define the traffic intensity ρ for the multiple-server case, it will be $\lambda/c\mu$ as Table 6.3 requires. When we use the divided arrival stream together with parallel single-server models, we will use the arrival rate λ/c divided by the service rate μ, which gives us exactly the same $\rho = \lambda/c\mu$. So we can compare models using the same numerical value of ρ. For example, when ρ is 0.9 for the case $c = 3$, we should use $\rho = 0.9$ for the case $c = 1$. We can directly compare W or W_q. However, if we want to compare L or L_q, we will have to remember that the divided model will contain only 1/c of the total number of customers in the whole system.

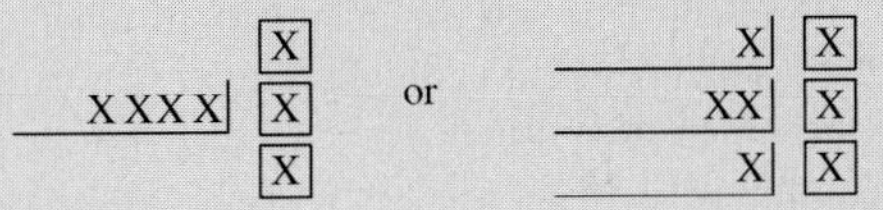

When we do this, it works out that for any value of ρ, and for any of the standard performance measures, it is always better to have a single line feeding multiple servers than to divide the arrival stream. Another way to state the conclusion is that pooling is always better than separation. One might suspect that this conclusion is somehow related to the peculiarities of the Markovian assumptions, but even more general models support the same conclusion. Hence, we have discovered a general principle of queueing system design. Other things being equal, it is preferable to design a service system so that just one line feeds all of the servers and the customer at the head of the line is directed to the first available server. We call this concept the principle of pooling, referring to the common pool of customers that feeds the servers.

In recent years, many banks, post offices, and airline ticket counters have adopted this principle. Perhaps you have had the experience of witnessing the change and seeing the interesting first-time reactions of customers to the unfamiliar new system. For the most part, however, the principle seems to be unknown to the average person. Certainly, one can find many opportunities to improve existing systems by applying the principle of pooling. As a design exercise, you may enjoy figuring out how to exploit the principle in supermarket checkouts. As you do so, be sure to think about side benefits and possible disadvantages, as well as the pure queueing aspects.

```
                    X              X  X
                    X              X  X
         XXXX       X    or        XX X
                    X              X  X
```

■ **FIGURE 6.18** Two alternative system designs

6.13 Which is Better: More Servers or Faster Servers?

EXAMPLE 6.4

Another interesting comparison is that of changing the number of servers versus changing the service rate. Specific situations are likely to involve other factors, but it makes sense to ask the question, "All other things being equal, will the system be better with more servers or faster servers?" Let us compare a system with two servers to a system with only one server operating at twice the speed of the first two servers. Use λ for the common arrival rate. For the two-server system, use μ as the service rate, which will mean that $\rho = \frac{\lambda}{2\mu}$. For the single server equivalent, we should use 2μ as the service rate (to represent twice as fast), which also translates to $\rho = \frac{\lambda}{2\mu}$. Therefore, we can use the same value of ρ in both equations and simply compare performance measures directly.

Now, what do we mean by "better?" If you are looking at the system from the outside, you may think that the better system is the one that has the smaller W—total time to pass through the system. However, if you are one of the customers, you may think that W_q—the delay encountered before being served—is what really matters.

Table 6.3 gives us L_q directly, so we will start there. You can see from Figure 6.17 that the value of L_q for two servers always lies below the curve for one server. Hence, the two-server system will always have a smaller average queue length. From Little's formula applied to the queue, that is,

$$L_q = \lambda W_q$$

we can conclude that the average time spent in the queue is also smaller for the two-server system. However, if we add in service time to consider the total time through the system, we have to add different mean service times for the separate cases. For the two server system, we will add $1/\mu$, but for the single-server system, we will add only $1/(2\mu)$ (because the single server works twice as fast.) To compare the two values of W, we will have to look at the formulas.

For the single-server system, W is given by

$$W = \frac{1}{\lambda}L = \frac{1}{\lambda}\left(\frac{\rho}{1-\rho}\right)$$

We got L formally when we looked at the M/M/1 model. Now, however, we need to use $\rho = \lambda/(2\mu)$ instead of the original $\rho = \lambda/\mu$ so that we can compare appropriate traffic intensities. So this equation for waiting time will now take the form

$$W = \frac{1}{\lambda}\left(\frac{\frac{\lambda}{2\mu}}{1-\frac{\lambda}{2\mu}}\right) = \frac{1}{2\mu - \lambda}$$

For the two-server system, we use (from Little's formula and Table 6.3),

$$W = W_q + \frac{1}{\mu} = \frac{1}{\lambda}(L_q) + \frac{1}{\mu} = \frac{1}{\lambda}\left(\frac{2\rho^3}{1-\rho^2}\right) + \frac{1}{\mu}$$

$$= \frac{1}{\lambda}\left[\left(\frac{2\rho^3}{1-\rho^2}\right) + \frac{\lambda}{\mu}\right] = \frac{1}{\lambda}\left[\left(\frac{2\rho^3}{1-\rho^2}\right) + 2\rho\right]$$

$$= \frac{1}{\lambda}\left[\frac{2\rho}{1-\rho^2}\right]$$

If you now substitute $\rho = \lambda/(2\mu)$ and simplify,

$$W = \frac{4\mu}{4\mu^2 - \lambda^2}$$

This expression can be factored to get a form that is easy to compare to the single-server result.

$$W = \left(\frac{4\mu}{2\mu + \lambda}\right)\left(\frac{1}{2\mu - \lambda}\right)$$

Notice that this expresses the mean waiting time for the two-server system as a multiple of the mean waiting time for the single-server system. The multiplier is

$$\left(\frac{4\mu}{2\mu + \lambda}\right)$$

But since we have the requirement that ρ must be less than 1, λ must be less than 2μ. Therefore, the denominator must be less than the numerator, which implies that the multiplier must be larger than 1, which finally leads to the conclusion that the W for the two-server system is always greater than that for the single-server system (under comparable conditions).

Our answer to the question, "Is it better to have more servers or faster servers?" depends upon what you want to achieve. If you want to minimize the delay time or the length of the queue, more servers are better. If you want to minimize the total time through the system or the total number of customers in the system, you are better off with the faster servers. In either case, the differences are small and it is unlikely that you can maintain all other factors exactly equivalent, so the particular circumstances should determine which design is better.

6.14 Case Study: A Grain Elevator

EXAMPLE 6.5

To illustrate the multiple-server queueing models and at the same time demonstrate their usefulness in a realistic situation, consider the following scenario. It concerns a grain elevator at a small town in one of the central plains states where wheat is the principal crop. During the harvest season, trucks loaded with wheat from the fields arrive at the elevator, where they must quickly deposit their loads and return to the fields for another load. The check-in process involves weighing the truck, drawing samples for moisture and contamination tests, and a few other details, before the load can be dumped through a grate.

The farmers are very concerned about getting their crop off the field and into the elevator quickly. Once the wheat is ripe, it is highly vulnerable to rain or wind. Any delay could threaten a significant portion of a farmer's income for the whole year. Of course, all of the fields in a region ripen at about the same time, so it is not surprising that a traffic problem could develop at the elevator.

To keep the numbers simple, let us suppose that the average interarrival time for trucks is 6.67 minutes and the average service time is 6 minutes. Then, using the standard formula for the M/M/1 model, we could expect an average time at the elevator (W) of one hour per truck, which matches the actual experience of the drivers. This delay is considered by the farmers to be intolerable.

At a meeting of the farmers' cooperative, three suggestions are put forth:

1. By adding sideboards to the trucks (to permit slightly larger loads) the average interarrival time could be lengthened to ten minutes. Simultaneously, an extra worker at the check-in could shorten the average service time to four minutes. These relatively minor adjustments would cost an estimated $30,000.

2. Some farmers believe that, although the first proposal would help, it would not really solve the problem and that more significant changes are warranted. The addition of another complete check-in station, including scale, test equipment, and personnel, would essentially double the service capacity. Arriving trucks would join a single line, and the truck at the front of the line would move to the first available of the two check-in points. This change would cost about $400,000.

3. A few farmers believe that the entire grain elevator facility should be duplicated on the other side of town. In addition to doubling the service capacity, this proposal would split the arrivals into equal halves, since each farmer would attend the nearer of the two elevators. The cost of this proposal would be close to $1,000,000.

In discussing the merits of the various proposals, the farmers find the low cost of the first proposal to be attractive, but many question the effectiveness of such minor changes in dealing with such a serious delay problem. The third proposal seems to offer an ultimate fix, but the cost would be an enormous burden. The second proposal is favored by many who seek a compromise between cost and effectiveness. Basing your judgment on what has been presented thus far, which would you choose? Most people assess the situation as presented in Table 6.4.

Let us now do some quantitative analysis to evaluate the true effects of the various proposals. In the terminology of queueing theory, the first proposal involves changing the parameter values of the model, but no structural changes. The second proposal is for a two-server, single-queue system; the third is for a two-server, separated-queue system. We will use the simple Markov models to perform our analysis, despite the fact that it is highly unlikely that service times would be negative exponentially distributed. We may excuse this defect in assumptions by pointing out that we are seeking only comparative performance, not absolute predictions.

The three proposals are represented by Table 6.5. The analysis, using results from Table 6.3 and the relations of Section 6.4, is straightforward, except possibly in the case of the third proposal. The key to analyzing the third proposal is to realize that the system separates into two independent M/M/1 queueing systems and insofar as any individual truck is concerned, the service process is as if there were only one. Consequently, the analysis reduces to obtaining results for an ordinary M/M/1 with half the arrival rate of the original.

The results are startling, to say the least! The least expensive proposal turns out to yield the greatest benefits. It is unlikely that any of the farmers would realize from intuition alone what the analysis reveals. This example clearly illustrates why formal methods are needed when dealing with queueing phenomena. The human mind, which

■ **TABLE 6.4**
Proposed Improvements

Proposal	Improvement	Cost
Larger loads	minor	$30,000
Extra check-in station	better	$400,000
Complete extra elevator	best	$1,000,000

TABLE 6.5
Results for Grain Elevator Proposals

Proposal	Cost	Model	Arrival Rate	Service Rate	Waiting Time (W)
Existing	0	M/M/1	9 per hr.	10 per hr.	60 min.
1	$30K	M/M/1	6	15	6.67 min.
2	$400K	M/M/2	9	10	7.52 min.
3	$1,000K	M/M/1	4.5	10	10.91 min.

is naturally adept in many judgment situations, is just not very good at dealing with the kind of random behavior that accounts for queues.

If you remain puzzled, or perhaps even skeptical, about the outcome of this example, you should satisfy yourself that you understand how each of the suggested changes actually affects the queue. The comparison of the original system, proposal 1, and proposal 3 can be simply related to traffic intensity. The difference between proposal 2 and proposal 3 is explained by the pooling principle.

6.15 The M/M/c/c and M/M/∞ Models

Some simplification is achieved if we restrict attention to the "no waiting" case. For the M/M/c/c system, the transition diagram would extend only to the state c. The steady-state equations are

$$\lambda \pi_0 = \mu \pi_1$$
$$(\lambda + \mu)\pi_1 = \lambda \pi_0 + 2\mu \pi_2$$
$$(\lambda + 2\mu)\pi_2 = \lambda \pi_1 + 3\mu \pi_3$$
$$\cdots$$

Setting π_0 to the value 1 to determine an unnormalized solution, these equations can be solved in the usual iterative way for birth-death equations to yield

$$\hat{\pi}_j = \frac{1}{j!}\left(\frac{\lambda}{\mu}\right)^j$$

Defining $\rho = \frac{\lambda}{\mu}$, as in the M/M/1 queue,

$$\hat{\pi}_j = \frac{\rho^j}{j!}$$

This same equation works for both the M/M/c/c model and the M/M/∞ model. The only difference appears when we sum the series to find the normalizing constant G.

For the infinite capacity system (that is, the M/M/∞ model), we have

$$G = 1 + \rho + \frac{\rho^2}{2!} + \cdots$$

The infinite series that appears is equivalent to e^ρ, so the multiplier needed to force all of the probabilities to sum to 1 is

$$\frac{1}{G} = e^{-\rho}$$

When we substitute that back into the general solution, we get the result shown below.

Result For the M/M/∞ queueing system

$$\pi_j = \frac{\rho^j}{j!}e^{-\rho}$$

This distribution happens to be a recognizable one—the Poisson distribution. It has nice properties, a simple mean (ρ) and a simple variance (also ρ). Therefore, we can compute almost any of the performance measures we want for the unlimited number of servers system without difficulty. One way in which a real-life system could have an unlimited number of servers is for the arrivals to serve themselves.

For the finite case (that is, the M/M/c/c model), we have a finite series to sum, and get the equation

$$G = 1 + \rho + \frac{\rho^2}{2!} + \cdots + \frac{\rho^c}{c!}$$

There is no clean expression for this finite sum, but it might be helpful to observe that it is the first $c + 1$ terms of the infinite series expression of e^ρ. If c is very large relative to ρ, so that the remaining terms in the infinite series are small, we can say that π_0 is approximately $e^{-\rho}$, and we can use the Poisson distribution for the M/M/∞ as an approximation for the M/M/c/c.

More generally, when c is finite and the approximation is not appropriate, the steady-state distribution is as shown below.

Result For the M/M/c/c queueing system

$$\pi_j = \frac{\frac{\rho^j}{j!}}{\sum_{k=0}^{c} \frac{\rho^k}{k!}}$$

---

and this distribution has been called a truncated Poisson distribution. Section 6.18, "Numerical Calculations Using a Spreadsheet," shows how to compute this solution efficiently.

One term, in particular, is of special interest.

Result In the M/M/c/c queueing system, the probability that all servers are occupied or that the system is full is given by

$$\pi_c = \frac{\frac{\rho^c}{c!}}{\sum_{k=0}^{c} \frac{\rho^k}{k!}}$$

This formula, known variously as Erlang's lost call formula, the Erlang B-formula, or the first Erlang function, has been used extensively in the design of telephone systems. Tables and graphs are readily available, but today it is easy to build your own in a spreadsheet.

An example of such a chart, built in a few minutes in Microsoft Excel, is shown in Figure 6.19. The horizontal axis covers values of ρ, which in this case is defined as λ/μ, from 0 to 50. The vertical axis is the probability that at least one server is available, or $1 - \pi_c$. This value can be taken as a measure of performance. The various curves plotted are for different values of the number of servers, c, ranging from 5 to 40. By visually interpolating, one can use this graph to estimate the appropriate number of servers to provide for a given level of traffic intensity and any desired level of performance.

The mean number of occupied servers or customers in the system is given by

$$L = \rho(1 - \pi_c)$$

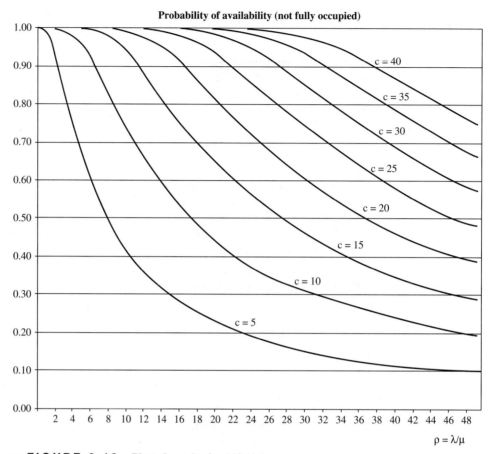

■ **FIGURE 6.19** Plotted results for M/M/c/c

This relation has a number of interesting verbal interpretations. The reader is encouraged to try to think of some.

Of course, all of these results were derived from a Markov model. In particular, service times were assumed to be negative exponentially distributed. Perhaps surprisingly, exactly the same results occur when the service time distribution is arbitrary. A. K. Erlang, who first obtained all of these results (for the negative exponential service time case) as early as 1917, conjectured that they would hold for arbitrary service times. The conjecture was confirmed, however, only as recently as 1969 (see the article by Takas cited in this chapter's "Recommended Reading"). As you might expect, the mathematics used is beyond the level of this text, but it is useful to know that the solution for the M/G/c/c system (including the case $c = \infty$) is the same as for the M/M/c/c system.

If you happen to be faced with the circumstances that allow you to use one of these two models (the $M/M/\infty$ or the M/M/c/c), you are in luck! You have extremely powerful results that do not depend upon the service time distribution or require any more data than the mean arrival and service rates.

6.16 Finite Sources

The next variation to be developed models the situation that occurs when the potential arrivals form a fixed, finite population. Such a situation would occur if, for example, the server were an

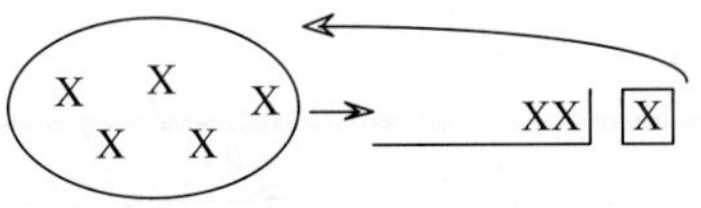

6.16 Finite Sources

[Figure 6.20 depicts an office copying machine scenario]

■ FIGURE 6.20 The finite source queue

office copying machine which only a dozen people were authorized to use. Figure 6.20 depicts the situation. It is assumed that when a customer completes service, he or she returns to the source.

The queueing nomenclature for this model is M/M/1/N/N. The second N refers to the number of items in the source, or the maximum number of arrivals. The first N represents the capacity of the system. Although the capacity could conceivably differ from the maximum number of arrivals, there is no point in considering a greater capacity—it would never be used—and it will be obvious how to treat a lesser capacity.

The difference between this model and previous ones is in the arrival rates. It is intuitively clear that when most of the customers are in the queue, the arrival rate should be lower than when most are in the source. Certainly, when all are the queue, the arrival rate should be zero.

Let λ represent the mean arrival rate *per customer*. This would be statistically estimated by recording the times spent in the source by each customer, that is, the durations of the intervals beginning when a customer returns to the source after completing service and ending when the same customer reappears at the queue requiring another service. Assuming these times fit a negative exponential distribution with a common mean for all customers, λ would be the reciprocal of the average time. Note that λ cannot be estimated by recording times between arrivals to the queue, because these times will be dependent on the number already in the queue.

If there is only one customer in the source, and the other $N-1$ are in the system, the arrival rate will obviously be λ. If there are two in the source, the arrival rate will be twice as great, or 2λ. The logic here is similar to that used to obtain the service transition rates when there were multiple servers. Continuing in this manner, we obtain the transition diagram shown in Figure 6.21. As in previous models, the state represents the number in the system (queue plus server), but it would also be possible in this case to use the number in the source as the indicator.

The steady-state equations are

$$N\lambda \pi_0 = \mu \pi_1$$

$$[(N-j)\lambda + \mu]\pi_j = (N-j+1)\lambda \pi_{j-1} + \mu \pi_{j+1} \quad \text{for } j = 1, 2, \ldots, N-1$$

$$\lambda \pi_{N-1} = \mu \pi_N$$

An unnormalized solution, obtained in the usual way, is

$$\hat{\pi}_j = \frac{N!}{(N-j)!}\left(\frac{\lambda}{\mu}\right)^j$$

Of course, we have to normalize these by dividing each by the sum of all of them. The expressions for π_0, L, and L_q are obtainable in a straightforward manner, but do not reduce to simple formulas that could provide additional insight.

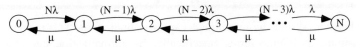

■ FIGURE 6.21 The transition diagram for the finite source queue

6.17 The Machine Repairmen Model

EXAMPLE 6.6

The finite source arrival rates can be combined with the multiple-server service rates to provide a useful model which is commonly called the machine interference model or the N machine, R repairmen model. In this application, the N machines form the source population, and the repairmen (who could, of course, be women) are the servers. The breakdown of a machine constitutes an arrival (not to imply, necessarily, that broken machines are physically moved to a repair shop). A repair restores the machine to operation and, in that sense, returns it to the source. Formally, the model is M/M/R/N/N.

The problem usually posed is, "Given the failure and repair rates, λ and μ, and a fixed number of machines, N, what is the optimal number of repairmen, R, to service the machines?" The method of solution is, of course, to model the various alternatives and to select that one which achieves the most advantageous trade-off between the cost of repairmen and the cost of idle machines. We illustrate this problem and its solution with a specific example.

Example: A job shop has four numerically controlled machine tools that are capable of operating on their own (that is, without a human operator) once they are set up with the proper cutting tools and all adjustments are made. Each setup requires the skills of an experienced machinist, and the time needed to complete a setup is negative exponentially distributed with a mean of 1/2 hour (implying that $\mu = 2$ per hour). When the setup is complete, the machinist starts the machine, which requires no further attention until it has finished its lot size and is ready for another setup. The lot size production times are negative exponentially distributed with a mean of 1 hour (which implies that $\lambda = 1$ per hour). The question is, "How many machinists should there be to tend the machines?" At opposite extremes, there could be one machinist tending all four machines or there could be one machinist for each of the machines. The optimal number to have obviously depends on a trade-off between the cost of machinists and the cost of idle machines. Of course, machinists are paid the same regardless of how much work they do, but the machines incur idle-time costs only when they are idle.

Assume that the cost of a machinist (including fringe benefits, overhead, and the like) is $20 per hour, and that the cost of an idle machine (including lost revenues and interest on the capital invested in the machine) is $60 per hour of idleness. Using the finite source model for each of the possible numbers of machinists, 1, 2, 3, or 4, we calculate the steady-state mean number of busy machines, B, from which we can find the mean number of idle machines over the long run. Finally, adding up the costs, we can select the number of machinists that produces the smallest total cost, which in this case is 2. These results are summarized in Table 6.6.

■ **TABLE 6.6**
Results for Machinist Problem

R, no. of machinists	Cost/hr. of machinists	Machine idleness	Cost of idleness/hr.	Total cost per hour
1	$20	2.19	$131.43	$151.43
2	$40	1.47	$88.28	$128.28
3	$60	1.34	$80.66	$140.66
4	$80	1.33	$80.0	$160.00

6.18 Numerical Calculations Using a Spreadsheet

When the system has unlimited capacity, finding the steady-state probabilities involves summing an infinite series. Specifically, the normalization step requires the infinite series sum. In those cases, if you want an algebraic solution you will have to rely upon the regularity of the equations in the higher numbered states to establish a pattern that you can work on to simplify. However, if the system capacity is limited or if you are content with approximating an infinite capacity system by a large finite capacity one (so that there are only a finite number of states), you can use a numerical technique to handle any of the variations that a birth-death process allows. This is a very practical and expedient approach when the equations get too messy for algebraic manipulation.

We have seen several variations that really affect only the transition rates, not the structure of the transition diagram. Multiple servers and finite sources just change some multipliers of the basic arrival rate, λ, and the basic service rate, μ. As long as the number of states is finite and within the capacity of a spreadsheet model (usually a large number), we can set up a template that can routinely compute the steady-state probabilities for any birth-death queueing model. Figure 6.22 shows one way to set this up; once you have the idea, you can construct your own version that may compute many other measures.

The boldfaced entries are fixed parts of the template. The data that describe any particular model are in the second and third columns and in the cells that are named "lambda" and "mu." All other entries in the spreadsheet contain formulas that will recompute automatically when any of the data change. By adjusting the entries in the arrival rate multiplier column we can handle finite sources or change the capacity of the system. When we have an infinite source population, we use a multiplier of 1 (times lambda) for all states up to $N - 1$ and a multiplier of 0 for all higher numbers. (This will ensure that there are no arrivals accepted when full capacity is reached.) Of course, if you want to handle system capacities larger than ten, you will have to extend the spreadsheet. To handle a finite source population, you will have to adjust the multipliers accordingly. For example, to represent a source population of five, you would use five as the arrival rate multiplier for state 0, four as the multiplier for state 1, and so forth. Multiple servers can be represented similarly. The service rate multiplier for state i should be equal to the number of servers that are working when the system is in that state. For states greater than the capacity of the system, it will not matter what the service rate multiplier is, but we should not set it equal to zero if we want to avoid a "division by zero" message. Figure 6.22 is currently set up to compute the probabilities for the M/M/2/4/4 model, which is one of the cases needed for the machine repairman analysis.

	A	B	C	D	E	F
1		**Arrival Rate,**	**lambda**	1.0000		
2		**Service Rate,**	**mu**	2.0000		
3						
4		**Arrival Rate**	**Service Rate**	**Unnormalized**	**Normalized**	
5	**State**	**Multiplier**	**Multiplier**	**Solution**	**Solution**	**j*p(j)**
6	0	4	0	1.0000	0.1839	0.0000
7	1	3	1	2.0000	0.3678	0.3678
8	2	2	2	1.5000	0.2759	0.5517
9	3	1	2	0.7500	0.1379	0.4138
10	4	0	2	0.1875	0.0345	0.1379
11	5	0	2	0.0000	0.0000	0.0000
12	6	0	2	0.0000	0.0000	0.0000
13	7	0	2	0.0000	0.0000	0.0000
14	8	0	2	0.0000	0.0000	0.0000
15	9	0	2	0.0000	0.0000	0.0000
16	10	0	2	0.0000	0.0000	0.0000
17						
18			**sum:**	5.4375	**Mean, L:**	1.4713
19			**(to normalize)**			

■ **FIGURE 6.22** A spreadsheet for birth-death queueing models

The third column contains an unnormalized solution to the steady-state equations, which we obtain by setting the first value to 1 and then using the relation

$$\pi_j = \left(\frac{\lambda_{j-1}}{\mu_j}\right)\pi_{j-1} = \frac{\text{arrival multiplier }(j-1) \times \text{lambda}}{\text{service multiplier}(j) \times \text{mu}}\pi_{j-1}$$

This formula can be entered in cell D7 and copied to all other cells beneath it in the same column.

The fourth column contains the true normalized steady-state probabilities, which we can get from the unnormalized solution simply by dividing each by the sum of all of them. This scaling has the effect of forcing them to sum to 1. Once we have the steady-state probabilities, we can compute other quantities easily. The sixth column and the sum below it show the calculation of L, the steady-state mean number of customers in the system.

Incidentally, Figure 5.12 in Chapter 5 (page 159) represents the solution of the differential equations for the same M/M/2/4/4 model with the same λ and μ. Section 5.9 (page 157) uses this case to show how the differential equations can be solved.

Another way to implement the same calculation is by defining a function or subroutine in Visual Basic. For example, the code for a function that computes the value of L for any M/M/c model is shown below.

```
Function L(c, lambda, mu)
RHO = lambda/(c*mu)
If (c<1 Or RHO<0 Or RHO >= 1) Then         'trap errors on input
L = 0
Exit Function
Else
PI = 1#                    'set unnormalized Pi zero to 1
G = 1#                     'initialize normalization constant
L = 0#                     'initialize unnormalized L
For i = 1 To c             'loop through the initial terms, 1 to c
    PI = (c*RHO/i)*PI
    G = G + PI
    L = L + i*PI
Next i
i = c
Do Until PI < 1e-20        'continue until term is negligible
    i = i + 1
    PI = RHO*PI
    G = G + PI
    L = L + i*PI
Loop
    L = L/G                'normalize the final L
End If
End Function
```

After a few statements to trap inappropriate parameters (always a good idea), the basic logic of the calculation is to loop through the first c equations solving for unnormalized state probabilities and adding each to the normalization constant G. At the same time, an unnormalized version of L is accumulated. After that loop is completed, the same kind of calculation is continued for as long as the calculated state probabilities are significant, but using the iterative equation that applies for states larger than c. At the end, L is normalized and returned. Similar logic will work for L_q, W, W_q, and so forth.

6.19 Queue Discipline Variations

Up to this point we have been tacitly assuming that customers are served in order of arrival. Most systems serving human customers are at least intended to operate by this rule, simply because anything else would be perceived as unfair. Occasionally, however, practical constraints prevent strict adherence to a FCFS rule. Or sometimes the arrivals are not human customers for whom the order of service is an issue. For example, they might be workpieces which are to be operated on by a machine. In such cases, there is no obvious incentive to hold to the first come-first served (FCFS) discipline. In fact, if the workpieces are stacked on one another, the last come-first served (LCFS) discipline is a natural one. If they are spread out in no particular arrangement, the random selection for service (RSS) discipline may be appropriate.

At first glance, it may appear that differences of this kind would immediately invalidate all previous results; however, such is not the case. A bit of reflection about the logic that went into the construction of the transition diagrams will convince the reader that the steady-state distribution of the number in the system is the same for any queue discipline that does not depend on service times. That is, the server could choose the next customer arbitrarily (as long as his choice is not biased by a prediction of the customer's service times) and the transition rates would be the same. The corresponding steady-state equations, and therefore the solution, would also remain the same. Since the distributions are the same, the mean (as well as other moments) of the number in the system or of the number in the queue would be the same. In other words, the L's and the L_q's that have been calculated for the FCFS cases are also applicable for the LCFS and RSS cases.

Furthermore, because the L = RW and L_q = RW_q (Little's formula) relations apply for any queue discipline, the mean waiting times in the system and in the queue are the same for all three cases. On the average, it will take just as long to pass through a first come-first served system as it will through a last come-first served system—a fact that is certainly not intuitively obvious.

What measures are affected by the queue discipline? The primary one is the distribution of waiting times. Just because the mean waiting times are unaffected by the discipline there is no cause to infer that, say, variances are also unaffected. In fact, it can be shown that, among the three disciplines mentioned so far, the FCFS rule will produce the smallest variance in waiting times, the LCFS rule will produce the largest, and the RSS rule will produce an intermediate value. In a sense, the risk of a very long wait is minimized if customers are served in the order of their arrival. By the same token, the chance of a very short wait is also minimized by the same policy. Thus the policy has the effect of tending to equalize waiting times, relative to other policies. It is in this sense, but in this sense only, that the first come-first served policy is fairer than other methods. Despite virtually universal acceptance of and insistence on the FCFS rule when a queue consists of people, there is no quantitative indication that average waiting times are any better under this rule.

One way to reduce average waiting times for selected customers is to use a priority (PR) discipline. There are many possible priority schemes, at least some of which can be modeled as Markov systems. The interested reader should consult a specialized textbook on queueing theory, such as references 2, 5, 7, and 11 in this chapter's "Recommended Reading" section.

Some of the more difficult priority schemes to model happen to yield some of the most useful results. Recall that the arguments leading to the conclusion that mean waiting times were unaffected by the queue discipline contained the qualification, "provided that the discipline is independent of service times." It may be correctly inferred that queue disciplines that are dependent on service times will influence the mean waiting times, W and W_q. This fact, in turn, suggests that wise selection of a service-time dependent queue discipline can reduce mean waiting time. Providing express service for customers whose predicted service times are short would be one way to implement such an idea.

In many industrial applications of queueing theory, the arrivals are not customers but jobs. Processing (service) times, or at least estimates of them, are available for use in scheduling. It is possible to imagine many seemingly reasonable scheduling concepts, such as, "Give highest priority to the job with the greatest work content," or "Give highest priority to the job with the nearest due date." Out of all of these rules of thumb, one has been shown to produce surprisingly good results (that is, small mean waiting times) under widely varying circumstances. This rule, which is briefly stated as the shortest processing time (SPT) rule, attributes highest priority to jobs requiring the least amount of time to complete. The rule can be applied weakly or applied strictly—even to the extent of interrupting or preempting a job being serviced whenever a job that could be finished in less time comes along. A method that separated arrivals into long and short jobs and then gave priority to the short ones would implement the SPT rule in a weak way. Even this method can achieve dramatic reductions in mean waiting time. In fact, almost any way of implementing the SPT idea will perform with surprising efficacy. For a more complete discussion, demonstrating comparative results for various queueing disciplines, see reference 1 in Chapter 8's "Recommended Reading" section.

6.20 Non-Markovian Queues

Aside from a few more general results that have been cited without proof, all of the queueing theory in this chapter has been based on Markov models. A remarkably rich variety of useful models was obtainable merely by suitably adjusting the transition rates. In each case, however, the probability distributions involved were negative exponential. This assumption, although certainly convenient, is not always realistic. Service-time distributions, in particular, frequently lack the "forgetfulness" property that is characteristic of the negative exponential family. Thus there is a practical need for models that do not rely on strict Markov assumptions.

As was mentioned in passing, two models permit easy generalization. The $M/G/\infty$ and the $M/G/c/c$ models have the same solutions as the $M/M/\infty$ and the $M/M/c/c$, respectively. Although the proof of these equivalencies are beyond the scope of this book, it is useful to know the simple fact that the form of the service-time distribution does not influence the steady-state results in these two cases.

The Erlang family of distributions, a subclass of the Gamma family, is representable in terms of Markov processes if a certain trick is employed. The trick is based on the fact that an Erlang-k distributed random variable is equivalent to the sum of k negative exponentially distributed random variables. To show how the trick would work in a specific example, suppose that we want to model an $M/E_3/1$ system. Each service time, which is actually Erlang-3 distributed, can then be thought of as consisting of three negative exponentially distributed "stages." A supplementary variable will have to be incorporated into the state definition to keep track of which stage the service process is in, but once this is done, the process becomes Markovian. The states would be designated as follows: "0" means the system is empty, "(i, j)" means the system contains i customers and the server is currently in stage j.

The transition diagram for the $M/E_3/1$ system would look like Figure 6.23. In this diagram, each arrow directed to the right represents a possible arrival transition, and each one directed down or to the upper left represents a service stage completion. The times between any such transitions are, of course, negative exponential. Although the resulting process is not of the birth-death type, as were most of the queueing models, it is a continuous time Markov process.

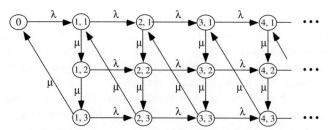

■ **FIGURE 6.23** The transition diagram for M/E$_3$/1

The same idea can be exploited to model systems with Erlang-distributed inter-arrival times. With two supplementary variables, it is even possible to model the system in which both the arrival process and service process are of the Erlang type. Although in principle no difficulties occur, in practice the steady-state solution is tedious to obtain and the expressions are rather nasty. Solutions for the M/E$_k$/1 and E$_k$/M/1 systems are derived in (see the T. L. Saaty text cited in this chapter's "Recommended Reading" for the full cite).

Another device is commonly used when either the service times or interarrival times are arbitrarily distributed but the other is negative exponentially distributed. The idea is known as embedding a Markov chain, and is generally attributed to D. G. Kendall. The basic concept is to view the process only at selected moments, ignoring the dynamic behavior of the process at intermittent times. The moments at which the process is viewed—the embedding points—are chosen so that the discrete time Markov assumption will hold. That is, given the state at one of these points, enough information is available to predict the state at the next point, and any additional information is superfluous.

6.21   The M/G/1 Model

In the case of the M/G/1 system, we may embed a Markov chain at the moments of service completions. At these times, and at these times only, it will be sufficient to know the number in the system to predict the future. At any other times, one would also have to know something such as elapsed service time on the customer currently in service. Because the arrival process is assumed to be negative exponential, it does not matter how long it has been since the last arrival. The same idea can be used to model the GI/M/1 queue, embedding at the moment when arrivals occur.

The embedded Markov chain method leads to a very useful and important result for the M/G/1 queue. It turns out that for this system, the four measures L, L$_q$, W, and W$_q$ depend on no more than the mean and variance of the service-time distribution. The formula for L$_q$, known as the Pollaczek-Khintchine formula, is shown below.

> **Result** In the M/G/1 queueing system
> $$L_q = \frac{\rho^2 + \lambda^2 V(S)}{2(1 - \rho)}$$

The formulas for L, W, and W$_q$ can be obtained from this one by the methods of Sections 6.4 and 6.5. It is convenient to have such a powerful result. As mentioned earlier, it is frequently the case that service time distributions are clearly not negative exponential. The Pollaczek-Khintchine formula covers arbitrary service-time distributions. Moreover, it does not even require that the form of the distribution be identified; the mean and variance of the distribution are sufficient. Of course, the statistical problem of obtaining good estimates of the mean and variance from empirical data is much easier than the problem of fitting an appropriate distribution to the same data. Because L, L_q, W, and W_q are independent of any queue discipline which does not depend on service times, the Pollaczek-Khintchine formula is not limited to the FCFS case. The restrictions to remember are that it applies only to single-server, infinite-capacity systems with Poisson arrivals.

One quick application of the formula yields an interesting insight. Consider the M/D/1 system, in which service times are constant. For this system, $V(S) = 0$, so

$$L_q = \frac{\rho^2}{2(1-\rho)}$$

which is exactly half of the analogous result for the M/M/1 system,

$$L_q = \frac{\rho^2}{(1-\rho)}$$

We see that the average number in the queue would be reduced by exactly one-half if the variability were eliminated from the service times. In a sense, half of the queue can be attributed to service-time variability. The other half is the consequence of arrival-time variability.

6.22 Approximate Solutions for Other Models

We now have solutions or at least solution methods for all of the Markov queueing models, and for a few non-Markov cases. Specifically, we can handle the M/G/c/c and M/G/∞ cases (because the results are identical to the Markov versions) and the M/G/1 case (using the Pollaczek-Khintchine result). The glaring omissions are the M/G/c case, and any model with non-Poisson arrivals. Unfortunately, none of those models yield to the analytical methods developed in this book.

As a matter of practical necessity, two useful approximation formulas have been developed to cover the missing cases. Specifically, there was a great need for results that go beyond the simple Markov models. We can use the following formulas to obtain results for arbitrary arrival and service processes, as long as they do not deviate too far from "ordinary" behavior. The formulas for L_q will be shown; it is easy to derive the corresponding results for L, W, and W_q using the relations given in Sections 6.4 and 6.5.

Both formulas make use of the coefficients of variation of the interarrival and service-time distributions. The coefficient of variation is defined at the ratio of the standard deviation to the mean, and it serves as a measure of the relative variability of a distribution. The negative exponential distribution has a coefficient of variation of 1 (because the standard deviation equals the mean), and most distributions have values between 0 and 1. But there are some highly variable distributions that have values greater than 1.

We have just seen from the Pollaczek-Khintchine formula that half of the queue can be attributed to variability in service time. That observation might help to make the following approximation plausible.

> **Result** Kingman's approximation: For the GI/G/1 queueing system
>
> $$L_q = \left(\frac{C_a^2 + C_s^2}{2}\right)\left(\frac{\rho^2}{1-\rho}\right)$$
>
> where $\rho = \lambda/(c\mu)$ and where C_a and C_b are the coefficients of variation of the interarrival and service-time distributions.

The next formula extends those results to multiple server cases. It will seem reasonable if you compare it to the approximation presented earlier for the M/M/c case in Section 6.11.

> **Result** Whitt's approximation: For the GI/G/c
> $$L_q = \left(\frac{C_a^2 + C_s^2}{2}\right)\left(\frac{\rho^{\sqrt{2(c+1)}}}{(1-\rho)}\right)$$
> where $\rho = \lambda/(c\mu)$ and where C_a and C_b are the coefficients of variation of the interarrival and service-time distributions.

These are clean results and easy to remember if you think of the first term as the average of the squared coefficients of variation of the arrival and service-time distributions, which either reduces or enlarges the second term, which is the exact formula for L_q for the M/M/1 model in the first formula and the approximate formula for L_q for the M/M/c model in the second case. They are plausible approximations, and they reduce to the exact ones for the M/M/1, M/G/1, and M/M/c. They give reasonably accurate estimates for ordinary situations, but the GI and G assumptions cover a lot of possibilities, including some that may deviate far from the typical. Of course, any model is only an approximation of reality, and one should always be on guard against faulty assumptions. As long as you do not expect precision in the numbers, the formulas will provide both reasonable quantitative results and, even more importantly, good insight into what factors are causing the queue.

6.23 Conclusion

It would be nice to be able to claim that most important queueing models have now been covered, and that, having completed this chapter, you should be equipped to handle the majority of situations you are likely to encounter. Regrettably, such is not the case. Real-life problems typically involve complexities not even touched on here. Often, the only practical analysis tool turns out to be simulation. On the other hand, there are many insights to be gained into even very complicated situations from studying the simpler models, and one should certainly attempt to develop this kind of understanding before resorting to simulation.

6.24 Recommended Reading

It is hoped that this introduction has provided some motivation to study the subject more thoroughly. The textbooks referenced as 1, 2, 3, 4, 5, 9, 10, 11, 14 or 15 each provide a good entry into the standard methods of queueing theory. For an emphasis on applications, complete with detailed case studies, see 7 or 8.

1. Cooper, R. B., *Introduction to Queueing Theory*, 3rd ed. CEEP Press, Washington, DC, 1990.
2. Cox, D. R., and W. L. Smith, *Queues*. Wiley, New York, 1961.
3. Giffin, W. C., *Queueing: Basic Theory and Applications*. Grid, Columbus, OH, 1978.
4. Gross, D., and C. M. Harris, *Fundamentals of Queueing Theory*, 2nd ed. Wiley, New York, 1985.
5. Kao, Edward, *An Introduction to Stochastic Processes*. Brooks/Cole, New York, 1996.
6. Khintchine, A. Y., *Mathematical Methods in the Theory of Queueing*. Hafner Publishing Co., New York, 1960.

7. Kleinrock, L., *Queueing Systems*, vol. II. Wiley, New York, 1975.
8. Kleinrock, L., *Queueing Systems*, vol. II. Wiley, New York, 1976.
9. Lee, A. M., *Applied Queueing Theory*. St. Martin's Press, New York, 1966.
10. Newell, G. F., *Applications of Queueing Theory*. Chapman and Hall, London, 1982.
11. Saaty, T. L., *Elements of Queueing Theory*. McGraw-Hill, New York, 1961.
12. Stidham, S. Jr., "A Last Word on $L = \lambda W$," *Operations Research* 22(1974), 417–421.
13. Takacs, L., "On Erlang's Formula," *Am. Math. Stat.* 40(1969), 71–78.
14. White, J. A., J. W. Schmidt, and G. K. Bennett, *Analysis of Queueing Systems*. Academic Press, New York, 1975.
15. Wolff, Ronald W., *Stochastic Modeling and the Theory of Queues*. Prentice-Hall, Englewood Cliffs, NJ, 1989.

Chapter 6 Problems

1. Set up the transition diagrams for the following queueing system models:
   a. M/M/2/4
   b. M/M/3/3
   c. M/M/1/3/3
   d. M/M/3/3/3

2. Write and solve the steady-state equations for the models in problem 1.

3. The following questions make use of the relations in Sections 6.4 and 6.5. No particular assumptions are made about the distribution of interarrival or service times.
   a. If a factory maintains an average in-process inventory equivalent to 300 work orders, or jobs, and an average job spends six weeks in the factory, what is the production rate of the factory in units of jobs per year?
   b. If an average of 3,000 people per day pass through an airport lobby, and the average time spent in the lobby is an hour and a half, what is the average number of people occupying the lobby?
   c. If a queueing system has three servers, the average number of customers in the system is 6.4, and the average number in the queue is 4.0, what is the average utilization of each server?

4. What is the capacity of the queue (that is, the number of customers that the queue can hold) in a M/M/3/4/5/FCFS system?

5. Consider an office complex in which ten employees share the use of two fax machines. The machines are together and are of the same brand and model. Name an appropriate queueing model using Kendall's notation.

6. The text provides formulas for L for the M/M/1 and M/M/1/N queueing systems. Find the formulas for W and W_q for both of these systems. Plot the results as functions of ρ to see how the mean waiting times behave as ρ approaches 1.

7. To model a single-server, infinite-capacity queueing system in which customers are "discouraged" from entering when the queue grows long, let the arrival rate be dependent on the state in the following way: when there are j customers in the system, let the arrival rate be $\frac{\lambda}{j+1}$. Make the usual Markov and stationarity assumptions.
   a. Show the steady-state equations and solve them if you can. Be sure to note any convergence requirement.
   b. Explain in words what the parameter λ represents in this model, and how it could be measured. (Note: Discouraged customers are never seen.)
   c. Show an expression for W_q, the steady-state mean time spent in the queue. Be as precise as you can.

8. Consider a barbershop with two chairs, two barbers, and no room for customers to wait. Say that the state of the system is the number of customers in the shop: 0, 1, or 2. If there is an empty chair when a customer arrives, he enters the shop and his haircut begins. If both chairs are occupied when he arrives, he does not enter the shop. As soon as a customer's haircut is completed, he instantaneously leaves the shop. The barbers do not assist one another when there is only one customer in the shop. On the average, a customer arrives every ten minutes, and each haircut takes an average of fifteen minutes.
   a. Set up the transition diagram and steady-state equations.
   b. Solve the steady-state equations for the distribution of the number of customers in the shop.
   c. Calculate the expected number of busy barbers.
   d. Calculate the expected number of customers turned away per hour.
   e. Notice in part d above that some potential customers are turned away.

If the shop had another barber, it might be able to profit from additional paying customers. On the other hand, the extra barber would have to be paid. Describe how you might determine whether it would be worthwhile for the shop to hire another barber.

9. A parking lot for a small shopping center has spaces for 100 cars. Assume that cars arrive according to a Poisson process, and that the durations of shopping trips are negative exponentially distributed. Also assume that a car will not wait for a parking space if the lot is full.

a. Identify an appropriate queueing model.
b. Write the steady-state equations and give the solution.
c. Interpret, in this context, the steady-state probabilities, L, W, L_q, and W_q. That is, explain what these mean in terms of the parking lot.
d. Explain how you would determine whether spaces for more cars would be worth having.
e. Explain what, if anything, could be done to model the situation if the shopping trip durations were not negative exponential.

10. Consider an airport with two runways, one of which is used solely for takeoffs and the other solely for landings. Assume that a plane, whether landing or taking off, will occupy a runway for an average of two minutes, where by "occupy" we mean "prevent use by any other plane." Delay on the ground, although unpleasant for passengers, does not pose a safety hazard; delay in the air, on the other hand, is of serious concern. Suppose that FAA regulations specify that the mean delay in the air (that is, W_q) must not exceed ten minutes. Assume that planes arrive according to a Poisson process, and construct a model from the information given above to answer the following questions.

a. What is the maximum tolerable load on the airport in terms of the mean number of planes that can arrive per hour?
b. Would it make much difference if you were told that the standard deviation of the time that a plane will occupy a runway is one minute?
c. What practical suggestion can you make as to what might be done to increase the load capacity of the airport, short of building new runways?

11. An executive must establish a secretarial staff. She has been allocated enough money to hire either two class A secretaries or three class B secretaries. In this company, secretaries are classified and paid according to their work efficiency. One class A secretary can complete the same job in an average of two-thirds the time that a class B secretary can, and is paid accordingly. Thus, at first glance, the two alternatives seem equal. Develop a Markovian queueing model (where the secretaries are the servers and the jobs submitted to them form a single queue with a FCFS discipline) to evaluate which, if either, alternative is preferable to the other.

12. A new hospital is to be built in a particular location. Restricting attention to just one section of the hospital, the maternity wing, imagine that the following information has been obtained:

- For the population that the hospital is intended to serve, we may expect an average of twelve deliveries per day.
- Recovery rooms are no problem. If necessary, rooms outside the maternity wing can be used. Delivery rooms are also not a problem because they are used only briefly for the actual delivery. The potential bottleneck is the availability of labor rooms.
- The average time that a labor room will be occupied by a patient is three and a half hours, but of course there is considerable variability in this time. Preparation of a room for the next patient can be accomplished in half an hour, if necessary.
- If all the labor rooms are occupied, an arriving maternity patient will be directed to another hospital. This is done very reluctantly, but state laws necessitate such action.

a. Provide a formula for the fraction of maternity patients who are turned away, as a function of the number of labor rooms designed into the hospital.
b. Provide a formula for the occupancy rate of the labor rooms, that is, the long-term average fraction occupied. (This figure is to be multiplied by the room rate to predict annual income from the labor rooms.)
c. Formulate a reasonable optimization problem, the solution to which would indicate the appropriate number of labor rooms to provide. (Introduce whatever cost parameters and the like that you believe are needed.)
d. Suppose that additional data were to reveal that the standard deviation of the labor room occupancy time is one hour. What effect would this information have on your analysis?

13. There is a problem with excessive delays at a self-service copying machine. It is a good machine, but recent increases in demand have strained its ability to keep up. The rated speed of the machine is fifteen copies per minute, but this does not allow for the time the customer spends aligning the material, checking the darkness, and so on. A time study reveals that the actual service times of customers, allowing for all of the variations that occur, are negative exponentially distributed with a mean of two minutes. Of this, a quarter (or a half minute) is machine time and the rest is human time.

The company representative offers two proposals. One is to replace the machine with a more expensive, faster machine that has a rated capacity of forty-five copies per minute. The other is to introduce another machine that is identical to the original one. The cost of the present machine is, say, k_1 dollars per hour. The faster machine would cost k_2 dollars per hour. The two-machine system would cost $2k_1$. Waiting time is valued at k_3 dollars per hour. The demand rate is twenty-four customers per hour.

Using Markovian queueing models, show how to evaluate the proposals to find the most cost effective alternative.

14. At a particular exit ramp of a turnpike, there is a single tollbooth. It was originally thought that there would be insufficient traffic at this point to justify more than one booth, but the traffic has increased recently. There are now (during rush hour, which is the period of concern) an average of 210 cars per hours, and the average service time is fifteen seconds. Two proposals have been made. The first is to add another toll booth identical to the first (assume the traffic would split evenly and at random between the two). The second is to add an automatic (that is, no human operator) booth that accepts exact change only. That booth would have an average service time of only five seconds, but only a third of the arrivals (selected at random) would have exact change. The performance measure of concern is the average delay incurred by the people who exit at this ramp.

a. Using an appropriate queueing model, estimate the delay under the present system, the first proposal, and the second proposal.

b. Recognizing that the service times are probably not negative exponential but are certainly not constant either, consider whether the results of part a should be modified.

c. Taking into account any other factors that may have been neglected, what do you recommend?

15. A large automobile dealer's lot attracts customers who arrive according to a Poisson process with a rate of thirty customers per hour (in peak times, which is all that concerns us here). Each customer, which could be an individual or a family, will look around for an average of twenty minutes whether or not there is a salesperson assisting. The dealer would like to have enough salespeople to handle the average number of customers present. How many is that?

16. In a two-bay car wash, where cars arrive at the rate $\lambda = 9$ cars per hour and the average duration of a car wash is twelve minutes, and using the formulas from Table 6.3 on page 205:

$$\pi_0 = \frac{1-\rho}{1+\rho} \qquad L_q = \frac{2\rho^3}{1-\rho^2}$$

what is the steady-state probability that an arriving customer will find at least one of the bays occupied?

17. Automobile accidents in a certain district occur according to a Poisson process with a rate of 2.4 per 24-hours. The probability that an accident involves an injury requiring ambulance service is 0.2. This district has only one ambulance, which takes an average of thirty minutes per run. What is the probability that the ambulance is busy when it is needed?

18. A single-server, infinite-capacity queue has Poisson arrivals with a rate of $\lambda = 10$ customers per hour. The service time distribution is approximately log-normal with a mean of four minutes and standard deviation of one minute. What is the steady-state average queue length?

19. An M/G/5/N system is observed for a long period of time. Statistical estimates (not model results) indicate that the long-term average number of customers in the system is 200, and the average time in the system is five hours. It is known that there are fifty arrival attempts per hour. What is the fraction of arriving customers that are not admitted?

CHAPTER 7

Networks of Queues

CHAPTER CONTENTS

7.1. Open Networks of Markovian Queues 226
7.2. An Example Open Network 227
7.3. Extensions 228
7.4. Closed Networks 229
7.5. A Preliminary Example 229
7.6. Relative Arrival Rates 230
7.7. Unnormalized Solutions for Individual Stations 232
7.8. Assembling the Pieces of the Solution 234
7.9. Calculating the Normalization Constant 235
7.10. Performance Measures for Closed Networks 237
7.11. Creating a Closed Model 239
7.12. Case Study 242
7.13. Extensions 247
7.14. Approximate Methods 247
7.15. Recommended Reading 248

In many of the practical situations you will encounter, queues occur as part of larger networks rather than isolated systems. For example, a factory ordinarily contains dozens of queues, linked together by the logical sequence of the production process. Such systems often exhibit complicated behavior resulting not only from the direct interaction of arrival and service processes, but such additional factors as branching, merging, and looping of the traffic streams. Each subsystem consisting of a single queue and the servers that go with it is called a *station*. Logically, we should probably call the system a network of stations, but the traditional terminology calls it a network of queues. This chapter explains two practical approaches for modeling networks of queues. They cannot accomplish everything one might like, but they do cover a wide range of applications.

7.1 Open Networks of Markovian Queues

Perhaps the most obvious approach to dealing with networks of queues is to separate them into stations, each of which has only one queue, and then analyze each station individually using the methods of the previous chapter. This would permit use of the rich variety of models for single queues, while immediately extending the range of applications to systems of arbitrary size. Unfortunately, there are two problems with this approach: (1) it does not always work, in the sense of yielding acceptably accurate results; and (2) even when it is technically correct, it neglects the interactions among the stations, which are often the most critical aspect of network behavior. Nevertheless, the difficulty of modeling even a single queue accurately, with all of the variations it can accommodate, suggests that some kind of decomposition strategy is desirable.

There is a class of queueing networks for which the decomposition strategy works very effectively. Provided that all of the required conditions are satisfied, each station will act exactly as if it were an independent Markovian queue of the M/M/c type. (Strictly speaking, the internal traffic streams are not always Poisson processes in these networks, but the results come out the same regardless.)

Definition An open product form queueing network is any network of queueing stations satisfying the following conditions:

1. All external arrivals to the network must occur in independent Poisson streams. (There may be several different streams entering at different points.)
2. All service times must be negative exponentially distributed with rates that depend at most upon the local queue. (Having queue-dependent service rates means that we can handle multiple servers and several other variations.)
3. All queues must have unlimited capacity.
4. Any branching of the internal traffic streams must be Markovian, with probabilities that are independent of everything except the position in the network. In other words, after completing service at subsystem i, a customer would go next to subsystem j with probability p_{ij}.

Open product form-networks of queues are sometimes called Jackson networks for James R. Jackson, the mathematician who discovered that the decomposition works in this case. The *open* part of the description refers to the fact that arrivals from the outside are accepted, and the *product form* description refers to the idea that the network factors into independent stations. Literally, the joint distribution of the state at each station is equal to the product of the marginal distributions for each station taken separately. In practical terms, this means that you can simply analyze each station separately, ignoring all of the others.

Those specific assumptions are critical. You cannot, for example, decompose a queueing network in which something special happens when a finite queue fills up. If customers back up in such a way that they affect previous servers (a behavior called *blocking*) or go somewhere else when the queue is full (called *overflow*), the independence of the separate stations is spoiled. Also, the routing cannot depend upon the states of the local queues, which means, for example, that a customer cannot choose to go to the shortest line. Such variations are common in real-life queueing networks, so it would be nice to be able to analyze them, but the simple approach just does not work. Although there are some more advanced techniques to handle these variations, they are beyond the scope of this book. Check the references in this chapter's "Recommended Reading" section if you need more advanced capabilities.

7.2 An Example Open Network

EXAMPLE 7.1

To illustrate the decomposition concretely, we will use the specific example shown in Figure 7.1. We assume that the external arrivals coming in from the left to station 1 and station 3 are Poisson processes, that all queues have unlimited capacity, that all service times are negative exponential, and that the branching switches after stations 2 and 5 are probabilistic. Provided that those required conditions are met, the analysis of each station amounts to straightforward application of the M/M/c results. For example, the station labeled "1" in the upper left is an ordinary M/M/3 queue.

The only step that may not be obvious is the calculation of the net *effective* arrival rate to each station, since it comprises (possibly) a direct external arrival stream and one or more internal traffic streams coming from other stations. You can see that station 1 has only an external input and so will behave just like an ordinary M/M/3 queue, which we know how to analyze. Station 2 receives its arrivals as the output of station 1. Now, under the required assumptions, it is known that the output process of an M/M/c will be a Poisson process with the same rate as the arrival process. The *timing* of the arrivals and departures is not the same because it has been disrupted by the service processes of station 1. However, in equilibrium, the mean *rate* of departures from station 1 has to be the same as the *rate* of arrivals (otherwise, one would be gaining on the other). So the arrival rate to station 2 is the same as the arrival rate to station 1. After station 2, there is a branching switch that sends some customers to station 4 and some to station 5, according to a probability. You could probably figure out branching and merging on your own—they are just what you would expect if the traffic streams are all Poisson processes. However, looping presents a different problem. We could work out special cases, but instead will jump to the fully general case in which any kind of probabilistic branching, merging, or looping is represented.

Before explaining that, we have a slight conflict in notation to deal with. It would be tempting to use the Greek letter λ for a parameter describing arrivals to the system, but if we do that we will not be able to use it for the internal effective arrival rates for the individual stations. Since all of our M/M/c results use λ, we prefer to reserve that symbol for the internal rates. That means we will have to use something else for the rates of the arrivals to the system. We will use α_j (a lowercase Greek alpha) for the given external arrival rate to queue j. Of course, α_j is zero if there are no direct external arrivals to queue j. Then we can use λ_j for the effective total arrival rate to queue j from all sources—something we will have to calculate.

In the most general case, where each of k subsystems has its own external arrival stream (at rate α_j) and every subsystem is connected to every other (with p_{ij} representing the probability of branching from subsystem i to subsystem j), the effective arrival rates λ_j are given by the solution to the so-called *traffic equations*:

$$\lambda_1 = p_{11}\lambda_1 + p_{21}\lambda_2 + \cdots + p_{K1}\lambda_K + \alpha_1$$
$$\lambda_2 = p_{12}\lambda_1 + p_{22}\lambda_2 + \cdots + p_{K2}\lambda_K + \alpha_2$$
$$\cdots$$
$$\lambda_K = p_{1K}\lambda_1 + p_{2K}\lambda_2 + \cdots + p_{KK}\lambda_K + \alpha_K$$

You can construct these equations by thinking, "The total rate of arrivals to station j (the left-hand side) is equal to the sum of the output rate from every station i times the fraction

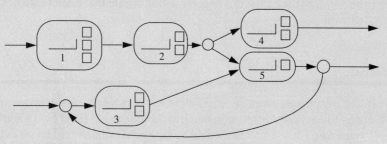

■ FIGURE 7.1   An example open network

that comes to station j from i, plus the rate of arrivals to j from the outside." For example, the structure of the open network in Figure 7.1 would be represented by the traffic equations:

$$\lambda_1 = 0\lambda_1 + 0\lambda_2 + 0\lambda_3 + 0\lambda_4 + 0\lambda_5 + \alpha_1$$
$$\lambda_2 = 1\lambda_1 + 0\lambda_2 + 0\lambda_3 + 0\lambda_4 + 0\lambda_5 + 0$$
$$\lambda_3 = 0\lambda_1 + 0\lambda_2 + 0\lambda_3 + 0\lambda_4 + p_{53}\lambda_5 + \alpha_3$$
$$\lambda_4 = 0\lambda_1 + p_{24}\lambda_2 + 0\lambda_3 + 0\lambda_4 + 0\lambda_5 + 0$$
$$\lambda_5 = 0\lambda_1 + 1\lambda_2 + p_{35}\lambda_3 + 0\lambda_4 + 0\lambda_5 + 0$$

To be even more concrete, let us suppose that the branching after station 2 sends two-thirds of the customers to station 4 and one-third to station 5. After station 5, four-fifths of the customers depart, and one-fifth are sent back to station 3. There are input streams entering at stations 1 and 3, so suppose that α_1 is 3 and α_3 is 2. With these assumptions about numerical values, we can solve explicitly for the five effective arrival rates. The equations become

$$\lambda_1 = 3$$
$$\lambda_2 = \lambda_1$$
$$\lambda_3 = 0.2\lambda_5 + 2$$
$$\lambda_4 = 0.67\lambda_2$$
$$\lambda_5 = 0.8\lambda_2 + \lambda_3$$

By simple substitution, the final solution is

$$\lambda_1 = 3 \quad \lambda_2 = 3 \quad \lambda_3 = 3.1 \quad \lambda_4 = 2 \quad \lambda_5 = 5.5$$

That gives us the effective arrival rates to use in the standard queueing formulas for each station separately. For example, since station 3 has two servers, we use the M/M/2 formulas with an arrival rate of 3.1.

You can also think of the problem of calculating the effective arrival rates in the following way. Ignoring how long anything takes and thinking of the stations as states, the routing of customers through the network is a Markov chain. The matrix of coefficients is the transient submatrix of an absorbing Markov chain (where absorption is equivalent to the departure of a customer from the network), and will be nonsingular in all practical situations. Hence, there is always a unique solution. In matrix form, the equations are

$$\boldsymbol{\lambda} = \mathbf{Q}\boldsymbol{\lambda} + \boldsymbol{\alpha}$$

where $\mathbf{Q}$ is the transient portion of the transition probability matrix and $\boldsymbol{\lambda}$ and $\boldsymbol{\alpha}$ are the appropriately defined vectors. It solves to

$$\boldsymbol{\lambda} = (\mathbf{I} - \mathbf{Q})^{-1}\boldsymbol{\alpha}$$

There are some implicit stability requirements for the effective arrival rates, which amount to the corresponding requirements for the separate subsystems. That is, the internal arrival rates must not exceed the subsystem processing rates, or $\lambda_j < c_j\mu_j$ for every j. If this constraint is violated for any station, the server(s) at that station will not be able to keep up with the arrivals, resulting in a queue that would grow endlessly over time.

The steady-state probabilities for each station are essentially independent, which also implies that the mean queue lengths, utilizations, and all of the other performance measures for each station can be computed separately. Once you know how to get the individual station arrival rates, the complete analysis of the entire network is easy. That is not to say that the results are obvious; sometimes queueing network behavior is surprising. One of the problems at the end of this chapter will illustrate that point in what is actually a very simple network.

7.3 Extensions

The facts that network decomposition works (when the assumptions are met) and is relatively easy to carry out create a temptation to use the same method even when the assumptions are not quite met. For example, we might be facing a situation in which we know that the service time distributions are something other than negative exponential, or perhaps the routing of customers is not exactly probabilistic in the way required. Can we get away with using the

Markov formulation anyway? Will the results be "close" to correct? The answer, unfortunately, is no. Experiments have shown that the results from the product form open network are quite sensitive to the assumptions and will give misleading information if the assumptions are not met. That is not too surprising if you realize that the entire decomposition approach neglects any effects of dependencies among stations. In other words, you have to pretend that no station has any effect on any other. That is quite a severe restriction.

Many scholars have attempted to extend the decomposition approach to include more complicated routing schemes or more general arrival and service-time distributions, while still seeking exact results. Unfortunately, in most cases whatever success they had came at a price of much more difficult mathematics. In a few cases, it is possible to incorporate some minor variations before losing the essential property of independence of the subsystems. Some of these variations are not very useful, but they do clarify just how far you can stretch the open network methodology. One of the most useful extensions is to multiple classes of customers, each with its own service time and routing properties. Of course, the notation becomes much more complicated, but the essential ideas remain the same and the computations are reasonably practical. If you want to learn more, consult some of the references listed in this chapter's "Recommended Reading" section.

7.4 Closed Networks

There is another class of computationally manageable networks of queues called *closed networks having a product form solution*. Conceptually, these networks differ from the open type by maintaining a captured population of customers; there are neither arrivals nor departures. In a way, this model is a network version of the finite source model treated in Chapter 6. Although this class is somewhat more difficult to handle, it has the distinct advantage of representing some dependency among the subsystems. That is, the closed network acts as a coherent system, not just a collection of unrelated components. For this reason, the class is very useful in applications, even when the open network would seem the more natural choice.

The assumptions required for a closed network to be of product form (which in this case means that the network "almost" factors) are the same as those required for the open network, except that there are no arrivals, so no assumption is required for them. However, the results for closed networks are generally less sensitive to deviations from the assumptions than are the results for open networks. Because the conditions will rarely be met precisely, this insensitivity property provides another reason to employ the closed network methods in practical applications. There is even a way to "fake" an open network model so that these advantages can be achieved; this trick is explained in Section 7.11, "Creating a Closed Model." As mentioned, one must pay a computational penalty to deal with the closed network case, but the methods are very effective at solving even large networks. Because the methods are iterative in nature, a computer will be required for all but the most trivial cases.

7.5 A Preliminary Example

EXAMPLE 7.2

We will start with a very small example. This will be far from the general case, but it will establish some ideas before the complications occur. Suppose that the network contains just two single-server stations, as in Figure 7.2.

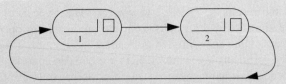

■ **FIGURE 7.2** A small closed network

If there are always four customers circulating around this system, then there will be five possible states, indicated by (i, j), where i represents the number at station one, j represents the number at station two, and i + j always equals 4. The five states are specifically (4,0), (3,1), (2,2), (1,3), and (0,4). When a service is completed at station one, the transition is of the form (i, j) to (i − 1, j + 1) and when a service is completed at station 2, the transition is of the form (i, j) to (i + 1, j − 1). That is, the sum of the indices remains constant while one increases by 1 and the other decreases by 1. The transition diagram will look like Figure 7.3 (the commas in the state designators have been suppressed).

This is a birth-death process—one of the classic Markov process models that is easy to solve. The equations are

$$0 = -\mu_1 \pi_{40} + \mu_2 \pi_{31}$$
$$0 = \mu_1 \pi_{40} - (\mu_1 + \mu_2)\pi_{31} + \mu_2 \pi_{22}$$
$$0 = \mu_1 \pi_{31} - (\mu_1 + \mu_2)\pi_{22} + \mu_2 \pi_{13}$$
$$0 = \mu_1 \pi_{22} - (\mu_1 + \mu_2)\pi_{13} + \mu_2 \pi_{04}$$
$$0 = \mu_1 \pi_{13} - \mu_2 \pi_{04}$$

We can get an unnormalized solution (one that satisfies all of the steady-state equations but may not sum to 1) by setting any one of the unknowns to an arbitrary value. It will be convenient to let

$$\hat{\pi}_{40} = \left(\frac{1}{\mu_1}\right)^4$$

■ **FIGURE 7.3** The transition diagram for the small closed network

Solving for the rest of the unnormalized probabilities is routine. You may verify for yourself that the unnormalized solution is given by:

$$\hat{\pi}_{ij} = \left(\frac{1}{\mu_1}\right)^i \left(\frac{1}{\mu_2}\right)^j$$

Here we use a caret mark over the unnormalized values to distinguish them from the final, corrected values. We still have to divide all of these by the normalization constant G in order to force them to sum to 1, but that will be the final step.

Examine the form of this solution. It is the product of two factors, the first of which uses only the parameter of the first station and the second of which uses only the parameter of the second station. In that sense, the solution factors. Furthermore, each term taken separately, namely

$$\left(\frac{1}{\mu_1}\right)^i \text{ and } \left(\frac{1}{\mu_2}\right)^j$$

is an unnormalized solution to the M/M/1 equations for the two stations taken separately (with arrival rates of 1). In other words, leaving aside the issue of normalizing, we can get the probability for any state by solving the pieces separately and multiplying the (unnormalized) marginal probabilities to get the (unnormalized) joint probability. This seems to suggest that the stations are independent, but that is not quite true. The normalization constant G contains the information about the dependencies. We still have to compute that value, and it does not factor. On the other hand, it is easy enough to compute because it is just the sum of all of the unnormalized probabilities.

For the closed network case, we say that a solution has a product form when, *apart from the normalization*, the joint probability is the product of the separate terms. Notice that the actual probabilities do not factor because G does not, but the unnormalized forms do.

This example was too simple to be much of a network, particularly with regard to the routing. However, it will give you an idea of where we are headed as we work through the general case. First, we have to introduce some definitions and notation. Be forewarned that the need to deal with a lot of parameters necessitates some messy looking expressions. They will be easier to understand if you keep in mind that we are looking for an unnormalized solution that is a product of terms that relate to just one station at a time. We gradually will build up to full complexity.

7.6 Relative Arrival Rates

We do not have external arrivals in the closed network model; all arrivals come from other stations. Therefore, we have to compute values for the internal arrival rates to replace the λ parameter for each station. In the most general case, the movements are governed by a Markov chain. That is, the probability that a customer finishing service at station i goes next to station j is p_{ij}, a transition probability that depends on nothing else. We will call the matrix $\mathbf{P}$ of transition probabilities describing the flow in a closed network the *routing* matrix. The states of this discrete time Markov chain correspond to the stations, and under normal circumstances the process will be ergodic. That means that we can calculate the steady-state probabilities for the routing matrix.

We are ultimately seeking steady-state probabilities for the entire queueing network, which is a continuous time Markov process. They are not the same as the steady-state probabilities we get from the discrete time chain, which only capture the routing information. In order to avoid confusion, let us reserve the usual symbol π for the larger queueing problem and use η for the routing chain. That is, η_i is the steady-state probability that station i is visited, or the fraction of visits, over the long run, that go to station i rather than somewhere else. We will only need these briefly.

Consider the interpretation of those probabilities. They do not contain any information about time, but they collectively represent how the visits are distributed over the stations. If you counted many (technically, an infinite number of) movements from one station to another, the fraction of them that went to a particular station would be represented by that kind of steady-state probability. Now that probability could not be an arrival rate in the sense of a Poisson process, because it does not have dimensional units of events per unit time. However, it measures the *relative* arrival rate. That is, if one of these probabilities (say, η_1) has twice the value of another (say, η_2) then whatever the real time units are, the arrivals will be coming twice as frequently to station 1 as to station 2. That will be close enough to actual arrival rates to serve our purposes, because we are only seeking unnormalized solutions. By the end of our development, everything will be corrected and we will be able to compute the actual arrival rates.

We can compute the η_i's, just as we would for any Markov chain, from the discrete time steady-state equations (refer to Chapter 3, Section 3.10 if you need reminding):

$$\eta_1 = p_{11}\eta_1 + p_{21}\eta_2 + p_{31}\eta_3 + \cdots$$
$$\eta_2 = p_{12}\eta_1 + p_{22}\eta_2 + p_{32}\eta_3 + \cdots$$
$$\eta_3 = p_{13}\eta_1 + p_{23}\eta_2 + p_{33}\eta_3 + \cdots$$
$$\cdots$$
$$\sum_{\text{all j}} \eta_j = 1$$

In the context of closed queueing network models, these are called the traffic equations.

It will work to just solve these in the usual way, but there is a further simplification that might save some time and be more convenient. We are seeking only *unnormalized* solutions for the separate stations, so we do not need the normalized η_i's; it will be just as valid to use unnormalized versions. That is, we can ignore the requirement that they sum to 1, set any one of them to any convenient value, and solve for the others using the traffic equations.

EXAMPLE 7.3

To illustrate these calculations concretely, we will use the routing matrix:

$$\mathbf{P} = \begin{bmatrix} 0 & 3/4 & 1/4 \\ 9/10 & 0 & 1/10 \\ 1 & 0 & 0 \end{bmatrix}$$

In words, three-fourths of the customers departing station 1 go next to station 2, one-fourth go to station 3; from station 2, nine-tenths go back to station 1, one-tenth go on to station 3; and from station 3, all customers go back to station 1. The steady-state equations (including the normalizing equation) would be:

$$\eta_1 = 0\eta_1 + 0.9\eta_2 + 1\eta_3$$
$$\eta_2 = 0.75\eta_1 + 0\eta_2 + 0\eta_3$$
$$\eta_3 = 0.25\eta_1 + 0.1\eta_2 + 0\eta_3$$
$$1 = \eta_1 + \eta_2 + \eta_3$$

> When you solve those algebraically in fractions (in order to avoid the round-off error that might occur in decimal form), you get
>
> $$\eta_1 = \frac{40}{83}, \ \eta_2 = \frac{30}{83}, \ \text{and} \ \eta_3 = \frac{13}{83}$$
>
> We could use those values, which are the true steady-state values for the routing matrix, for our relative arrival rates.
>
> Alternatively, since we need only relative values, you may scale them any way you like. For example, you could get rid of the fractions by multiplying by 83. These values would work just as well:
>
> $$\hat{\eta}_1 = 40, \ \hat{\eta}_2 = 30, \ \text{and} \ \hat{\eta}_3 = 13$$

Another way to get valid relative arrival rates is to estimate them directly. If you are working from data describing the movements (such as a process plan for a manufactured product or count statistics), you can use that data to estimate the relative frequencies of visits to the different stations. That approach would be both simpler and more direct than using the same data to construct a transition matrix and then solve equations. Or sometimes the structure of the network alone will suffice to tell you the relative rates, as in the case of a cyclical network, where all of the rates would have to be equal. Figure 7.2 illustrates a case like that.

This entire explanation of the relative arrival rates has been somewhat informal, in the hope that it will help you to make sense of them. You may be assured, however, that the theory is backed up by rigorous proof.

7.7 Unnormalized Solutions for Individual Stations

Once we have the effective arrival rates for each station, we can begin to assemble the parameters we need for the entire network. Let us take a moment to review some results for an individual station. A single station with c servers will have steady-state equations of the form (taken from Chapter 6, Section 6.11).

$$\lambda \pi_0 = \mu \pi_1$$
$$(\lambda + j\mu)\pi_j = \lambda \pi_{j-1} + (j+1)\mu \pi_{j+1} \quad \text{for } j < c$$
$$(\lambda + c\mu)\pi_j = \lambda \pi_{j-1} + c\mu \pi_{j+1} \quad \text{for } j \geq c$$

If we arbitrarily set $\hat{\pi}_0$ to 1 in order to specify an unnormalized solution, and then solve the rest of the equations, we get an unnormalized solution of the form

$$\hat{\pi}_j = \frac{1}{j!}\left(\frac{\lambda}{\mu}\right)^j \quad \text{for } j \leq c$$

and

$$\hat{\pi}_j = \frac{1}{c!c^{j-c}}\left(\frac{\lambda}{\mu}\right)^j \quad \text{for } j > c$$

The first formula applies to states less than or equal to the number of servers, c; the second applies when all of the servers become occupied and subsequent arrivals have to wait. There are two special cases for which the formulas simplify. For a single-server case, these expressions reduce to just

$$\hat{\pi}_j = \left(\frac{\lambda}{\mu}\right)^j \quad \text{for all } j$$

7.7 Unnormalized Solutions for Individual Stations

and if there are enough servers to handle all of the customers in the network, the second form (for j > c) is not needed. However, in the general case, we need both forms to cover all of the states.

They will simplify a little if we define the traffic intensity ρ as λ/μ:

$$\hat{\pi}_j = \frac{\rho^j}{j!} \quad \text{for} \quad j \leq c$$

and

$$\hat{\pi}_j = \frac{\rho^j}{c!c^{j-c}} \quad \text{for} \quad j > c$$

This function may look awkward in symbolic form, but it is easy enough to compute when you have specific numbers. If you want to use a spreadsheet, you should take advantage of the iterative form that expresses each one in the sequence in terms of the previous one:

$$\hat{\pi}_j = \frac{\rho}{j} \hat{\pi}_{j-1} \quad \text{for } j \leq c$$

and

$$\hat{\pi}_j = \frac{\rho}{c} \hat{\pi}_{j-1} \quad \text{for } j > c$$

You will also have to use $\hat{\pi}_0 = 1$ to start the sequence. A more compact way to express the iteration formula is

$$\hat{\pi}_j = \frac{\rho}{\min\{j, c\}} \hat{\pi}_{j-1}$$

Figure 7.4 shows a portion of a Microsoft Excel spreadsheet with the iteration formula showing. After establishing cells for ρ and c (B1 and B2 in the figure), and a column for the state

	A	B
1	ρ =	1.5
2	c =	4
3		
4	state	unnormalized probability
5	0	1.0
6	1	=($B$1/MIN(A6,$B$2))*B6
7	2	=($B$1/MIN(A7,$B$2))*B7
8	3	=($B$1/MIN(A8,$B$2))*B8
9	4	=($B$1/MIN(A9,$B$2))*B9
10	5	=($B$1/MIN(A10,$B$2))*B10
11	6	=($B$1/MIN(A11,$B$2))*B11
12	7	=($B$1/MIN(A12,$B$2))*B12
13	8	=($B$1/MIN(A13,$B$2))*B13
14	9	=($B$1/MIN(A14,$B$2))*B14
15	10	=($B$1/MIN(A15,$B$2))*B15
16	11	=($B$1/MIN(A16,$B$2))*B16
17	12	=($B$1/MIN(A17,$B$2))*B17

■ **FIGURE 7.4** Spreadsheet formula for unnormalized solution

numbers, you put the initial value for $\hat{\pi}_0$ (usually just 1.0) in a starting cell (B5). Then, in the cell below, you enter the formula "= ($B$1/MIN(A6, $B$2))*B5" to express $\hat{\pi}_1$. Once you have that formula entered, you can drag it down the column to copy it into the other cells. You should take it far enough so the values have become so close to zero that they are negligible.

Of course, once you have a column of unnormalized probabilities, you can normalize them by dividing each by the sum of all of them. However, we will not have to do that for any of the closed network results that follow. The most we would want (and in most cases we will not even need these) would be the unnormalized values. We usually are more interested in summary performance measures such as mean waiting times and queue lengths than detailed probabilities, and we will soon discover that we can get those without computing the probabilities.

7.8 Assembling the Pieces of the Solution

We now have the form of the unnormalized solution that we want to use, but we will have to change the notation a bit to accommodate the network situation. In order to keep track of which information goes with which station, we need to add subscripts to each parameter. However, up until now, we have used a subscript to indicate the state. So, to begin, we have to move the state index out of the subscript to make room for a different index that represents the station. That is, we start by converting π_j to $\pi(j)$. Then we add a subscript to each parameter so that we know to which station it belongs. Therefore, c_i will represent the number of servers at station i and μ_i will be the mean service rate at station i. The relative arrival rate $\hat{\eta}_i$ is already appropriately subscripted. We will also want a version of the traffic intensity, so we will define $\hat{\rho}_i = \hat{\eta}_i/\mu_i$. The caret marks indicate that these are relative values; any consistent scaling of them will work just as well. We will call them *relative workloads*, rather than traffic intensities, because the true traffic intensities will have to be computed.

Once we have the relative workload, we will have no further need for the relative arrival rates (the $\hat{\eta}_i$'s); they were needed only to compute the $\hat{\rho}_i$'s. It is recommended that you complete that task before addressing the job of combining stations. For purposes of the next steps, each station is fully described by its number of servers, c_i, and its relative workload $\hat{\rho}_i$.

For a closed network with M stations, we can define the network state as an M-dimensional vector,

$$\mathbf{j} = (j_1, j_2, \ldots, j_M)$$

where j_i represents the local number of customers at station i, and the sum of all terms remains constant. We will use N for the total number of customers in the system (at all times). The entire state space, J, consists of all vectors $\mathbf{j}$ satisfying

$$j_i \geq 0 \quad \text{for all i}$$

and

$$\sum_{i=1}^{M} j_i = N$$

Those requirements taken together imply that $j_i \leq N$ for every i.

Every transition will cause one of the components of the network state vector to increase by one and another to decrease by one. With the assumptions of a continuous time Markov process, the steady-state probabilities are determined by linear equations that are easy to conceptualize but very messy to write out. Fortunately, we do not have to solve these equations directly. The key result is that an unnormalized solution to the steady-state equations is given by a simple product of the unnormalized solutions to the individual stations. That is,

$$\hat{\pi}(\mathbf{j}) = \hat{\pi}_1(j_1)\hat{\pi}_2(j_2)\cdots\hat{\pi}_M(j_M)$$

Of course, you will still have to normalize by dividing these products by G, the sum of all of them, so the actual probability for a particular network state **j** is

$$\pi(\mathbf{j}) = \frac{1}{G}[\hat{\pi}_1(j_1)\hat{\pi}_2(j_2)\cdots\hat{\pi}_M(j_M)]$$

or, more compactly,

$$\pi(\mathbf{j}) = \frac{1}{G}\prod_{i=1}^{M}\hat{\pi}_i(j_i)$$

This is the product form solution for the closed network. The normalization constant G does not factor as it does for the open network. We will have to compute it for the network as a whole. In fact, that is the essential difference between the solutions for the open and closed cases.

7.9 Calculating the Normalization Constant

In principle, it is easy to declare how to get the normalization constant. Simply use the normalization equation (based on the fact that the probabilities have to sum to one),

$$\sum_{\text{all states}} \frac{1}{G}\hat{\pi}(\mathbf{j}) = 1$$

and solve for G,

$$G = \sum_{\text{all states}} \hat{\pi}(\mathbf{j})$$

So G is just the sum of all of the unnormalized steady-state probabilities. As a practical matter, however, that approach will be prohibitively difficult for all but the smallest networks. The difficulty is tied to the number of states. The number of possible states in a system with M stations and N customers is exactly

$$\binom{M+N-1}{M-1} \quad \text{or} \quad \frac{(M+N-1)!}{(M-1)!(N)!}$$

For even small values of M and N, this number can be very large. For example, a modest-sized system of ten stations and thirty customers would possess 211,915,132 states. Of course, the probability of finding the system in any one particular state would be tiny.

There is a much more efficient way to find G that involves building it up in stages, station by station. In order to describe that approach, we need to extend some notation. Start by defining a state vector for the first m stations (where $m \leq M$)

$$(j_1, j_2, \cdots j_m)$$

in which each j_i is nonnegative and they all sum to $n \leq N$. Further, denote the set of all such states J(m, n). This is, in effect, the state space for a related process involving only the first m stations and n customers. Then define

$$G(m, n) = \sum_{n \varepsilon J(m,n)} \prod_{i=1}^{m} \hat{\pi}_i(j_i)$$

which is the normalizing constant for that reduced state space. In this notation, the original G that we were seeking—the normalization for the entire network—will be G(M, N). That is, it will be the final term when all of the stations and all of the customers have been accounted for.

We can develop an iterative formula for G(m, n) as follows:

$$G(m, n) = \sum_{n \varepsilon J(m,n)} \hat{\pi}_m(j_m) \prod_{i=1}^{m-1} \hat{\pi}_i(j_i)$$

$$G(m, n) = \sum_{j_m=0}^{n} \hat{\pi}_m(j_m) \sum_{n \varepsilon J(m-1, n-j_m)} \prod_{i=1}^{m-1} \hat{\pi}_i(j_i)$$

$$G(m, n) = \sum_{j_m=0}^{n} \hat{\pi}_m(j_m) G(m-1, n-j_m)$$

This expresses the normalizing constant for any m and n in terms of the normalizing constants for the system with one less station and the same or less number of customers. In order to initialize this iteration properly, we have to define

$$G(1, n) = \hat{\pi}_1(j_1)$$

(which is not really a normalizing constant, but is defined as it needs to be for the iteration to work). We start with that function of n, and then generate G(2, n) for all n from 0 to N, then G(3, n), and so forth.

Table 7.1 provides a way to visualize how the calculations can be organized. If you set up a table (either on paper or in a spreadsheet) with the data for each station across the top, columns for each station, and a row for each value of n from 0 to N, you will work down the first column filling in the values for G(1, n) (which are obtainable from just the workload and number of servers at station 1). Then, using the values from the first column, you fill in the second column working from top to bottom. The specific equation you will use is

$$G(2, n) = \sum_{j=0}^{n} \hat{\pi}_2(j_2) G(1, n-j)$$

which involves only the values from the first column and above row n—the various $G(1, n-j)$—and the unnormalized solution for the second station, obtainable from the workloads and number of stations at the top of the second column. When the second column

■ **TABLE 7.1**
Computation of G(m,n)

Station	1	2	3	4
$\hat{\rho}_i$				
c_i				
0	1			
1				
2				
3				
4				
5				

is full, proceed to the third, and so forth. Two of the problems will give you a chance to work out examples with easy numbers.

To save memory in a computer implementation, you can keep all of the information in a single-column array—overwriting previous values as you add the next station—provided that you order the calculations so that information is not destroyed before it is needed. That requires that the loop run backward, calculating G(m, N) first, then G(m, N − 1), and so forth down to G(m, 0).

What all of this achieves is simply to add up all of the unnormalized joint probabilities, but cleverly organizes the addition to minimize the number of steps. The number of additions is of the order of MN^2, which for ten stations and twenty customers is 4,000—a sizable number, but far less than the almost 212 million that would be required by the direct approach. By the way, there is no need to be concerned about convergence in any of these sums. There are no infinite series to sum (because the number of states is always finite for a closed model) so the sums always converge (although one could run into overflow problems if the numbers grow too large).

You would not want to attempt these calculations by hand for anything but the tiniest illustrations, and the formulas are a bit complicated to put into a spreadsheet, so Appendix II provides some Visual Basic© code to implement a primitive version of the algorithm.

7.10 Performance Measures for Closed Networks

We now have, in principle, all that we need to analyze any closed network of product form. Any individual steady-state probability can be obtained by multiplying together unnormalized solutions for the individual stations, computing the normalization constant, and dividing the product by the sum. However, for a network of significant size, the number of possible states will be huge, so the probability for any specific individual state will be tiny. Such numbers are not very useful. What we really want, usually, are more aggregate performance measures, such as the mean queue lengths or waiting times. Happily, some of these can be calculated without additional effort, as part of the calculation of the normalization constant G. In fact, for most practical applications, there will never be a reason to look at the steady-state probabilities themselves; just calculating G may give you all you need.

For example, let B_i represent the steady-state mean number of busy servers at station i. Using the marginal distribution, it can be shown that

$$B_i = \hat{\rho}_i \left[\frac{G(M, N-1)}{G(M, N)} \right]$$

That is, we just multiply the unnormalized intensity for the station, $\hat{\rho}_i$, by a constant that is the normalization constant for N − 1 customers in the system (which was already calculated) divided by the normalization constant for N customers.

That ratio of normalization constants appears in multiple places, so let us give it a name. We will call it the *Pace* of the system and use the full name in expressions in order to avoid excessive notation. The definition is

$$\text{Pace} = \frac{G(M, N-1)}{G(M, N)}$$

If you carefully trace the calculations, you can discover that the dimensions of this constant are events per unit time, so it is something like an arrival rate or service rate. It measures how fast the customers are circulating.

Let λ_i be the throughput rate for station i, which is to say, the steady-state mean number of customers served per unit time. This is the true rate of arrivals to the station, taking into account all of the network behavior (unlike the unnormalized values $\hat{\eta}_i$). We know that each of the c_i

servers will process at the rate of μ_i when busy, and by the above formula the mean number of busy servers at station i will be $B_i = \hat{\rho}_i$Pace. Hence the throughput rate must be

$$\lambda_i = \mu_i \hat{\rho}_i \text{Pace}$$

But, since $\hat{\rho}_i = \hat{\eta}_i / \mu_\iota$,

$$\lambda_i = \hat{\eta}_i \text{Pace}$$

In addition to giving us a formula for the local throughputs, this result provides two ways to interpret the Pace. Most immediately, it is the "correction" that must be applied to the unnormalized arrival rates to obtain the true arrival rates. Second, in many applications, there is a particular station where the processing begins or ends. In such cases, we could scale the $\hat{\eta}_i$ for either the first or last station to have the value 1. Then the throughput of that station would equal the Pace, which would also be the throughput of the overall system. Of course, the calculation of G, and therefore the Pace, would have to use the same scaling. It is not necessary to use that particular scaling, but it does provide a simple interpretation for the Pace.

When the production rate is the same as the Pace, we can use Little's formula to find the overall time to pass through the system. The mean number in the system is N and the throughput rate is the Pace, so Little's formula translates to

$$N = (\text{Pace})(W)$$

So W, the expected total time to pass through the network, is N/Pace. Another easily obtained performance measure is the utilization, or the fraction of time each server is busy, defined by $U_i = B_i/c_i$,

$$U_i = \left(\frac{\hat{\rho}_i}{c_i}\right)[\text{Pace}]$$

If a station happens to have more than enough servers to handle all customers (that is, $c_i \geq N$), then there will never be a queue and the steady-state mean number of customers at the station will equal the mean number of busy servers. For this case only,

$$L_i = \hat{\rho}_i[\text{Pace}]$$

On the other extreme, if the station happens to have a single server, it can be shown that

$$L_i = \frac{1}{G(M,N)} \sum_{k=1}^{N} (\hat{\rho}_i)^k G(M, N-k)$$

Those two special cases provide expressions for L_i that require nothing more than the calculation of $G(M, N)$. However, the general case, where the number of servers is more than one but less than N does require some additional work.

To obtain those results, we need an expression for the marginal steady-state distribution for the number of customers at a particular station (without specifying the numbers at other stations). We will denote that by the notation

$$\pi_i(k) = P(n_i = k)$$

Of course, we already have the joint distribution, so we can use the normal rule for extracting the marginal distribution from the joint. You can look this up in Chapter 1, Section 1.6, if necessary. We have to sum all of the joint probabilities where n_i (the number of customers at station i) is held constant at j_i and the other j_k vary over all of their possible values,

$$\pi_i(j_i) = \sum \frac{1}{G(M,N)} \hat{\pi}(\mathbf{j})$$

The sum is over all states where the particular value for j_i is k and the other values range from 0 to N, keeping the sum of all values equal to N. If you pull the constant $1/G(M, N)$ out of the sum and factor the fixed term out of the product, you get

$$\pi_i(j_i) = \frac{1}{G(M, N)} \hat{\pi}_i(j_i) \sum \prod_{\substack{k=1 \\ k \neq i}}^{M} \hat{\pi}_k(j_k)$$

where now the sum runs over all combinations of values where the various j_k sum to $N - k$. For the sake of simplifying the notation, let us define

$$H_i(n) = \sum \prod_{\substack{k=1 \\ k \neq i}}^{M} \hat{\pi}_k(j_k)$$

where, again, the sum runs over all combinations of nonnegative values of the j_k where the sum totals n. That sum is, in effect, the normalization constant for a system that is missing station i and with n customers. In other words, we can compute the sum in the same way that we computed G, only we omit station i (or, equivalently, set the intensity of station i to zero). Then, for our marginal probability we have the expression

$$\pi_i(j_i) = \frac{1}{G(M, N)} \hat{\pi}_i(j_i) H(N - j_i)$$

That may not seem very simple, but it is directly computable. Moreover, it is a useful intermediate result that enables us to develop some easier expressions. In particular, the general formula for the steady-state mean number of customers at station i is (by simply using the definition of the mean)

$$L_i = \frac{1}{G(M, N)} \sum_{j_i=1}^{N} j_i [\hat{\pi}_i(j_i) G_i(M, N - j_i)]$$

Computing these for each station is a significant job that you would not want to attempt without a computer. However, a straightforward approach is represented in the Visual Basic code in Appendix II.

After all this development, the work of computing all of these performance measures comes down to two extended calculations (that are best implemented in code) and easy transformations of the results from those two. Table 7.2 summarizes the formulas.

7.11 Creating a Closed Model

Constructing a closed queueing network model is usually easy. You must first identify the stations and customers. The number of servers for each station is a simple count; the mean service times can be estimated as averages. The routing information can be represented by a Markov chain (where the stations are the states), or in some cases a more direct estimation of the relative arrival rates can be made. The case study in the next section will illustrate each of these steps.

The closed network model does not require any information about external arrival rates. However, it does depend heavily upon the value of N, the total number of customers in the system. In many applications, the value to use is not immediately apparent. The production rate will increase monotonically with N but will approach a maximum value, as shown in

■ **TABLE 7.2**
Closed Queueing Network Formulas

Name	Notation	Formula
Normalization constant	$G(M,N)$	computer code
Pace	Pace	$\dfrac{G(M, N-1)}{G(M, N)}$
Throughput rate (sometimes—see text)	λ	Pace
Mean time in system (sometimes)	W	N/Pace
Mean number of busy servers at station i	B_i	$\hat{\rho}_i \text{Pace}$
Utilization of servers at station i	U_i	$\left(\dfrac{\hat{\rho}_i}{c_i}\right)\text{Pace}$
Throughput at station i	λ_i	$\hat{\eta}_i \text{Pace}$
Mean number of customers at station i	L_i	computer code
Mean wait time at station i (per visit, including service time)	W_i	$\dfrac{L_i}{\lambda_i}$
Mean queue length at station i	Lq_i	$L_i - \dfrac{\lambda_i}{\mu_i}$
Mean delay at station i (not including service time)	Wq_i	$W_i - \dfrac{1}{\mu_i}$

Figure 7.5. That implies that the throughput, the station utilizations, and the number of busy servers all approach saturation limits as N increases. At the same time, the waiting times and delays increase without limit, as shown in Figure 7.6. Therefore, if one has the ability to control N, there is a trade-off to consider. One can get more complete use of the server resources by increasing N, but only at the expense of increasing waiting times. Unless you have a way to evaluate the importance of one versus the other, there is no particular optimum value. Typically, however, a reasonable balance will occur when N is around three times the total number of servers in the system. At least, that is a convenient rule of thumb for initial evaluation, and you may vary N to see what happens.

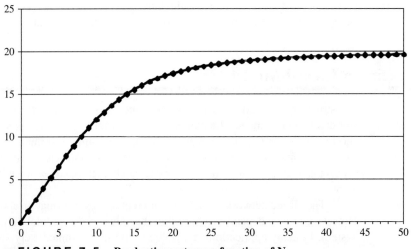

■ **FIGURE 7.5** Production rate as a function of N

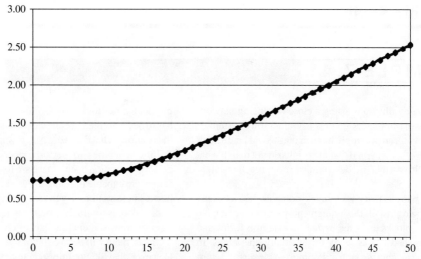
■ **FIGURE 7.6**   Flow time (in hours) as a function of N

Sometimes you may think that an open network model is the most natural choice (because there are in fact arrivals and departures), but you would like to take advantage of the robustness of the closed network model. There is an easy way to convert an open model to closed, as shown in Figure 7.7; simply define a dummy station to represent the outside world. That is, you imagine that customers, instead of leaving forever, "recycle" to the dummy station where they await another entry. The dummy server is set up to release customers into the system at a rate that corresponds to what would otherwise be the external arrival rate. In other words, the dummy service times are not really doing anything but spacing the arrivals and the dummy queue is just holding the potential arrivals. For purposes of analysis, we treat the entire system (including the dummy station) as a closed network.

Why would you want to do that (as opposed to just using the easier open network approach)? When you employ this trick, the variation in the number of customers in the system is bounded. It is not a fixed value, as in the purely closed model, nor can it grow very large, as in the open model, but ranges from zero to an upper limit of N. (You can make N a large number if you want to.) In many cases, that is genuinely more realistic than allowing it to range from zero to infinity, which would be what the open network model would do. More importantly, it is known that the results from the closed model are much more robust than the open model. That is, they seem to be close to accurate even when the assumption about negative exponential service times is violated. This characteristic is not yet fully understood, but many researchers have corroborated the finding. Although one should always be careful about trusting a model that is based on weak assumptions, the closed model seems to permit considerable latitude.

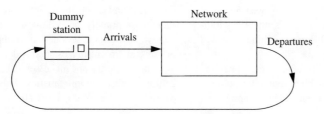
■ **FIGURE 7.7**   **Employing a dummy station**

7.12 Case Study

EXAMPLE 7.4

This case study illustrates using a closed queueing network model to assist in the design of a manufacturing system. It shows that you can accomplish much that would be difficult to do in any other way, and that the results do not match what more primitive (that is, deterministic) approaches might suggest.

The scenario is that a motorcycle company has to select, purchase, and install the equipment to machine engine blocks. The blocks enter the system as cast iron forms and leave as finished blocks ready for assembly operations. In between, the machines rough mill some outside surfaces, turn the piston chambers, drill several holes, and fine mill a few slots. There also is some cleaning and inspection involved, and the workpieces have to be transported. We have a little bit of information about processing requirements (but not much), and we have a ballpark estimate of the required production rate. However, we cannot fine-tune the design to any particular data, because any values we use are likely to be revised before the plant goes into operation. Because of the uncertainty in exact values (both in advance of purchasing the equipment and over the long run as production requirements change), the company wants a "robust" design that will continue to function well even as the variations occur.

Although it is a nontrivial task in itself to identify the right types of machines to purchase, we will assume that this has already been done. The question, for the moment is, "How many of each type of machine should be ordered?" Later, the exact layout of the system will need to be specified, but for now it will suffice to determine the quantities. In queueing network terms, we need to find the number of servers required for each station in order to meet the requirements. The most expensive machine is a heavy-duty mill, used for face milling surfaces of the casting. We will say each one costs $300,000. We need at least one boring machine to bore and hone the cylinders, at a cost of $200,000 each. A lighter duty mill costing $150,000 will mill some slots. Most of the time will be taken in drilling and tapping various holes, using drilling machines that cost $75,000 each. All of the machines have automated tool changers, so that they can perform all of the operations required without stopping for setups. Finally, we need some fixturing machines that have precision alignment capabilities; each of these cost $10,000. Each machine type will comprise a station.

Of course, we want to pool servers of the same type rather than have separate queues for machines that are capable of doing the same work. The machines should be grouped—not physically, but logically—so that when a certain operation has to be performed, it can be done on any of the machines of the station. That is, we do not want to dedicate certain drilling machines to perform operation 3 and others to perform operation 6, unless we are forced to by other factors. If they are capable of doing the work, they should be pooled into a single station. (By the way, this principle is not widely understood, is routinely violated in real-life factories, and is an easy improvement to make, often with dramatic results.)

Thus, there will be five stations where work is performed. However, they do not fully account for all of the time that is spent in the system, even with their queues. We need some way to deal with the time consumed in transporting workpieces from one station to another. We could ignore that, of course, by making the assumption that it is inconsequential, but a better way is to include a sixth station for the transport servers. If the parts are moved manually, or perhaps by automated carts, we would have a limited number of servers; if there is a conveyor system, we would make the number of servers greater than or equal to the number of pieces in the system. Let us assume the latter.

The "customers" in this application are individual workpieces. If the system were required to produce in batches (say, ten at a time), we would regard the entire batch as one customer. One of the problems explores this possibility. For the present example, we will assume that workpieces are mounted on fixture pallets (of which there are a fixed number) and that they are moved individually from machine to machine by an automated material handling system. Although raw castings enter the system and leave as finished engine blocks, the constant number of pallets justifies the use of a closed model. That is, as soon as a block has finished its operations, it is removed from its fixture and immediately replaced by a raw casting. The pallets will never enter or leave the system, so will remain constant in number.

Table 7.3 shows a typical process plan for a particular block of the kind that the system will be required to produce. It is not necessarily the only process plan—just representative in the sense that the routing and times are typical. We could go through a more elaborate procedure of averaging the times and visits from all of the process plans (weighting each appropriately according to the quantities of each relative to of total production volume). However, under the present circumstances of considerable uncertainty about the data, it would not make sense to spend a lot of effort getting more accurate statistics. This one typical process plan will suffice for our illustration.

TABLE 7.3
Typical Process Plan

Operation	Description	Station	Time (minutes)
1	Attach workpiece to fixture	fixture	2.7
2	Mill base	mill	6.0
3	Drill locator holes	drill	3.0
4	Refixture (turn over)	fixture	3.0
5	Bore and hone cylinders	bore	4.2
6	Drill various holes	drill	3.2
7	Mill slots	light mill	4.0
8	Detach from fixture	fixture	2.0

We number the stations in the order they are first visited, except for the transport station, which we will put last. For each of these, we need the number of servers, a relative arrival rate, and a mean service time. The number of servers for each station are the decision variables that we have to find values for (by test and refinement). The relative arrival rates and mean service times will be multiplied to get the relative workloads—the ρ_i's that appear in the formulas. So we might begin by setting up a table in a spreadsheet with all of this data, something like in Table 7.4. The last column will be computed from the previous two, but we will have to supply all of the other values.

The mean service times for the mill, the bore, and the light mill are easy because each of these stations is only visited once. The times can be copied directly from the process plan. The drill is visited twice (operations 3 and 6), so we will have to average the two times for drill's mean service time. The fixture station is visited three times, so the time there will have to averaged over the three operations. The transport station is unique. There is no explicit mention of it in the process plan, but implicitly, there is a movement of the workpiece after every operation. If we had a layout and could estimate the speed and distance of each trip, we might be able to get a good estimate of the mean trip time. However, we do not have a layout, so we have to be content with using a reasonable value. Let's say that two minutes is a reasonable number with which to start. If it turns out that this number is critical to the results, we will explore variations; if not, it will not matter very much what value we use.

Now for the relative arrival rates, we can proceed in two different ways (which lead to the same values). We could construct the appropriate Markov chain and solve for some unnormalized steady-state probabilities. That might be the preferred approach if the process plan were very complicated, or if there were many to combine. To show what the transition matrix would look like, we construct a count matrix as follows:

$$\begin{bmatrix} 0 & 0 & 0 & 0 & 0 & 3 \\ 0 & 0 & 0 & 0 & 0 & 1 \\ 0 & 0 & 0 & 0 & 0 & 2 \\ 0 & 0 & 0 & 0 & 0 & 1 \\ 0 & 0 & 0 & 0 & 0 & 1 \\ 3 & 1 & 2 & 1 & 1 & 0 \end{bmatrix}$$

The first row says that, in one complete tour of the process plan, the workpiece left the fixture station 3 times, always going next to the transporter. The last row counts the number of times the transporter delivers to the mill, the drill, and so

TABLE 7.4
Empty Data Table

Station	Station name	Number of servers (c_i)	Relative arrival rate ($\hat{\eta}_i$)	Mean service time ($\frac{1}{\mu_i}$)	Relative workload ($\hat{\rho}_i$)
1	Fixture				
2	Mill				
3	Drill				
4	Bore				
5	Light mill				
6	Transport				

forth. When we normalize each row to get the transition probabilities, we get:

$$\begin{bmatrix} 0 & 0 & 0 & 0 & 0 & 1 \\ 0 & 0 & 0 & 0 & 0 & 1 \\ 0 & 0 & 0 & 0 & 0 & 1 \\ 0 & 0 & 0 & 0 & 0 & 1 \\ 0 & 0 & 0 & 0 & 0 & 1 \\ \frac{3}{8} & \frac{1}{8} & \frac{2}{8} & \frac{1}{8} & \frac{1}{8} & 0 \end{bmatrix}$$

Finally, the steady-state probabilities for this Markov chain are

$$\eta_1 = \frac{3}{16}, \eta_2 = \frac{1}{16}, \eta_3 = \frac{1}{8}, \eta_4 = \frac{1}{16}, \eta_5 = \frac{1}{16}, \eta_6 = \frac{1}{2}$$

and any positive multiple of these values will work for our relative arrival rates.

The more direct approach is to calculate visit frequencies directly. Eight operations and the accompanying transport steps imply that there will be sixteen separate visits to stations by a workpiece passing through the system. Of these, eight will be to the transport station, three to the fixture station, and so forth. The fraction of the total that goes to station i provides the same results as the steady-state probabilities. Furthermore, since we can use unnormalized values, we might as well get rid of the fractions by multiplying each by sixteen. We will use

$$\hat{\eta}_1 = 3, \hat{\eta}_2 = 1, \hat{\eta}_3 = 2, \hat{\eta}_4 = 1, \hat{\eta}_5 = 1, \hat{\eta}_6 = 8$$

You may notice that this solution matches the number of visits to each station, so that when we multiply these relative arrival rates by the mean service times, we are actually making the relative workloads equal to total service time at each station. Hence, we could have saved all this trouble by using those values directly. Of course, we did not know that we could do that until we worked out the algebra. Now that we do know it, it will be convenient hereafter to simply use the total processing times as the relative workloads.

The table of required data now looks like that shown in Table 7.5.

The next step is to specify the number of servers for each station. For the transport station, we said that we would assume a conveyor-type material handling system, which would permit simultaneous movements of many parts. We can represent that by providing many servers—say, one hundred. That way, we know that workpieces will never have to wait to be moved, but we are still including the time that the physical transfer requires. Next, we need initial values for each of the machine types. These are the decision variables for our problem, so we will be changing them to find the best combination of values to meet the production requirement and minimize cost.

The way this planning is usually done in practice (in the absence of any queueing knowledge) is to base everything on deterministic capacity calculations. The required production rate of twenty completed parts per hour translates to one every three minutes on average. That inverse of the rate is commonly called the cycle time. In order to keep up, each station will have to be capable of finishing one workpiece in that time. So, for example, if a particular operation takes eleven minutes, there will have to be at least four machines working in parallel to ensure that the rate can be met. (You take the operation time, divide it by the cycle time, and round up.) The only flaw in that logic is that it does not take account of the time lost to variability, which is to say the queueing behavior. The variations cannot "average out" because, while it is possible to fall behind, it is not possible to get ahead. Planners who have sufficient experience know that they have to include an "allowance" to account for some unavoidable time losses. Still, that is usually applied at the end, to provide a cushion. We can do better. A queueing network model can quantify the effects of the variability so that the planning decisions can be based on more realistic and accurate calculations of capacity. However, just so that we can see what happens, we will use the cycle-time calculations to provide initial values for the number of machines of each type.

The data table for the first option, as shown in Table 7.6, is now sufficient to permit the queueing analysis, but we want to add one more factor: the cost of each machine.

■ **TABLE 7.5**
Partially Completed Data Table

Station	Station name	Number of servers (c_i)	Relative arrival rate ($\hat{\eta}_i$)	Mean service time ($\frac{1}{\mu_i}$)	Relative workload ($\hat{\rho}_i$)
1	Fixture		3	2.9	8.7
2	Mill		1	6.0	6.0
3	Drill		2	3.1	6.2
4	Bore		1	4.2	4.2
5	Light mill		1	4.0	4.0
6	Transport		8	2.0	16.0

TABLE 7.6
Input Data for First Option

Station number	Station name	Number of servers (c_i)	Relative arrival rate ($\hat{\eta}_i$)	Service time ($\frac{1}{\mu_i}$)	Relative workload ($\hat{\rho}_i$)	Cost per machine (thousands)
1	Fixture	3	3	2.9	8.7	$ 20
2	Mill	2	1	6.0	6.0	200
3	Drill	3	2	3.1	6.2	75
4	Bore	2	1	4.2	4.2	180
5	Light mill	2	1	4.0	4.0	150
6	Transport	100	8	2.0	16.0	0

We are looking for the lowest cost system that can meet the requirements, so we will want to compare the total system cost for each option. To do that, we need the last column in Table 7.6, where costs are expressed in thousands of dollars. Although there will certainly be some fixed costs (such as the cost of the material handling equipment), it will suffice for our purposes to compare the linear cost functions. That is, we will just multiply the cost of each machine type by the number of that type and total.

We need just one more parameter value: N, the number of customers/workpieces/pallets that are allowed in the system. This is a management choice. A high value will increase production, but will create a lot of in-process inventory and long production flow times. Figures 7.5 and 7.6, shown earlier, actually plot the functions of N for the data in this table. We can experiment with different values, but a reasonable starting choice would be three times the number of machines, or thirty-six. Alternatively, we could just put in a larger value, like fifty, and plot graphs similar to Figures 7.5 and 7.6 to let management judge how to balance the trade-off. A third option is to use the code to search for the minimum N that achieves our desired production goal of twenty finished blocks per hour. Pallets and fixtures cost much less than machines, but they are not cheap. We will assume that each one required will add $5,000 to the cost of the system.

The production rate of the system—the rate at which finished blocks are produced—can be related to the throughput rate of station 2, 4, or 5, because each of these stations is visited exactly once in the process plan. Alternatively and equivalently, it could be the throughput of the fixture station divided by three, or the drill station divided by two, or the transport station divided by eight (because these divisors match the number of times the station is visited per completed part).

The goal is to achieve a production rate of at least twenty blocks per hour at the minimum cost. If we complete the calculations with the number of servers suggested by the deterministic analysis, we find that we cannot meet the goal. As Figure 7.5 shows, the production rate approaches twenty as the number of pallets increases, but it never quite gets there. We will have to add at least one machine to the system to increase the production capacity. It might make sense to add one of the most heavily utilized type (the bottleneck station), which in this case would be a boring machine. On the other hand, boring machines are expensive, so there might be a more cost-effective choice.

There are actually many combinations of numbers of servers that will meet the goal of producing at least twenty blocks per hour. As long as we have code to handle the calculations, it is easy to explore many alternatives. Starting from the initial solution suggested, which we will identify by the vector (3, 2, 3, 2, 2), (where we are simply listing the number of machines of each type in order), we can consider all perturbations in the neighborhood of that vector. That is, we can increase each number in turn, then in pairs, and so forth, until we have exhausted every reasonable possibility. We do not have to consider combinations with fewer machines, because we know that we do not have enough to start. Table 7.7 shows are some calculated results for the first round of changes, where we increase the number of machines at each station one-by-one.

As you can see, only one of the changes—option 2—increases the capacity sufficiently to meet the goal of twenty blocks per hour. That is because the bottleneck is at the first station, and increasing the resources elsewhere cannot overcome that limitation. The higher cost of some of those options tells you clearly that you do not always get what you pay for. You must add resources where they can do some good.

Option 2 in Table 7.7 is the only one yet considered that can be considered a feasible design. We will use this as the stepping off point for another round of possible changes. If we add one machine to each of the stations now, and calculate the corresponding effects, we get Table 7.8.

All of these possibilities can achieve the production goal of twenty blocks per hour. Option 6 is interesting because it is actually cheaper than option 2, despite having one more fixturing machine. The cost savings occurs because, although the fixturing machine costs $20,000, its presence

TABLE 7.7
Results of Analysis, First Round

Option	Configuration vector	Pallets required	Production rate	Flowtime (hours)	Cost (thousands)
0	3, 2, 3, 2, 2	36	19.352	1.86	$1,525
1	4, 2, 3, 2, 2	36	19.948	0.80	1,545
2	3, 3, 3, 2, 2	26	20.079	0.93	1,675
3	3, 2, 4, 2, 2	36	19.388	1.86	1,600
4	3, 2, 3, 3, 2	36	19.401	1.86	1,705
5	3, 2, 3, 2, 3	36	19.391	1.86	1,675

TABLE 7.8
Results of Analysis, Second Round

Option	Configuration vector	Pallets required	Production rate	Flowtime (hours)	Cost (thousands)
2	3, 3, 3, 2, 2	26	20.079	0.93	$1,675
6	4, 3, 3, 2, 2	19	20.060	0.95	1,660
7	3, 4, 3, 2, 2	25	20.082	1.24	1,870
8	3, 3, 4, 2, 2	25	20.107	1.24	1,745
9	3, 3, 3, 3, 2	24	20.027	1.20	1,845
10	3, 3, 3, 2, 3	25	20.116	1.24	1,820

means seven fewer pallets, which saves $35,000. The other options on this list are inferior in terms of both cost and flowtime, so option 6 is now the primary contender. Let us take a closer look at that option. As Table 7.9 shows, the fixturing station is still the most heavily utilized.

The numbers seem reasonably balanced, at least to the extent that no one stands out enough to suggest a problem. But we have not yet determined that this option will be sound if it turns out that some of our rough estimates for the input were in error. In particular, what if the workload on the busiest station were underestimated? Would the system still be able to meet its production goals? Or would additional equipment have to be purchased? We can run the code again to see what happens if the workload on the fixturing station (the busiest station) turns out to be 10 percent greater than expected. After doing that analysis, we find that the system will still be capable of producing twenty blocks per hour, but only if we increase the number of pallets by one. Errors in the estimates of other station workloads would have less effect, so we can be confident that option 6 will work.

This case study is an excellent example of how queueing network models can support the design of a production system. They can tell you things that you could not otherwise predict, and can help to avoid serious and long-lasting consequences of mistakes. Very little data is needed, and even that can be quite rough.

TABLE 7.9
Performance Details, Option 6

Station	Number of machines	Mean number busy	Utilization per machine	Mean no. at station	Mean no. in queue
Fixture	4	2.91	0.73	3.67	0.76
Mill	3	2.01	0.67	2.64	0.63
Drill	3	2.07	0.69	2.81	0.73
Bore	2	1.40	0.70	2.38	0.98
Light mill	2	1.34	0.67	2.15	0.81

7.13 Extensions

The closed queueing network model is capable of much more than was explained here. There are many "tricks" to deal with things such as different types of customers (with different routing and different service times), server failures, batch processing (with setup times), assembly operations, different kinds of material handling, and capacity expansion. Also, the computational algorithms can be considerably more sophisticated that those explained here. The analysis tools are mostly limited to steady-state results, so they are best suited for long-term issues such as facility planning or equipment replacement. They are not very helpful for, say, real-time control or scheduling.

A lot of experience and experimental testing has established that the closed queueing network model is surprisingly robust. That is, it seems to give results that are close to accurate even when the assumptions are not well met (unlike the open network model). That property seems to be related to the fact that having a limited number of customers restricts the variability of events. That is, arrivals slow down as queues fill up. As long as the system is large enough and the routing mixes the traffic streams, only the means of the service-time distributions seem to matter. Of course, we only calculated mean values, not variances or higher moments. Although they could in principle be calculated, the correct values are much more dependent upon the specific service time distributions and the Markov models are not likely to be reliable.

7.14 Approximate Methods

Every model is an approximation of reality. There is never a possibility of achieving perfect fidelity because you can only include so much and must ignore everything else. On the other hand, there is a difference between good models and bad models; we cannot excuse every assumption that we might find convenient. The queueing models we have seen so far (with two exceptions) approximate reality only in their assumptions. That is, the solutions are exact if the assumptions and input data are accepted as correct. We could, in principle, carry out the computations to as many decimal places as we want and still be exact. The advantage of such results, when you can obtain them, is that you can be certain that any flaws or errors can be attributed to the assumptions or the data, not from anything in the handling of the model. However, in many practical circumstances, the exact results are unobtainable. What can you do in those cases?

One approach is to resort to discrete event simulation, using any of the many software packages that are available for this purpose. This approach is more flexible than the analytical methods because you can make just about any assumptions you want and still be able to obtain results. The software packages make it relatively easy to program your model. However, simulation is not free of its own weaknesses. For one thing, the results you get are statistical estimates of the true values; all of the statistical issues associated with estimation apply. For example, it will take a very large sample size to get more than two significant digits of precision. Also, much of the simplest statistical theory depends upon having independent samples, but the observations taken from a simulation typically are correlated. The point is that it can take sophisticated statistical knowledge to do the job properly. Some beginners make the mistake of thinking that simulation is easier or safer than analytical methods; in fact, it is neither. Simulation is a powerful and wonderfully practical modeling tool, but it takes at least as much knowledge and skill to use it properly as any other modeling tool. You should definitely take a complete course on the subject so that you can add it to your arsenal.

There is another kind of approximation that does not depend upon statistical methods. The results are exact in the sense that the computations come out the same every time. It makes use, however, of formulas that are simplified (and therefore inexact) representations of the conditions assumed by the model formulation. Such results are derived from a deep understanding of the influences on the results, a good guess about the forms of the equations, and then a lot of experimental validation to ensure that they behave as they should. The downside is that you cannot be completely confident that they will always be close to what you would get if you could compute the exact results. Still, approximate results are better than no results. For most of the situations that are appropriate for queueing network models, and considering that the input data are usually imprecise anyway, accuracy within a few percent is usually acceptable.

We have already seen two examples of this kind of approximation; namely, the Kingman and Whitt approximations to the solutions of the GI/G/1 and GI/G/c cases, presented in Chapter 6, Section 6.22. For queueing networks, we can combine those approximations for the individual stations with the decomposition approach to the network of stations. Ward Whitt is regarded as the principal architect of this approach, and he has developed a computer package called QNA (Queueing Network Analyzer), as well as numerous refinements to improve the approximations in particular situations.

7.15 Recommended Reading

Although interest in queueing networks goes back to earliest days of queueing, the decomposition approach got its real start with the work of Jackson, 8, and Gordon and Newall, 7. Buzen, 3, was responsible for the breakthrough idea on how to compute the normalization constant for closed networks. Lavenburg and Reiser, 9, developed an alternative approach to the same class of problems which they called mean value analysis. Baskett et al., 1, summarized by 1975 the full extent of the capabilities for open and closed networks in a landmark paper that is known to all in the field. Further developments since then have filled out some additional details. Any of the more recent books, 2, 4, 5, 6, and 10, can guide you to more advanced work. The use of closed queueing network models to assist in manufacturing system design can be pursued in 11, 12, and 13. Whitt's approach to approximating network solutions can be found in 14.

1. Baskett, F., K. M. Chandy, R. R. Muntz, and F. G. Palacios, "Open, Closed and Mixed Networks of Queues with Different Classes of Customers," *J. ACM* 22 (1975), 248–260.
2. Bloch, G., S. Greiner, H. de Meer, and K. S. Trivedi. *Queueing Networks and Markov Chains*. John Wiley, New York, 1998.
3. Buzen, J. P. "Computational Algorithms for Closed Queueing Networks with Exponential Servers," *Comm. ACM* 16 (1973), 527–531.
4. Chao, X., M. Miyazawa, and M. Pinedo, *Queueing Networks: Customers, Signals, and Product Form Solutions*. John Wiley, New York, 1999.
5. Conway, A., and N. Georganas, *Queueing Networks—Exact Computational Algorithms*. MIT Press, Cambridge, 1989.
6. Gelenbe, E., and G. Pujolle, *Introduction to Queueing Networks*, 2nd ed. John Wiley, New York, 1998.
7. Gordon, W. I., and G. F. Newell, "Closed Queueing Systems with Exponential Servers," *Oper. Res.* 15 (1967), 154–165.
8. Jackson, R., "Jobshop-like Queueing Systems," *Management Science* 10 (1963), 131–142.

9. Lavenburg, S., and M. Reiser "Mean Value Analysis of Closed Multichain Queueing Networks," *Journal of the ACM* vol. 27, no. 2, April 1980.
10. Perros, H., *Queueing Networks with Blocking*. Oxford, New York, 1994.
11. Solberg, J., "Capacity Planning with a Stochastic Workflow Model," *AIIE Transactions, Special Issue on Current Research Directions in Production Planning and Control,* 1980.
12. Solberg, J., "A Mathematical Model of Computerized Manufacturing Systems," Proceedings, 4th International Conference on Production Research, Tokyo, August, 1977. (Also reprinted in *Production and Industrial Systems*, eds. R. Muramatzu and N. A. Dudley, Taylor & Francis, London, 1978, pp. 1265–1275.)
13. R. Suri, G. W. W. Diehl, S. D. Treville, and M. J. Tomsicek, "From CAN-Q to MPX: Evolution of Queueing Software for Manufacturing," *Interfaces* vol. 25, no. 5 (1995) 128–150.
14. W. Whitt, "The Queueing Network Analyzer," *The Bell System Technical Journal* vol. 62, no. 9 (1983) 2779–2814.

Chapter 7 Problems

1. The first few problems compare alternative designs for some very simple networks of queues. Besides illustrating the theory, they expose some principles of network design; you should take note of the comparative results as well as obtaining them. To start, assume that we have a Poisson arrival process with $\lambda = 0.9$ customers per minute. That stream feeds two queues in series, each with a server that has $\mu = 1.0$ customers per minute, as shown in Figure 7.8.

■ **FIGURE 7.8** A simple open network

a. What is the expected total time to pass through this system?

b. Now imagine that the Poisson arrival process has $\lambda = 1.8$, but the arriving customers are randomly assigned (with probability 1/2) to either of two parallel subsystems, each like the one above. The network would be as shown in Figure 7.9. What is the expected total time to pass through this system?

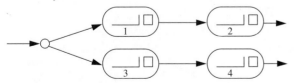

■ **FIGURE 7.9** A larger network

2. Next, suppose that the final queues for each of the parallel subsystems can be merged, as shown in Figure 7.10. Keeping all of the parameter values the same, what is the expected total time to pass through the system now?

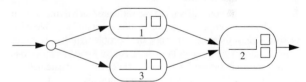

■ **FIGURE 7.10** A modified network

3. Now, suppose that the first two queues can also be merged, giving the system shown in Figure 7.11. Keeping all of the parameter values the same, what is the expected total time to pass through this variant of the network design?

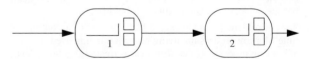

■ **FIGURE 7.11** A further modified network

4. Finally, compare these waiting times to what would result if you could merge all four servers into a single station with a shared queue—a simple M/M/4 system. In order to have comparable parameter values, the mean service time would have to be double the time for the two-stage versions, so use $\mu = 0.5$. (This calculation does not require any of the theory in this chapter, but it completes the comparison in an interesting way. Be sure to take note of the comparative results from the four different ways of serving the same traffic stream with the same resources.)

5. The next few questions tackle a more realistic design situation. In Chapter 2, a case study explored the formulation of a Markov chain model of a call center. In that model, time was represented by transfers; we could not answer any questions about how long calls lasted. In Chapter 4, we were able to use rewards to extend the model to incorporate real time and operator costs; however, we could not yet represent contention for resources and the additional queueing time that would result. Now, we have the tools to build a realistic model of a complete call center, including the number of operators of each type, and we can evaluate the quality of service in terms of the expected delays that customers encounter. In this case, we want to focus on the times spent waiting to speak to an operator, which are W_q's. The stations will be: 1 = Sales, with six operators; 2 = Billing, with four operators; 3 = Support, with four operators; and 4 = a single supervisor. Calls arrive according to a Poisson process at the rate of fifty per hour. The routing is as described by the Markov chain developed in Chapter 2's Section 2.8. The service times are as described in Chapter 4's problem 7, page 138.

    a. Find the effective arrival rates to the four operator stations. (Hint: The external arrivals are dividing into three streams by the first state in the Markov chain model for transfers.)

    b. Find the expected delays (W_q's) at each station. (Hint: The formulas for the L_q's are given in Chapter 6, Table 6.3, page 205. Remember to use the correct definition of ρ for each model.)

6. Extending the problem 5 above, suppose that one of the delay times is simply too long to be acceptable to customers, so another operator is to be hired. First figure out which type of operator is most needed, and then recalculate the delay times after that change is made.

7. Although you will want to use software (such as the code in Appendix II) to compute the normalization constant for problems of realistic size, you should work a few small problems by hand in order to understand what the algorithm does. An easy one to start with is the one shown in Figure 7.2. In order to keep the hand calculations simple, use workloads of two for each station and $N = 4$.

    a. Find the normalization constant G, using the table method described in Section 7.9.

    b. Using the table and the results in Table 7.2, find the utilization of both servers (which should be equal, because they are symmetric).

8. Repeat the same problem using two stations that have two servers each. Again, the utilizations should come out equal, but the calculations and results will be different.

9. Problem 15 in Chapter 5 had you model a continuous time Markov process involving recycling molds used in manufacturing process. At that time, it was treated as a birth-death process. However, it could also be treated as a two-station network of queues. The first station (machines) had two servers; the second station (workers) had three. The number of customers (molds) was eight. Assume that the service time at both stations is six minutes. (The numbers are not realistic, but will make the calculations relatively clean, so that you can do them by hand.)

    a. The goal was to keep the machines working as much as possible, so calculate the utilizations of the machines at station 1.

    b. One way to increase the utilization is to add more molds to the system. What would the result be if you added one more? (Hint: Just add a row to the table you used to calculate G.)

10. Continuing the problem 9 above, imagine that the management is concerned about how the productivity of the shop is affected by the absence of a worker. What would the utilization of the machines become if one of the three workers were absent?

11. To gain a better understanding of how to use a closed network with a dummy station to represent (approximately) an open network, consider a very small and simple example. For the open network, use an ordinary M/M/1 queue with $\lambda = 0.8$ and $\mu = 1$. All of the formulas in Section 6.7 will apply. To create an approximately equivalent closed network, we will need a dummy station that acts like the arrival process. That is, it will have one server that completes service at a rate that is equal to the arrival rate of the open network. We must also have enough customers circulating in the system to ensure that, almost always, the dummy server can continue to work. Whenever it goes idle, the simulated arrival process would be interrupted. We can measure how close we are to that condition (and hence how good the approximation is) by the utilization of the dummy server. So let us start with a value of $N = 10$ for the number of customers in the closed system. The internal relative arrival rates in the closed network will be equal so use 1 for the common value. The service rate (the reciprocal of the mean service time) of the dummy server must match the arrival rate of the open network, so use $\hat{\rho} = 1.25$ or 5/4 for the dummy. (The numbers are clean enough to work with by hand, but you may want to use a spreadsheet if the work is too tedious.)

    a. What is the utilization of the dummy server when $N = 10$?

    b. Compare the utilization of the real server in the original M/M/1 and in the closed network approximation.

    c. Compare the values of L for the real server in the original M/M/1 and in the closed network approximation.

12. Continuing the problem 11 above, repeat the same calculations with $N = 20$ to see how the approximations improve.

CHAPTER 8

Using the Transition Diagram to Compute

CHAPTER CONTENTS

8.1. An Example 252
8.2. Definitions 254
8.3. Steady-State Probabilities 258
8.4. How to Generate All In-Trees 259
8.5. Check Your Understanding 262
8.6. Generalization to Other Quantities 263
8.7. Mean First Passage Times 264
8.8. Results for Terminating Processes 265
8.9. How to Simplify the Arithmetic 266
8.10. How to Systematically Generate r-Forests 267
8.11. Summary of Results 267
8.12. How to Remember the Formulas 268
8.13. Advanced Topics 268
8.14. Recommended Reading 269

This chapter is optional, and although you can get along fine without knowing anything that is covered here, you might find the subject useful and even intriguing. The topic involves the graph theoretic expression of a number of quantities of interest directly from the transition diagram of a discrete or continuous time Markov process. That is, instead of solving sets of linear equations whose coefficients come from the transition matrix, you look at the transition diagram, recognize certain subgraphs, and write out the solution in terms related to those subgraphs. The formulas are presented here for the first time; you will not find them referenced anywhere in the standard literature.

The new methods possess a number of unique advantages.

- Identical formulas apply to both the discrete and continuous time cases, despite the fact that the parameters have different meaning and the equations are different. Thus, the approach achieves a unification of those two cases, allowing them to be treated in exactly the same way.
- The new methods use the transition diagram in its natural form. That is, not even minor modifications need be made to the diagram prior to applying the formulas.
- The subgraphs required are particularly simple, easily generated graphs. Small problems can be handled simply by looking at the transition diagram and writing out the answer. Computer programs can generate the graphs using a simple algorithm for larger problems.
- It is possible to avoid decimal calculations (in solving by hand), except for one division operation at the very end. In most cases, the formulas apply exactly the same if the transition probabilities or rates are scaled to convert them to integers. In a few, a very simple correction (multiplying by the scaling factor) will adjust the integer results.
- It is just as easy to get a symbolic solution as a numerical one.
- All terms appearing in the formulas are positive. In addition to simplifying the computation by eliminating sign difficulties, this property ensures that no cancellation of terms (which would imply redundant computational effort) can occur. In that sense, the method is efficient.
- The formulas are surprisingly "clean." They are easy to remember and apply.
- The formulas cover virtually every case of every quantity you might want to calculate for a discrete or continuous time Markov process.

There are also some limitations or disadvantages to be aware of:

- The process must be finite (that is, have a finite number of states).
- Although, in theory, the formulas are applicable to large problems, in practice one would want to use them only for fairly small ones.
- The computational complexity is no better (or worse) than solving linear equations by determinant and co-factor methods.

For the sake of brevity, primary attention is directed to those Markov chains and processes that are either ergodic or terminating. It is also assumed that the state at time zero is known with certainty. It should be obvious that the methods can be extended to cover other cases. Expressions will be given for steady-state probabilities and mean first passage times for ergodic processes, absorption probabilities, mean number of entries, and conditional mean first passage times for terminating processes. More advanced theory is also given at the end of the chapter just for completeness (in case you take a serious interest), but you can safely ignore that if you choose.

8.1 An Example

EXAMPLE 8.1

Before introducing a lot of new terminology, let us present a very simple case that will give a suggestion of how this method will work. Consider the following two-state Markov chain shown in Figure 8.1.

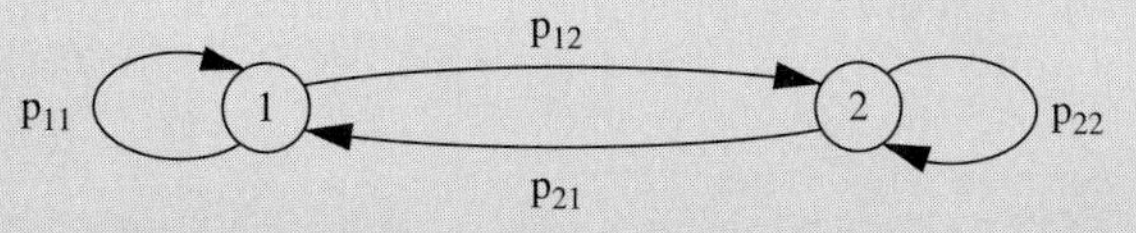

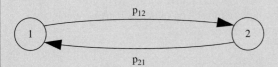

■ **FIGURE 8.1** A two-state Markov chain

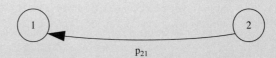

■ **FIGURE 8.3** The in-tree to point 1

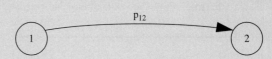

■ **FIGURE 8.2** The transition diagram with loops removed

■ **FIGURE 8.4** The in-tree to point 2

If you were given this transition diagram, you could of course set up the transition matrix, and from there establish the steady-state equations:

$$\pi_1 = p_{11}\pi_1 + p_{21}\pi_2$$
$$\pi_2 = p_{21}\pi_1 + p_{22}\pi_2$$

You would then drop either one of these, and replace it by the normalizing equation

$$1 = \pi_1 + \pi_2$$

You could also make use of the fact that the rows sums of **P** equal 1 to reduce the number of parameters. In particular,

$$p_{11} = 1 - p_{12}$$

After working out the algebra to express the solution in terms of p_{12} and p_{21}, you would get

$$\pi_1 = \frac{p_{21}}{p_{21} + p_{12}} \qquad \pi_2 = \frac{p_{12}}{p_{21} + p_{12}}$$

(You may want to verify that, just to establish the amount of effort involved.)

Now, instead of going through all of that, here is an alternative way that leads to the same results. Look at the transition diagram in Figure 8.2, and ignore the loops (whose weights are implicitly known when we know all of the other weights, because the sum of the weights of all outgoing arcs is known). So, without actually redrawing it, we are imagining the diagram in Figure 8.2.

Then, using the approach that this chapter will explain, we can obtain an *unnormalized* solution (one where the probabilities will be proportional to the correct values but may not add up to 1) by finding the weights of in-trees to each point. In this case, point 1 has the single in-tree shown in Figure 8.3. And point 2 has the in-tree shown in Figure 8.4. Neither of these would have to be drawn. Once you know what you are looking for, you can "see" them as subgraphs within the original transition diagram.

So, according to the theory, an unnormalized solution is given by:

$$\hat{\pi}_1 = p_{21} \qquad \hat{\pi}_2 = p_{12}$$

from the first and second in-tree, respectively. These unnormalized solutions, designated by the caret marks, must still be scaled (multiplied or divided by something) to force them to sum to 1. But that is easily done. If you divide an unnormalized value by the sum of all of them, you will achieve what you want. (Verify that to see what is going on.)

Our final solution, then, is

$$\pi_1 = \frac{p_{21}}{p_{21} + p_{12}} \qquad \pi_2 = \frac{p_{12}}{p_{21} + p_{12}}$$

which is just what we would have obtained by the algebraic method.

Now let's switch to a continuous time version of the same process. The transition diagram would have the same structure, but with rates instead of one-step transition probabilities, as shown in Figure 8.5. This time the equations are of somewhat different form:

$$0 = \lambda_{11}\pi_1 + \lambda_{21}\pi_2$$
$$0 = \lambda_{21}\pi_1 + \lambda_{22}\pi_2$$
$$1 = \pi_1 + \pi_2$$

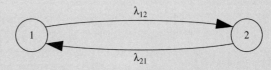

■ **FIGURE 8.5** A two-state continuous time Markov process

and, of course, the rows of Λ sum to 0 rather than 1. Nevertheless, once you work out the algebra, you get the solution

$$\pi_1 = \frac{\lambda_{21}}{\lambda_{21} + \lambda_{12}} \qquad \pi_2 = \frac{\lambda_{12}}{\lambda_{21} + \lambda_{12}}$$

which is just what you would get from applying the graph theoretic method. Hence, we have the remarkable conclusion that the graph theoretic method will produce the correct dsolution when it is applied in exactly the same way to either the discrete or continuous time version, despite the fact that the meanings of the coefficients and the forms of the equations are different.

This small example gives a suggestion of the possibilities to come. However, it is not at all obvious how to generalize from this two-state case to larger problems or how to compute other quantities of interest. For the general theory, we will have to back up to define some terminology.

8.2 Definitions

It will be easiest to work into the special definitions we need if we start with some terms from ordinary graph theory. Some of this is pertinent only indirectly to our needs, but may be helpful in clarifying the exact meaning of terms. Although there are quite a few terms to define, they are all quite easy to understand and remember.

> **Definition** A *graph* is a nonempty set of "points" and a set (possibly empty) of "lines" that are unordered pairs of points. If there is no distinguishable difference among the points, the graph is said to be *unlabeled.* We will always want to distinguish the points, so the graph will be *labeled,* even if we fail to mention it. The values associated with lines, if any, are called *weights.*

Although it is conventional to draw a graph in the obvious way (circles for points and line segments for the lines), the set theoretic formal definition makes it clear that we are talking about abstract relationships among objects, not any particular arrangement of points or lines on paper. The same graph could be drawn differently—that is, with different spacing or arrangement of the points—but if the same points are connected, it would be the same graph.

Figure 8.6 illustrates a labeled graph having four points and five lines, and Figure 8.7 shows another way to draw exactly the same graph. Notice that the lines are not necessarily straight line segments; they are just "connectors." Also, it may not be immediately obvious that two drawings represent the same graph.

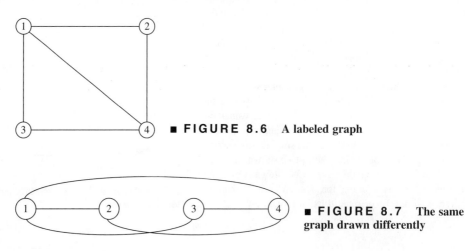

■ **FIGURE 8.6** A labeled graph

■ **FIGURE 8.7** The same graph drawn differently

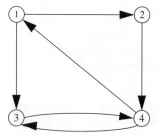

■ **FIGURE 8.8** A labeled directed graph

> **Definition** A *directed graph*, sometimes called a digraph, is a nonempty set of points and a set (possibly empty) of "arcs" that are *ordered* pairs of points.

The only difference between an ordinary graph and a directed graph is the ordering of the points in an arc. An arc is drawn as a line with an arrowhead. Since the order of two points i and j one way (i, j) is different from the reverse order (j, i), it is possible to have two arcs connecting the same two points (as long as they are in opposite directions). Figure 8.8 shows a directed graph with four points and six arcs.

A line or an arc from a point to itself is called a *loop*. The graphs shown above do not have any loops, and we will not be making any use of them (even though they might appear in transition diagrams).

In graph theoretic terminology, the transition diagrams that we use to describe either discrete or continuous time Markov processes are labeled, weighted directed graphs. There is an intimate relationship between labeled, weighted directed graphs and square matrices. Any n × n matrix can be represented by an n-point weighted, labeled directed graph, or vice versa. The point labels are taken from the rows and columns and the weights of the arcs correspond to the elements of the matrix. That is, we construct an arc from point i to point j if there is a nonzero element in the *i*th row and *j*th column, and give it a weight equal to whatever value is in that position. Conversely, if we are starting from the digraph, we construct a matrix having as many rows and columns as there are points, and identify the element in the *i*th row and *j*th column as the weight of the arc from i to j. If there is no such arc, give that matrix element the value 0.

> **Definition** The *adjacency matrix* of a digraph is a square matrix whose rows and columns correspond to the points and whose nonzero elements correspond to the arc weights.

The transition diagram of a discrete time Markov process is that network whose adjacency matrix is **P**, the one-step transition matrix. In the continuous time case, it is that network whose adjacency matrix is Λ, the matrix of instantaneous transition rates. In both cases, the points correspond to states and the arcs correspond to transitions. The weights associated with the arcs are transition probabilities in the discrete case and instantaneous transition rates in the continuous case. In practice, of course, the transition diagrams are constructed from first principles. Some people prefer to omit the loops, corresponding to the diagonal elements of Λ, in the continuous case. Because loops do not appear in any of the subgraphs used in the theorems, their inclusion or exclusion may be left as a matter of personal preference.

We will be using portions of the transition diagram to compute things, so we need a few precise terms for those portions.

> **Definition** A *subgraph* of a given graph is a graph that contains some of the points and lines of the original. A *spanning subgraph* is a subgraph that contains *all* of the points of the original. A spanning subgraph could contain any subset of the lines or arcs, including even the empty set, as long as all of the points are included.

It will be useful to have a term for the number of lines or arcs in a graph or subgraph. Although there is no standard term in conventional graph theory for this number, we will provide one for this chapter.

> **Definition** The *size* of a graph or subgraph is the number of lines or arcs contained in it.

After we have found subgraphs, we will need to associate values with them. In every case, we will simply multiply the weights of the arcs together.

> **Definition** The *weight* of a subgraph is the product of the weights of the lines or arcs contained in the subgraph. If a subgraph contains no lines or arcs, its weight is defined to be 1. We use w(G) to indicate the weight of subgraph G.

We discussed walk probabilities in Chapter 3. We will use the same terminology here.

> **Definition** A *walk* in a graph or directed graph is a subgraph containing a sequence of lines or arcs, respectively, that occur end to end. That is, each line or arc in the sequence must share an endpoint with the previous one in the sequence. In the case of directed graphs, the walk must occur in the direction of the arrow. A walk that ends at the point where it began is a *closed* walk.

It is natural to think of tracing a walk over a graph or directed graph by starting at a point and moving from point to point along the lines or arcs without lifting the pencil from the paper (and always moving in the directions of the arrows in the directed case). In a transition diagram, a walk would represent a possible realization of the process because the sequence of arcs would correspond to a possible sequence of transitions from state to state. A walk is not necessarily a subgraph of a transition diagram (because a walk could repeat arcs, but a subgraph could not); however, it might be (if no arcs are repeated).

> **Definition** A graph, digraph, or subgraph of either is *connected* if for every two points there exists a walk between them. A digraph is *strongly connected* if for every two points there are walks in both directions between them.

These are the graph theoretic terms for reachability and communication. If the transition diagram is strongly connected, it is irreducible because all states communicate.

> **Definition** A *path* is a walk in which no point is visited more than once, except possibly the first point in the case of a closed path that returns to where it started. A *cycle* in a graph or a directed graph is a closed path. A graph that contains no cycles is said to be *acyclic*.

Figure 8.9 illustrates cycles in graphs and directed graphs. The first two are subgraphs of the ordinary graph shown in Figure 8.6 and the last two are subgraphs of the directed graph in

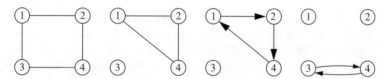

■ **FIGURE 8.9** Examples of cycles

Figure 8.8. We will not have to generate cycles; we simply have to recognize cycles when we see them so that we can eliminate subgraphs that contain them. None of the subgraphs we are looking for are allowed to contain cycles.

The following definition is one of several possible definitions of an important class of graphs or subgraphs.

> **Definition** A *tree* is a graph or spanning subgraph that is connected and acyclic.

Figure 8.10 shows some trees that are subgraphs of the undirected graph in Figure 8.6. Trees are a common category of graph. They are useful for many things, such as hierarchies, decision trees, searching, and other applications. There are many theorems about trees in graph theory, some of which will be mentioned here without proof. It will help your understanding to verify each of these statements using the examples given.

1. A tree is a minimal connected subgraph. ("Minimal" means that it has just enough lines to be connected; the removal of any line will disconnect the graph.)
2. A tree is a maximal acyclic subgraph. ("Maximal" means that the graph has the most number of lines it can have without creating a cycle; the addition of even one more line will complete a cycle.)
3. The size of a tree with n points will always be $n - 1$; that is, it will have $n - 1$ lines.
4. An acyclic graph or subgraph with n points and $n - 1$ lines is necessarily a tree.

We do not have any specific use in this chapter for undirected trees, but it is helpful to form a visual image of what a tree looks like, so that you can recognize one immediately when you see it. For our purposes in computing results for Markov processes, we need a directed version, which will have a similar appearance and some additional requirements on the directions of the arcs. From this point on, we will be speaking only of directed graphs.

> **Definition** An *in-tree to a point j* in a directed graph is an acyclic connected digraph or subgraph in which all walks end at j. That is, for every point in the in-tree, there is a walk that ends at the point j. The point where all the walks end is called the *root* of the in-tree. We will use T_j as a symbol to represent an in-tree with root j.

(By the way, there is another kind of directed tree called an out-tree, in which all of the walks originate at single point, but we will have no application for them here.)

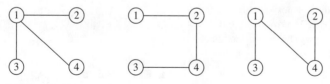

■ **FIGURE 8.10** Examples of trees

If you were to ignore the arrowheads, an in-tree would look like a tree in an undirected graph. That is, it would be connected, would have no cycles, and would have exactly n − 1 lines. One of the characteristics of in-trees that will help in their identification is that every point except the root will have exactly one outgoing arc. Later, we will discuss a systematic way to generate all of the in-trees.

8.3 Steady-State Probabilities

We now have enough terminology to state the theorem for computing steady-state probabilities. There will have to be some more terms to state the results for other quantities, but we will come to that soon. For now, make sure you understand how the in-trees give the steady-state results for both discrete and continuous processes. The following formula applies to both cases.

> **Result** In an ergodic Markov process, the steady-state probabilities are given by
>
> $$\pi_j = \frac{\sum w(T_j)}{\sum w(T)}$$
>
> The numerator sum is taken over all in-trees having root j; the denominator is taken over all in-trees in the diagram, regardless of root.

A good way to think of these graphs and subgraphs is to consider the original transition diagram to be a plumbing system with one-way pipes. The nodes are reservoirs holding a liquid (the one total unit of probability) that "flows" through the pipes (the transition arcs) in accordance with the transition probabilities or rates. An in-tree to point j would be a minimal drainage system, using pipes from the original system. That is, every reservoir would have to have one of its outgoing pipes used, and they would all have to flow either directly or indirectly to reach point j. Think of the weight of the in-tree as measuring the capacity of that one minimal piping system (the analogy is imperfect, because multiplying the weights does not make any physical sense). Add up the weights over all available in-trees to the point j to get the total flow to j. The total flow is the numerator term. The normalization amounts to finding the percentage of the total flow (the total to all points) that goes to point j. The exclusion of cycles, or equivalently, the requirement that the subgraph be connected, just ensures that every reservoir can send its fluid to the root at j.

EXAMPLE 8.2

You can try this out on the two-state example in Figure 8.1. There, the in-trees consisted of single arcs because there were only two points. The example in Figure 8.11 is only slightly larger, but gives a better indication of the general approach. The figure shows a small transition diagram of an ergodic Markov chain having three states: This has every possible transition, so any three-state Markov chain could be handled with the same in-trees that will be shown, though of course the weights would change if the transition probabilities were different.

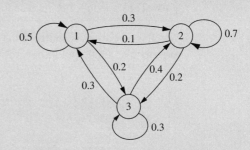

■ **FIGURE 8.11** A three-state Markov chain

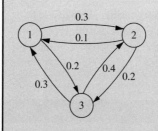

FIGURE 8.12 The diagram with loops removed

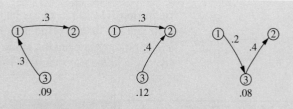

FIGURE 8.14 All in-trees to point 2

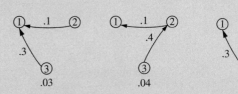

FIGURE 8.13 All in-trees to point 1

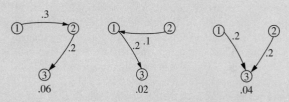

FIGURE 8.15 All in-trees to point 3

First, just to avoid confusion, we will suppress the loops and their values. (In actual practice, we would simply ignore them.) The diagram without the loops would look like Figure 8.12. Of course, this is no longer a proper transition diagram, but it is all that we need. (The loop weights could always be recovered, if needed, from the knowledge that each bundle's weights sum to 1.) All of the in-trees to point 1 (with weights written below) are shown in Figure 8.13. All of the in-trees to point 2 are shown in Figure 8.14 and all of the in-trees to point 3 are shown in Figure 8.15. Consequently, the steady-state probabilities are:

$$\pi_1 = \frac{0.13}{0.54} \qquad \pi_2 = \frac{0.29}{0.54} \qquad \pi_3 = \frac{0.12}{0.54}$$

Each of the numerators is obtained by summing the weights of the appropriate in-trees, and then the denominator for each (which is the same for all) is obtained by summing all of the numerators. From there, you can carry out the division to obtain the decimal equivalents of the fractions.

You can learn to do all of this in your head without actually drawing any of the subgraphs, as long as the transition diagram is small. You "see" the first in-tree to the first point, multiply the arc values and write down the product, then "see" the second in-tree, get the product and write it down, and so forth. When you have all of the in-trees to the first point, you move to the second point and repeat. When you have all of the numerator terms, you normalize by dividing each by the sum of all the numerators.

8.4 How to Generate All In-Trees

For very small transition diagrams, you can "see" whether you have everything generated, but as the diagram grows, it becomes difficult to be sure that you haven't missed something. Also, if you wanted to implement the algorithm on a computer, you could not rely upon visual methods. A systematic way to generate all of the forests of a specified order one-by-one in a prescribed order follows.

Definition The *bundle* of a point is the *set* of outward arcs from that point. We will use B_i to represent the bundle of point i. The *out-degree* of a point in a graph or subgraph is the *number* of arcs leaving the point. We will use δ_i to represent the out-degree of point i. Of course, δ_i is the cardinality of (number of elements in) the set B_i.

EXAMPLE 8.3

Figure 8.16 shows a directed graph of the sort that might portray the transitions in an ergodic process for either a discrete or continuous time case. (We are using unique symbols for the arc weights to help in identifying each individual arc.) This is also a slightly larger example with some asymmetries to show variation.

The out-degrees of the four points are 2, 2, 2, and 1, respectively. The bundle of point 1 is the set of two arcs shown in Figure 8.17. It will help later if you get used to the idea of perceiving the transition diagram as consisting of the collection of bundles. That is, when you look at a particular point, you want to "see" the outgoing arcs and ignore the arcs entering the point. The bundle is just a set, so it could be specified in any way that enables you to distinguish the members of the set. Here we have distinct weights for each arc, so we could use {a, b} to indicate the bundle, but that method will not work when the weights are numerical or use duplicate symbols. Perhaps the most obvious way is to name the arcs by their endpoints in order. So, for example, we can refer the bundle in Figure 8.17 as {(1,2), (1,3)}. A slightly more compact way is to name the point whose bundle you are considering and then (after a colon or some other separator) list the names of the points at the other ends of the arcs. For example, the bundle of point 1 shown above could be specified as 1: 2, 3. Yet another way, which is less compact but perhaps easier to relate to the digraph, is to use a row of the adjacency matrix. In that method, we have to ignore any 0 entries. For example, the adjacency matrix of the above

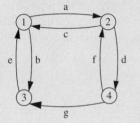

■ **FIGURE 8.16**
A directed graph

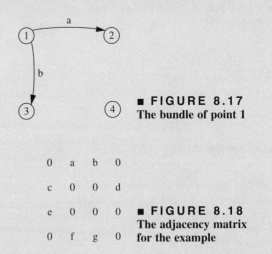

■ **FIGURE 8.17**
The bundle of point 1

0	a	b	0
c	0	0	d
e	0	0	0
0	f	g	0

■ **FIGURE 8.18**
The adjacency matrix for the example

digraph is shown in Figure 8.18 so we could indicate the bundle of the first point as (0, a, b, 0).

We can tell where the arcs are directed by their positions or column indices. The 0 entries are not arc weights, but simply placeholders to fill in empty cells. Note also that the out-degree of each point is equal to the number of nonzero entries in the corresponding row of the adjacency matrix.

For the algorithms that generate subgraphs, it is helpful (though not essential) to maintain a consistent order for the arcs in each bundle, whatever the representation may be. The most obvious order is to go from lowest to highest numbered destination points. For example, in the bundle for point 1, we would always list arc a ahead of arc b because the former goes to point 2 and the latter goes to point 3.

We will use the matrix row designation in our explanation of the algorithms, but a computer implementation would preferably use a more compact data structure. For typical transition diagrams, most of the entries in the adjacency matrix will be zeros, and there is no reason to consume storage for all those unused terms.

Definition A *1-graph* is a spanning subgraph of a directed graph in which every point but one (the root) has out-degree one. That is, there is exactly one outgoing arc from each point except the root, which has no outgoing arcs.

EXAMPLE 8.4

A 1-graph is a slight generalization of an in-tree, in that it could contain cycles. If we generate a 1-graph and subsequently verify that it contains no cycle, it will necessarily be an in-tree. We can use that logic to find all the in-trees. We generate all 1-graphs individually, discard any that contain cycles, and are left with all the in-trees. To generate the 1-graphs, fix a root and select one arc from each bundle of the other points. The following provides a more explicit description of the algorithm.

1. Mark out (or just ignore) the bundle of the point that is selected to be the root. (You can start with point 1 and step through each point in order, or any other order would do.) The matrix representation would look like Figure 8.19 when the first point is the selected root (because when it is a root, none of the arcs in its bundle can appear in its in-trees).

2. Put a "marker" on the first item in each remaining bundle (shown below as circles over the weights in Figure 8.20). This marks one arc in each bundle of each source and therefore identifies a 1-graph, which might or might not be an in-tree. We have to check to see if this 1-graph contains a cycle. If it does, reject it and move on. If it does not contain any cycles, it is an in-tree to the specified root, so calculate the weight and add it to the unnormalized (numerator) term.

3. To get the next in-tree, advance the last marker to the next arc in the bundle, if there is one, and repeat the cycle check as described in step 2. If there is no next arc (because it is the last in the bundle), go to step 4. In Figure 8.21, the marker for point 4 has been advanced to the next arc, generating the second 1-graph, which is an in-tree.

4. If you have already reached the end of the bundle, reset the marker for that bundle to the beginning of the list, back up to the previous bundle and advance the marker there. Reapply rule 4 whenever you reach the end of a bundle. In Figure 8.22, we reached the end of bundle 4, so we set its marker back to f, moved up to point 3 and tried to advance its marker. But it was already at the end of its bundle, so we backed up to point 2 and advanced that marker. The 1-graph this time contains a cycle, so it is not an in-tree.

5. Continue until all combinations of arcs from each of the bundles of the nonroot points have been exhausted. In the case of point 1, there is only one more combination, namely arcs d, e, g.

6. Repeat the entire five steps above for a different point as root. When each of the points has been a root, you have generated all possible in-trees in the digraph.

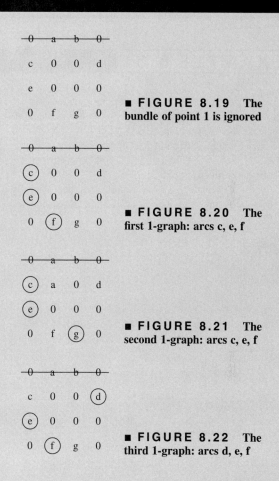

■ **FIGURE 8.19** The bundle of point 1 is ignored

■ **FIGURE 8.20** The first 1-graph: arcs c, e, f

■ **FIGURE 8.21** The second 1-graph: arcs c, e, f

■ **FIGURE 8.22** The third 1-graph: arcs d, e, f

This algorithm effectively "counts" through all of the combinations of one arc from each bundle in order. It is actually quite similar to ordinary counting where the digits advance from 0 to 9 and then reset, carrying over an increment to the next place. It is easier to do than it is to explain. Once you have the idea, you can confidently generate all the in-trees systematically and in a reproducible order.

There are some ways to be more efficient in the algorithm. If you discover, for example, that the first three points you consider have already formed a cycle, you do not have to generate all of the combinations that contain that set of arcs. You can just eliminate all of the 1-graphs that start out that way. A computer implementation that was to be used on large problems would certainly need to make use of any such shortcuts, because the number of 1-graphs grows rapidly with the size of the problem.

8.5 Check Your Understanding

EXAMPLE 8.5

The complete set of 1-graphs with point 1 as the root are shown in Figure 8.23. Of these, only the third contains a cycle, so the others are in-trees to point 1.

The 1-graphs with point 2 as root are shown in Figure 8.24. Of these, the third and fourth contain cycles, so the others are in-trees to point 2.

There are eight 1-graphs with point 3 as root. They are shown in Figure 8.25. Of these, the first three and the seventh have cycles, so there are only four in-trees remaining.

Finally, the 1-graphs with point 4 as root are shown in Figure 8.26. Of these, only the second is an in-tree; all the others have cycles.

Make sure that you can generate these on your own, and verify that you have found all of them.

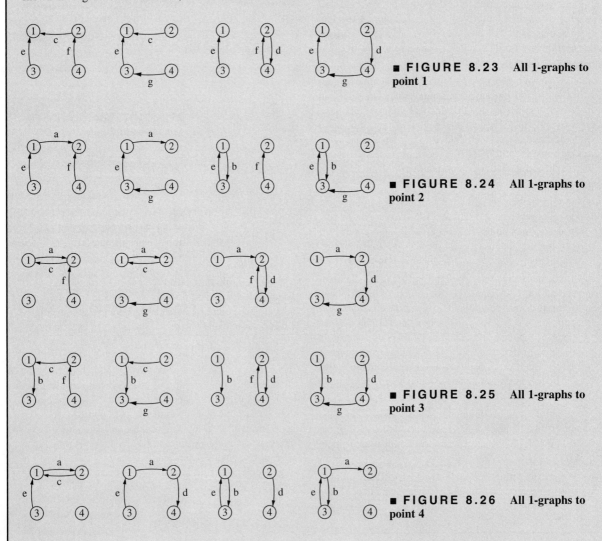

■ FIGURE 8.23 All 1-graphs to point 1

■ FIGURE 8.24 All 1-graphs to point 2

■ FIGURE 8.25 All 1-graphs to point 3

■ FIGURE 8.26 All 1-graphs to point 4

8.6 Generalization to Other Quantities

To generalize beyond steady-state probabilities, we have to define more terms. Once we have done that, all of the remaining results will come quickly.

> **Definition** An *r-forest* is an acyclic spanning subgraph in which r points have out-degree 0 and the remaining points have out-degree 1. The points of an r-forest that have out-degree 0 are called *roots*.

An r-forest with arbitrary roots is denoted F^r, and its weight will be given by $w(F^r)$. If we want to say that a particular point j is a root, we show the r-forest as F_j^r. When an r-forest must have a specific state i (which for our purposes will always be the initial state) connected to or in the same component as a particular root j we write F_{ij}^r. You can think of this notation as requiring the initial state to drain at the particular state j. Also, if we want the opposite effect, that is, to require that a specific state i *not* be connected to or in the same component as a particular root j, we write $F_{i/j}^r$. The slash between the subscripts can be thought of as separating i from j.

An r-graph with n points must be of size n − r; that is, it must have exactly n − r arcs. The proof follows directly from the definition. Visually, an r-forest (logically) looks like a collection of in-trees. That is, it will consist of a number of (in fact, exactly r) disconnected components, each of which satisfies the definition of an in-tree to its respective root, except for the fact that it does not span the entire original graph but only a subset of the points. A 1-forest automatically will be an in-tree.

An r-forest will never contain loops. A loop is a special case of a cycle, so it cannot appear in an r-forest. We use only r-forests in our formulas, which explains why the inclusion or exclusion of loops in the original transition diagram does not matter. Since any r-forest always has the same number of arcs, namely n − r, the product in each of the r-forests will always have the same number of terms. Since the individual values are always positive, the weight of any r-forest is always positive.

For purposes of generating r-forests, it also helps to define a generalization of 1-graphs.

> **Definition** An *r-graph* is a spanning subgraph in which r points have out-degree 0 and the remaining points have out-degree 1. The points of an r-graph that have out-degree 0 are called *roots*.

Comparing the two definitions, an r-forest is just an r-graph that has no cycles. You can generate an r-graph simply by selecting one arc from the bundle of each point that you want to be a source. The roots, of course, will have no arcs selected from their bundles. Pay no attention to how many arcs are entering the points; it does not matter. Once you have an r-graph, you can determine whether it is an r-forest by checking for cycles. Usually, it is quite easy to spot cycles visually. Consequently, it is easy to identify r-graphs that are not r-forests and eliminate them, leaving just the r-forests. To generate all r-forests, generate the r-graphs and discard the ones that have cycles.

We need to find *all* of the r-graphs having particular designated roots. How can we be sure that we have found them all? If we know which points are roots, the number of r-graphs to look for is shown in the following result.

> **Result** The number of r-graphs with specific roots equals
> $$\prod_{\text{all nonroots } i} \delta_i$$

Each combination of one arc from each nonroot bundle is a distinct r-graph, so the number of r-graphs is the number of such combinations. So, for example, if we wanted to be sure we had found all of the 1-graphs having point 1 as root (like the above two examples), we would know that there are $2 \times 1 \times 2 = 4$ (the product of the out-degrees of points 2, 3, and 4) subgraphs to find.

8.7 Mean First Passage Times

We now have all we need to express the graph theoretic version of the mean first passage times.

> **Result** In an ergodic Markov process, the mean first passage times are given by (for $i \neq j$)
> $$m_{ij} = \frac{\sum w(F^2_{i/j})}{\sum w(T_j)}$$

The denominator term matches the numerator of the steady-state probability, so you most likely already have that by the time you are using this formula. The numerator is a bit odd: It sums the weights of 2-forests in which j is a root, but the initial state i is *not* in the same component as j. (i might or might not be the other root of the 2-forest.)

EXAMPLE 8.6

For the mean first passage times, we need 2-forests. In a graph with only three points (such as the one we used in Figure 8.11), a 2-forest will have just one arc, because the two roots have out-degree 0, leaving only one point with out-degree 1. Hence, the entire set of 2-forests for that example is as shown in Figure 8.27.

From these we can pick out the ones satisfying the requirements for the various m_{ij}. For example, m_{12} requires all 2-forests in which 2 is a root and 1 is not in the same component. Of the six 2-forests shown in Figure 8.27, the ones that qualify are the first two in the top row and last one in the bottom row. Consequently,

$$m_{12} = \frac{0.9}{0.29} = 3.10$$

Similarly (in case you want to check your own calculations),

$$m_{13} = \frac{0.6}{0.12} = 5.00 \qquad m_{21} = \frac{0.9}{0.13} = 6.92$$

$$m_{23} = \frac{0.6}{0.12} = 5.00 \qquad m_{31} = \frac{0.7}{0.13} = 5.38$$

$$m_{32} = \frac{0.8}{0.29} = 2.76$$

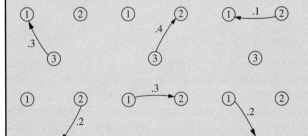

■ **FIGURE 8.27** All 2-forests for the three-state example

8.8 Results for Terminating Processes

We will present the remaining formulas without examples. All of these refer to terminating processes in which i is the initial state.

> **Result** In a terminating process with r absorbing states, the absorption probability (the probability that absorbing state j is ever reached) is given by
> $$a_{ij} = \frac{\sum w(F_{ij}^r)}{\sum w(F^r)}$$

The r absorbing states must be roots (since they only have loops, which cannot be used in an r-forest). Hence each r-forest has r components, one of which must contain the initial state i. Of the total flow through all of the them (represented by the denominator term), only a fraction of the r-forests will send the flow from the initial state i to the particular absorbing state j. The formula computes the fraction of the total flow that runs from i to j.

> **Result** In a terminating process with r absorbing states, the expected total time (or number of times—a count—when time is measured discretely) that transient state j is occupied is given by:
> $$e_{ij} = \frac{\sum w(F_{ij}^{r+1})}{\sum w(F^r)}$$

For the special case of i = j, the same formula applies, but the point i will always be contained in the same component as itself. Notice that the denominator in this formula is the same as for the absorption probabilities; you would not have to regenerate the r-forests if you have already done it for another formula. However, the numerator requires $(r+1)$-forests, which have not appeared before. Of the $(r+1)$ roots, r will be the absorbing points. The remaining one must be the transient state j. Of course, we want to sum the weights for only those $(r+1)$-forests in which i is connected to j.

> **Result** In a terminating process with r absorbing states, the expected duration of a process (the total time spent in transient states, measured as a count in discrete time) is given by:
> $$d_j = \frac{\sum_{\text{trans } j} \sum w(F_{ij}^{r+1})}{\sum w(F^r)}$$

Hit probabilities can be expressed for transient states j, except for the case where j is also the initial state.

> **Result** In a terminating process with r absorbing states, the probability that transient state j is ever reached, f_{ij}, is given by (for $i \neq j$)
> $$f_{ij} = \frac{\sum w(F_{ij}^{r+1})}{\sum w(F_j^{r+1})}$$

The denominator in this formula contains all of the $(r+1)$-forests in which the destination state j is a root; the numerator contains the subset of those same $(r+1)$-forests in which i connects to j.

> **Result** In a terminating process with r absorbing states, the conditional mean first passage time to absorbing state j given by
>
> $$m_{ij}^{(c)} = \frac{\sum w(F^{r+1})}{\sum w(F^r)} - \frac{\sum w(F_{ij}^{r+1})}{\sum w(F_{ij}^r)}$$

In the two cases that are probabilities, the numerator terms are restricted versions of the denominator terms; or, if you prefer, the denominator terms are taken over a larger set that includes the numerator terms. A practical way to organize the computations is to find all of the subgraphs on the kind needed—the r-forests or whatever—to determine the denominator, then pick out the ones that meet the requirements to form the numerator. In the other cases (such as m_{ij} and e_{ij}), the numerator and denominator terms involve forests of different order. In those cases, you have to generate all of the forests of each order, r and r + 1. If you need to find several different measures, such as e_{ij} and a_{ij}, you can generate the required subgraphs once and put their weights together in different ways to get what you need.

To do a complete analysis of a terminating process, the order in which you should carry out the computations is shown below:

1. Find all of the r-forests. The roots will necessarily be the absorbing states. The sum of the weights of all r-forests appear in several denominator terms, but only needs to be calculated once.
2. Out of the collection of r-forests, add the weights of the ones where i is in the same component as j to get the numerators of the a_{ij}.
3. Find all of the (r + 1)-forests. All but one of the roots must be absorbing states; the other will move through the transient states one at a time.
4. Depending upon connections (i appearing in the component that has root j), the weights add up in different combinations to give the numerator terms required for e_{ij}, f_{ij}, and $m_{ij}^{(c)}$.

8.9 How to Simplify the Arithmetic

If you are solving a very small problem by hand, there is a nice trick to avoid decimal calculations. The first step is to scale all of the transition probabilities or rates to integer values by multiplying all of them by some common power of 10. For example, if you have a value like 0.125 (three decimal digits), multiply everything by 1,000. You will then have integer values for the weights of the arcs. When you want the weight of an in-tree or r-forest, just multiply the integer values. Those weights will not be the same as the original, obviously, but because every in-tree or r-forest of a specified size has the same number of arcs, they will all be scaled by the same multiple. When you add weights of subgraphs of the same size, you will add terms that have been scaled in the same way. We could "undo" the scaling at that point, but it turns out that there is no need to do so.

In the formulas for steady-state probabilities, absorption probabilities, and hit probabilities, the numerator and denominator terms contain subgraphs of the same size. That is, whatever scaling was applied to the numerator is exactly the same as that applied to the denominator. Therefore, the common scale factor cancels out, and you will get the exact result without having to do anything more than divide the integer-valued weights. That means that you can multiply and add terms without any need to contend with decimal numbers until the final division step.

In the cases of mean first passage times, mean times spent in transient states, mean durations, and conditional mean first passage times, the numerator terms and denominator terms differ in size. In particular, each denominator term will contain one more arc than each numerator term in each of these formulas. Therefore, the final result will have to multiplied by the common scaling factor to restore the balance between numerator and denominator. As long as you remember to make that correction, you may again work entirely in integers up until the final division. You can practice this with the problems at the end of the chapter.

8.10 How to Systematically Generate r-Forests

The algorithm for generating all r-forests is essentially the same as that for finding in-trees, except that we cross out more rows to account for multiple roots.

1. Fix the roots by marking out the bundles of the points that you want to be roots. Absorbing states *must* be roots, others *may* be.
2. Put a "marker" on the first item in each remaining bundle. This marks one arc in each bundle of each source and therefore identifies an r-graph, which may or may not be an r-forest. We have to check to see if this r-graph contains a cycle. If it does, reject it and move on. If it does not contain any cycles, it is an r-forest.
3. To get the next r-graph, advance the last marker to the next arc in the bundle and check for cycles. Continue step 3 until you reach the last arc of the bundle.
4. If you have already reached the end of the bundle, reset the marker to the beginning of the list, back up to the previous bundle, and advance the marker there. Reapply rule 4 whenever you reach the end of a bundle.
5. Advance to the next combination of roots.

8.11 Summary of Results

Name	Notation	Type of model	Formula
Steady-state probability	π_j	ergodic	$\pi_j = \dfrac{\sum w(T_j)}{\sum w(T)}$
Mean first passage time	m_{ij}	ergodic	$m_{ij} = \dfrac{\sum w(F^2_{i/j})}{\sum w(T_j)}$
Mean time accumulated in a transient state j	e_{ij}	terminating	$e_{ij} = \dfrac{\sum w(F^{r+1}_{ij})}{\sum w(F^r)}$
Expected duration	d_j	terminating	$d_j = \dfrac{\sum\limits_{\text{trans } j} \sum w(F^{r+1}_{ij})}{\sum w(F^r)}$
Absorption probability	a_{ij}	terminating	$a_{ij} = \dfrac{\sum w(F^r_{ij})}{\sum w(F^r)}$
Probability a transient state is ever visited	f_{ij}	terminating	$f_{ij} = \dfrac{\sum w(F^{r+1}_{ij})}{\sum w(F^{r+1}_j)}$
Conditional mean first passage time	$m^{(c)}_{ij}$	terminating	$m^{(c)}_{ij} = \dfrac{\sum w(F^{r+1})}{\sum w(F^r)} - \dfrac{\sum w(F^{r+1}_{ij})}{\sum w(F^r_{ij})}$

8.12 How to Remember the Formulas

The formulas look a lot alike. It may seem that it will be hard to keep track of the slight distinctions from one quantity to another. But after you have worked with them a while, they begin to make sense individually. There are consistent patterns that, once assimilated, tell you what to do. Here are some of these patterns:

1. When you are finding a *probability*—π_j, a_{ij}, or f_{ij}—the denominators are just normalizing constants. In other words, the numerators and denominators involve the same family of subgraphs, with the numerators containing ones that satisfy certain conditions and the denominators containing all of them. It makes sense that the numerators and denominators would have the same kinds of forests; probabilities are dimensionless units, so whatever is in the numerator should cancel whatever is in the denominator.
2. When you are finding mean *times*—m_{ij}, e_{ij}, d_i, or $m_{ij}^{(c)}$—the numerators involve one more root than the denominators. This also makes sense from the perspective of dimensional units. The numerator will have one less transition rate in it than the denominator. When you work out what that implies, you find that the result has units of time.
3. When the process is *terminating* with r absorbing states, those r states must be roots. Hence the lowest possible ordered forests will be r-forests. You will use those for the probabilities and $(r+1)$-forests for the numerators of the mean times.
4. When the process is *ergodic* there are no states that must be roots. Hence, the lowest-ordered forests will be trees. You will use just trees for steady-state probabilities and 2-forests for the mean first passage time numerators.
5. Mean first passage times and conditional mean first passage times are more similar than they first appear. If you look closely, you will see that the formula for the mean first passage time in an ergodic process is a special case (namely, $r = 1$) of the formula for the conditional mean first passage time in a terminating process.
6. The slight difference between a_{ij} and f_{ij} is logical. Both are probabilities, so the order of the forests in numerator and denominator should match. Both occur in terminating processes, so the r absorbing states must be roots. For a_{ij}, the state j is one of those roots, so r-forests will be enough. For f_{ij}, the state j is a transient state and you will want it to be a root, so you will need to generate $(r+1)$-forests.

8.13 Advanced Topics

There are more advanced graph theoretic formulas that express the Laplace transform or zeta transform of transition probability functions. In fact, virtually everything that one might want to calculate in a Markov process can be put into graph theoretic form using the r-forests. The fundamental reason that it works is that the equations are always linear and the two matrices **P** and **Λ** have the required properties. The proofs of the theorems are not at all obvious (they involve relating the r-forests to cofactors). However, the results are remarkably clean and useful, and do not require understanding the proofs to be able to use them. Once you learn to visualize the subgraphs, you can directly relate solutions to the transition diagram without passing through any intermediate mathematics.

8.14 Recommended Reading

Although you will not find the results given in this chapter in any other book, you may be interested in other graph theoretic approaches. R. Howard's method uses signal flow graphs. The original theorem linking the steady-state probabilities to in-trees is in reference 3.

1. Harary, F., *Graph Theory*. Addison-Wesley, Reading, MA, 1969.
2. Howard, R., *Dynamic Probabilistic Systems*, vols. I and II. Wiley, NY, 1971.
3. Solberg, J., "A Graph Theoretic Formula for the Steady-State Distribution of Finite Markov Processes," 21 *Man. Sci.* 1,040–1,048 (1975).

Chapter 8 Problems

If you want to practice, you can use the methods of this chapter on many of the problems from Chapters 2, 3, 5 and 6. The answers, obviously, should come out the same. In at least some of the cases, the graph theoretic approach will be dramatically easier than algebraic methods. Here are a few easy ones that will help you practice finding the right subgraphs.

1. The fishing problem formulated in Chapter 2, problem 1 (page 68), was an ergodic Markov chain with three states, much like the example in Figure 8.11, but with different transition probabilities.
   a. Find the steady-state probabilities, using the graph theoretic method.
   b. Find m_{31} by the graph theoretic method.

2. The air quality problem, whose transition matrix can be found in the answers to Chapter 2, problem 3 (page 68) is a four-state ergodic Markov chain. Figure 8.28 shows the transition diagram (with loop omitted, because they are not used here).
   a. Find the steady-state probabilities, using the graph theoretic method.
   b. Find m_{41}, the mean first passage time from state 4 to state 1.

3. Figure 8.29 shows the transition diagram for a small terminating process (with loops removed), which we can use to illustrate several calculations.
   a. Find the absorption probability a_{14}.
   b. Find the expected time accumulated in transient state 2, e_{12}.
   c. Find the hit probability f_{12}.
   d. Find the conditional mean first passage time $m_{14}^{(c)}$.

4. Repeat the same calculations as in problem 3 above after multiplying each of the transition probabilities by 10. (The answers should be the same.)

5. The transition diagram for the hospital problem can be found in the answers to Chapter 2, problem 13 (page 70). Figure 8.30 shows it with the loops removed.
   a. Find the absorption probability a_{15}.
   b. Find the expected number of times state 2 is entered, starting from state 1, e_{12}.
   c. Find the value of the hit probability f_{12}.

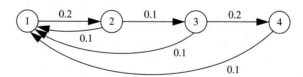

■ **FIGURE 8.28** The air quality problem

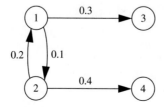

■ **FIGURE 8.29** A terminating process

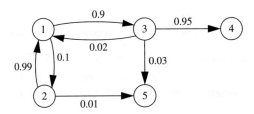

■ **FIGURE 8.30** The hospital problem

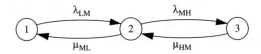

■ **FIGURE 8.31** The traffic problem

6. An easy continuous time Markov model on which to try the techniques is provided in the answers to Chapter 6, problem 1 (page 222). Figure 8.31 demonstrates that it is just as easy to get a symbolic solution as a numerical one using the graph theoretic approach.

  a. Find the steady-state probabilities in terms of λ and μ symbols.
  b. Find the symbolic solution for m_{21}, the mean first passage time from medium to light traffic.

7. Most of the Markov queueing models are birth-death processes, for which the subgraphs are easy to identify. However, many of them involve infinite state spaces, for which the graph theoretic formulas given would not apply. A small finite queueing problem would be an M/M/2/4 model. Find the steady-state probabilities in symbolic form for this model (that is, in terms of λ and μ).

APPENDIX 1

Visual Basic Code for Closed Queueing Networks

This appendix contains all of the essential Visual Basic code to implement a Microsoft Excel spreadsheet analyzer for closed queueing networks having product form solutions. This simplified code is far from optimal; there are more sophisticated procedures that consume less time and memory and will handle larger networks. Also, any computer code that will be reused over time should have error trapping statements to avoid problems with inappropriate input data (such as zero or noninteger numbers of servers). However, it is relatively complete and certainly adequate for small problems.

If you have never used Excel macros, you should consult a manual on how to do it. Briefly, you will open a module in the Visual Basic editor (under the "tools" menu), type in the code, debug any typing errors, and save the module. In addition, you will need to name some cells and arrays in the worksheet to which the module is attached so that the code knows where to find the input data and where to place the calculated results. You may arrange those special locations on the worksheet in any way you want, so you could customize an application with additional information or alter the appearance for your own purposes.

The code assumes that there are two named *column* arrays somewhere on the active worksheet called "servers" and "workloads," which contain the number of servers for each station (the c_i) and the relative workloads for each station (the ρ_i). There must also be a named cell somewhere on the active worksheet called "customers" from which the value of N is taken. The number of stations, M, does not have to be explicitly defined; it is inferred from the length of the arrays. When the macro named "Main" is run, the code will compute all of output and hold it in hidden variables. In order to see those values, you must have defined cells or arrays on the same active worksheet. The names must be (exactly) L, B, U, and W, for the expected number of customers at the station, the expected number of busy servers, the utilization, and the expected total time at the station, respectively. If you have a cell named Pace, the value of that variable will be placed in it, so any of the performance measures that depend upon it could be calculated from that. These named areas may be arranged on the worksheet in any place or order, but the length of the arrays should match the number of stations. If there is no named destination for a particular result, an error trap will simply skip over that output. You may therefore design your own output format without altering the code.

If you save and close a file with the macro attached, you will get a warning when you try to open it again. Excel will ask if you want to enable macros, because you should never enable a file that contains them unless you know they are deliberate (computer viruses are sometimes implemented in macros). In this case, they are intended to be there, so you will have to allow them in order to call the program. Once the file is open, you may change variable values at will and rerun the macro to see the effect of changes. You may want to save a template file with empty input and output fields so that you can easily start a new analysis without clearing out old data.

This first section of code establishes some conventions and declares some global variables.

```
Option Base 0    'sets all arrays to start at 0.
Option Explicit  'requires all variables to be declared.
Dim M As Integer, N As Integer, i as Integer
Dim c() As Integer
Dim r() As Double
Dim G() As Double
Dim H() As Double
Dim L() As Double
Dim Scalefactor as Double
```

The next portion is the main program (called, logically, Main). It is the only macro that appears in the list of macros under the tools menu; all of the others are hidden subroutines. To see or edit them, you will have to open the Visual Basic editor.

```
Public Sub Main()
   GetData     'reads the data from the worksheet into variables
   ScaleRW     'scales relative workloads for best performance
   Gcalc       'calculates the normalization constant
   LCalc       'calculates the L's and W's
   PutResults  'places results back into the worksheet
End Sub
```

The following subroutine gets the input data from the active worksheet and puts it into array variables that are hidden from the worksheet, but used in the Visual Basic calculations.

```
Private Sub GetData()
    Dim i As Integer
    M = ActiveSheet.Range("Servers").Count
    N = ActiveSheet.Range("Customers").Value
    'dimension arrays that use M
    ReDim c(M)
    ReDim r(M)
    ReDim L(M)
    'dimension arrays that use N
    ReDim G(N)
    ReDim H(N)
    'read the data into variables
    For i = 1 To M
        c(i) = ActiveSheet.Range("Servers").Cells(i, 1).Value
        r(i) = ActiveSheet.Range("Workloads").Cells(i, 1).Value
Next i
End Sub
```

If the relative workloads are numbers that are either too large or too small, there is a risk of overflow or underflow in the calculations when N is large, because they are (in effect) being multiplied by themselves many times. The following subroutine scales the relative workload values to minimize the likelihood of that occurring. It may still happen if N is extremely large, but the scaling applied here will accomplish as much as a personal computer is capable of doing.

```
Private Sub ScaleRW()
   Scalefactor = 0#
   For i = 1 To M
     If c(i) > 0 Then
        If r(i) / c(i) > Scalefactor Then Scalefactor = r(i) / c(i)
     End If
   Next i
   If Scalefactor = 0# Then Scalefactor = 1#
   For i = 1 To M
     r(i) = r(i) / Scalefactor
   Next i
End Sub
```

The next subroutine computes all of the terms of the the normalization constant array G(M,n) and leaves them in a one-dimensional array variable called G. Although it essentially implements the algorithm explained in the text, it contains a number of computational refinements to speed the calculations and save memory. It is not recommended that you attempt to modify this section of the code unless you really understand what it is doing.

```
Private Sub Gcalc()
   Dim i As Integer, j As Integer, k As Integer
   Dim MinNs As Integer
   Dim temp As Double
   G(0) = 1#
   For i = 1 To M         'do for each station
      If r(i) > 0 Then
         If c(i) > 1 Then
            'multiple server case
            If N > c(i) Then
               For j = N To c(i) + 1 Step -1
                  temp = 1#
                  For k = 1 To j - c(i)
                     temp = G(k) + (r(i) / c(i)) * temp
                  Next k
                  For k = j - c(i) + 1 To j
                     temp = G(k) + (r(i) / (j + 1 - k)) * temp
                  Next k
                  G(j) = temp
               Next j
            End If
            If N < c(i) Then MinNs = N Else MinNs = c(i)
            For j = MinNs To 1 Step -1
               temp = 1#
               For k = 1 To j
                  temp = G(k) + (r(i) / (j + 1 - k)) * temp
               Next k
               G(j) = temp
            Next j
         Else                        'single server case
            For j = 1 To N
               G(j) = G(j - 1) * r(i) + G(j)
            Next j
         End If
      End If
   Next i
End Sub
```

The next subroutine calculates L for each station and keeps the results in a (hidden) global array for later use.

```
Private Sub LCalc()
   Dim j As Integer, k As Integer
   Dim prod As Double, ra As Double, sum As Double
   For i = 1 To M
      If c(i) >= N Then
         L(i) = r(i) * G(N - 1) / G(N)
      Else
      prod = 1#
      For k = 1 To c(i)
           prod = r(i) * prod / k
      Next k
      HCalc
      sum = 0#
      ra = r(i) / c(i)
      For k = 1 To N - c(i)
         sum = sum + k * (ra ^ k) * H(N - c(i) - k)
      Next k
      L(i) = (r(i) * G(N - 1) + prod * sum) / G(N)
      End If
   Next i
End Sub
```

The following subroutine computes all of the H function from 1 to $N - s$, for a given station. It is called by LCALC when it is needed, but not otherwise.

```
Private Sub HCalc()
   Dim jkc As Integer, j As Integer, k As Integer
   Dim temp as Double
   H(0) = 1#
   For j = 1 To N - c(i)
      temp = 0#
      For k = 1 To j
         If (j - k + 2 < c(i)) Then jkc = (j - k + 2) Else jkc = c(i)
         temp = H(k - 1) + (r(i) / jkc) * temp
      Next k
      H(j) = G(j) - r(i) * temp
   Next j
End Sub
```

The following subroutine is the one that writes the calculated results into named cells or arrays in the worksheet.

```
Private Sub PutResults()
   WriteL
   WriteB
   WriteU
   WriteW
   WritePace
End Sub
```

This function writes the expected number of customers at each station into an array on the active worksheet named "L" if there is one; otherwise, it does nothing. The first station must correspond to the first row in the array, the second to the second, and so forth.

```
Function WriteL()
  On Error GoTo Skip
  For i = 1 To M
      ActiveSheet.Range("L").Cells(i, 1).Value = L(i)
  Next i
Skip:
End Function
```

This function below writes the expected number of customers at each station into a column array on the active worksheet named "B" if there is one.

```
Function WriteB()
  On Error GoTo Skip
  Dim B As Double
  For i = 1 To M
      B = r(i) * G(N - 1) / G(N)
      ActiveSheet.Range("B").Cells(i, 1).Value = B
  Next i
Skip:
End Function
```

The next function writes the station utilizations into a column array on the active worksheet named "U" if there is one.

```
Function WriteU()
  On Error GoTo Skip
  Dim U As Double
  For i = 1 To M
      U = (r(i) * G(N - 1) / G(N)) / c(i)
      ActiveSheet.Range("U").Cells(i, 1).Value = U
  Next i
Skip:
End Function
```

The next function writes the expected number of customers at each station into an array on the active worksheet named "W" if there is one.

```
Function WriteW()
  On Error GoTo Skip
  Dim wait As Double
  wait = N / ((G(N - 1) / G(N)))
  ActiveSheet.Range("W").Value = wait
Skip:
End Function
```

This last function writes the value of the Pace into a single cell on the active worksheet named "Pace" if there is such a named cell.

```
Function WritePace()
   On Error GoTo Skip
   Dim Pace As Double
   Pace = (G(N - 1) / G(N)) / Scalefactor
   ActiveSheet.Range("Pace").Value = Pace
Skip:
End Function
```

APPENDIX 2

Answers to Selected Problems

Chapter 1: Probability Review

1. a. The student is not in engineering. $\overline{A}$
   b. The student is in engineering and in this class. $A \cap B$
   c. The student is not in engineering but is in this class. $\overline{A} \cap B$
   d. The student is not in engineering and is not in this class. $\overline{A} \cap \overline{B}$
   e. The student is either in engineering or is in this class. $A \cup B$
   f. The student is either in engineering or in this class, but not both.
      Several forms will work, including $(A \cap \overline{B}) \cup (\overline{A} \cap B)$ or $(A \cup B) \cap \overline{(A \cap B)}$ or any equivalent expression. (The point of this problem is to understand the correspondence between English language words ("not," "and," "but," and so forth and their set theoretic expression.)

2. a. The show is liked by critics, the public, and advertisers. $A \cap B \cap C$
   b. Critics do not like the show, but it is popular with the public and advertisers. $\overline{A} \cap B \cap C$
   c. Critics and advertisers seem to like the show, but the public does not care for it. $A \cap \overline{B} \cap C$
   d. None of the three target audiences likes the show. $\overline{A} \cap \overline{B} \cap \overline{C}$
      (This problem extends the lessons of the first problem to combinations of three events. Alternate, equivalent expressions could accomplish the same thing.)

3. a. You bet on a single horse and care whether you win or lose. $S = \{\text{win, lose}\}$
   b. You care which of the five horses wins. $S = \{A, B, C, D, E\}$

Chapter 1: Probability Review

c. You care about which horses come in first, second, and third.   S = {ABC, ABD, ABE, ACB, ACD, ACE, ⋯}
(There are 5 × 4 × 3 = 60 possibilities.)
(The point of this problem is to demonstrate that the same experiment can have different sample spaces, depending upon what your interested is.)

4. a. $P(A) = P(B)$ — Yes, equally likely case is an example.
b. $P(A) = 2P(B)$ — Yes, P(A) = 1/2, P(B) = 1/4.
c. $P(A) = 1 - P(B)$ — Yes, any case where P(3) = 0, so P(A) + P(B) = 1.
d. $P(A) + P(B) > 1$ — Yes, equally likely case is an example.
e. $P(A) - P(B) < 0$ — Yes, any case where P(A) < P(B).
f. $P(A) - P(B) > 1$ — No, because P(A) > 1 + P(B), cannot be larger than 1.
(This problem and the next test your understanding of the basic laws of probability, applied in a general way without specifying any particular distribution.)

5. a. A∩B — A∩B = {0, 2}, cannot infer any value.
b. B∪C — B∪C = {0, 1, 2, 3, 4, 5} = S, so $P(B∪C) = 1$.
c. $\overline{A}$ — $\overline{A}$ = {3, 4, 5}, $P(\overline{A}) = 1 - P(A)$.
d. B∩$\overline{C}$ — Set is the same as B, so $P(B∩\overline{C})$ is $P(B)$.
e. (A∩B)∪C — Set is, {4} cannot infer any value.

6. $P(S) = 1$ from axiom 2.
$P(S∪\varphi) = 1$ by set theory (when you take the union of any set with the empty set, you leave the original set unchanged.)
Since S and φ are mutually exclusive, using axiom 3 gives $P(S) + P(\varphi) = 1$.
Now subtract the first equation to leave $P(\varphi) = 0$, which proves relation 4.
(This problem and the next are about using the axioms, not about proving theorems.)

7. Again, start from $P(S) = 1$ from axiom 2. Then from set theory, substitute S = A∪$\overline{A}$ to get $P(A∪\overline{A}) = 1$. Since A and its complement $\overline{A}$ are mutually exclusive, we can use axiom 3 to get $P(A) + P(\overline{A}) = 1$. Rearrangement gives the result, $P(\overline{A}) = 1 - P(A)$.

8. a. Rain today, rain tomorrow. — Most likely dependent, because the time is close enough to expect some carry-over dependence.

b. Rain today, rain one month from today. — Most likely independent, because the time is so long.

c. Rain one year ago today, rain today. — Almost certainly independent.

d. Receiving grade of A in introductory probability course, receiving same grade in this course (for the same person). — Dependent, because the courses are related.

e. Receiving grade of A in freshman-level physics, receiving same grade in this course. — Could be argued either way.

(This problem makes the point that deciding whether events are dependent or independent usually involves thinking about the events, not checking whether some equation is satisfied. In modeling applications, that "call for judgement" is common.)

9. When events are mutually exclusive, they are definitely dependent; in fact, that relationship is a strong kind of dependency. All other combinations are possible. Here is a table of the various combinations.

	Independent	Dependent
Mutually exclusive	impossible	possible
Not mutually exclusive	possible	possible

10. No. The relation is not transitive. A and C could be closely related (in fact, identical), while neither has anything to do with B.

11. We are looking for $P(A \cup B)$, but the union includes the intersection, or the possibility of getting offers from both companies. The probability that at least one rejects is $P(\overline{A} \cup \overline{B})$, given as 0.8. But the set can be rewritten as $\overline{A \cap B}$, so $P(A \cap B) = 1 - P(\overline{A \cap B}) = 1 - 0.8 = 0.2$. Finally, $P(A \cup B) = P(A) + P(B) - P(A \cap B) = 0.6 + 0.5 - 0.2 = 0.9$.

12. Let A be the event that the selected person is right-handed, and let B be the event that he or she owns a dog. We want $P\{A \cap \overline{B}\}$. We know two facts: $P(\overline{A}) = 0.1$ and $P(B|A) = 0.4$. We can deduce from the first that $P(A) = 0.9$ (since a person must be either right- or left-handed). From the second, we can deduce that $P(\overline{B}|A) = 0.6$ (since a person must either own or not own a dog). Using the relation between joint and conditional probabilities, we can put these two deduced facts together to get the desired result

$$P(A \cap \overline{B}) = P(\overline{B}|A)P(A) = (0.6)(0.9) = 0.54$$

13. Start with the definition

$$P(A|B) = \frac{P(A \cap B)}{P(B)} \text{ which we assume is } > P(A)$$

Hence, $P(A \cap B) > P(A)P(B)$. Substitute $P(A) = 1 - P(\overline{A})$ to get $P(A \cap B) > [1 - P(\overline{A})]P(B)$ or $P(A \cap B) > P(B) - P(\overline{A})P(B)$.
Now expand the first P(B) on the right hand side as $P(B) = P(B \cap A) + P(B \cap \overline{A})$. Then substitute to get

$$P(A \cap B) > P(B \cap A) + P(B \cap \overline{A}) - P(\overline{A})P(B)$$

which means

$$P(\overline{A})P(B) > P(B \cap \overline{A})$$

or

$$P(B) > \frac{P(B \cap \overline{A})}{P(\overline{A})}$$

But the right-hand side is $P(B|\overline{A})$, so the relation is proved.
(This was a fairly complicated proof, so you can be forgiven if you needed help. Again, it was given as an problem in using the rules of probability, rather than constructing proofs.)

14. $P\{Y = 6|X = 2\} = 0.2$ (obtained by looking up the joint probability $P\{Y = 6, X = 2\} = 0.02$ and dividing by the marginal probability $P\{X = 2\}$ which you get by adding the terms in the second column = 0.10).

15. No, the two random variables are dependent. In order for the two random variables to be independent, *every* cell in the matrix must be the product of the marginals. Some cells satisfy the requirement, but not all. (Note that independence of random variables is not as easy to satisfy as independence of events.)

16. The marginal distribution of Y is obtained by summing across rows:

$$P\{Y = 2\} = 0.5, \ P\{Y = 4\} = 0.2, \ P\{Y = 6\} = 0.2, \ P\{Y = 8\} = 0.1$$

17. The easiest way to compute is to use $COV(X, Y) = E(XY) - E(X)E(Y)$. We get $E(X)$ and $E(Y)$ from the separate marginal distributions:

$$E(X) = (1)(0.2) + (2)(0.1) + (3)(0.3) + (4)(0.2) + (5)(0.2) = 3.1$$
$$E(Y) = (2)(0.5) + (4)(0.2) + (6)(0.2) + (8)(0.1) = 3.8$$

To get $E(XY)$, we need to compute each product-pair, multiply each by the corresponding joint probability, and add

$$\begin{aligned}E(XY) =\ &(2)(1)(0.10) + (2)(2)(0.05) + (2)(3)(0.15) + (2)(4)(0.10) + (2)(5)(0.10) \\&+ (4)(1)(0.04) + (4)(2)(0.02) + (4)(3)(0.06) + (4)(4)(0.04) + (4)(5)(0.04) \\&+ (6)(1)(0.04) + (6)(2)(0.02) + (6)(3)(0.06) + (6)(4)(0.06) + (6)(5)(0.02) \\&+ (8)(1)(0.02) + (8)(2)(0.01) + (8)(3)(0.03) + (8)(4)(0.00) + (8)(5)(0.04) \\=\ &11.82\end{aligned}$$

Finally, collecting terms,

$$COV(X, Y) = E(XY) - E(X)E(Y) = 11.82 - (3.1)(3.8) = 0.04$$

18. To evaluate the unknown parameter a, we make use the fact that the integral of the density function over the entire range must equal 1. Set up the integral, set it equal to 1, and solve for a:

$$\int_1^a 2x\, dx = 1$$
$$x^2 \big|_1^a = 1$$
$$a^2 - 1 = 1$$
$$a^2 = 2$$
$$a = \sqrt{2}$$

19. The problem contains more information than is needed for this question. The expectation of a sum is the sum of the expectations, even if the random variables are dependent. Hence,

$$E(W) = E(X + Y + 4) = E(X) + E(Y) + 4 = 10 + 8 + 4 = 22$$

20. Variance is more complicated. W is the sum of three terms. The constant term, 4, drops out, according the rule given on page 17. That leaves $V(W) = V(X + Y)$. If X and Y were independent, the variance of the sum would be the sum of the variances, but that is not the case here. Instead, we have to use the more general form given on page 18, $V(X + Y) = V(X) + V(Y) + 2COV(X, Y)$. Hence, we need the covariance, which is given most conveniently by $COV(X, Y) = E(XY) - E(X)E(Y)$. Putting all the required data together,

$$V(X, Y) = 9 + 4 + 2(84 - 10 * 8) = 21$$

21. The correlation coefficient is the covariance divided by two standard deviations,

$$\rho = \frac{COV(X, Y)}{\sigma_X \sigma_Y} = \frac{4}{(3)(2)} = \frac{2}{3}$$

22. "Success" and "failure" are arbitrary words used to describe the two outcomes. If the probability of success is 0.4, the probability of failure must be 0.6. Then let that be

considered the probability of success (of the event we are interested in, which is really "failure"), so the expectation will be np or $(10)(0.6) = 6$.

23. This question can be easily answered in either of two ways. You can resort to the definition,

$$E(X) = (10)(0.6) + (5)(0.4) = 8$$

or you can use the linear transformation of the ordinary Bernoulli random variable Y which takes on the values 1 and 0, for success and failure, respectively. To get X from Y, we have to figure out that $X = 5Y + 5$. (When $Y = 1$, X will equal 10 and when $Y = 0$, X will equal 5.) Then, using $E(Y) = p$, the probability of success,

$$E(X) = E(5Y + 5) = (5)(0.6) + 5 = 8$$

24. The negative exponential distribution is the continuous time analog of the geometric.

25. The variance of a sum of independent random variables is just the sum of the variances (regardless of the type of distribution). Furthermore, the variance of any Poisson random variable is the same as the mean, which in turn is equal to the parameter of the distribution. Hence,

$$V(X + Y) = 3 + 2 = 5$$

26. First, you use the linear transformation rules for expectations,

$$E(Y) = E(2X + 3) = 2E(X) + 3$$

Then you plug in the known mean for a binomial random variable = np

$$E(Y) = 2(20)(0.4) + 3 = 19$$

27. If odds against an event are expressed as a:b, then the probability of the event is $\frac{b}{a+b}$. So odds of 5:1 translates to a probability of $\frac{1}{6}$. If the probability of an event is expressed as the fraction $\frac{x}{y}$, then the odds against it would be $(y - x) : x$. So a probability of 0.25 translates to odds of 3:1.

28. The addition rule for mutually exclusive events becomes: "If the odds against event A are a:b and the odds against B are c:d, and A and B are mutually exclusive, then the odds against either A or B are $(ac - bd) : (bc + 2bd + ad)$." Others are similarly messy.

Chapter 2: Markov Chain Formulation

1. Time units: States 1 = good, State 2 = medium, and State 3 = bad. The transition matrix is

$$\mathbf{P} = \begin{bmatrix} 0.5 & 0.3 & 0.2 \\ 0.2 & 0.6 & 0.2 \\ 0.3 & 0.2 & 0.5 \end{bmatrix}$$

3. Time units: days. State 1 = green, State 2 = yellow, State 3 = orange, and State 4 = red. Normalize the count data to convert to probabilities.

$$\mathbf{P} = \begin{bmatrix} 0.8 & 0.2 & 0 & 0 \\ 0.1 & 0.8 & 0.1 & 0 \\ 0.1 & 0 & 0.7 & 0.2 \\ 0.1 & 0 & 0 & 0.9 \end{bmatrix}$$

5. Time units: years. State 1 = corn, State 2 = soy, State 3 = other, and State 4 = fallow. Notice that the data matrix has time running in reverse order; that is, the earlier time is in columns and the later time is in rows. Therefore, normalize the *transpose* of the data matrix.

$$\mathbf{P} = \begin{bmatrix} \frac{2}{3} & \frac{1}{9} & \frac{1}{9} & \frac{1}{9} \\ \frac{1}{4} & \frac{3}{10} & \frac{1}{4} & \frac{1}{5} \\ \frac{5}{26} & \frac{3}{13} & \frac{1}{2} & \frac{1}{13} \\ \frac{5}{16} & \frac{1}{4} & \frac{7}{16} & 0 \end{bmatrix} = \begin{bmatrix} 0.667 & 0.111 & 0.111 & 0.111 \\ 0.25 & 0.3 & 0.25 & 0.2 \\ 0.192 & 0.231 & 0.5 & 0.077 \\ 0.312 & 0.25 & 0.438 & 0 \end{bmatrix}$$

7. Normalize the data matrix,

$$\mathbf{P} = \begin{bmatrix} \frac{1}{30} & \frac{4}{15} & \frac{3}{5} & \frac{1}{10} \\ \frac{1}{3} & 0 & \frac{2}{15} & \frac{8}{15} \\ \frac{1}{5} & \frac{4}{15} & \frac{2}{15} & \frac{2}{5} \\ \frac{1}{10} & \frac{3}{5} & \frac{1}{5} & \frac{1}{10} \end{bmatrix} = \begin{bmatrix} 0.033 & 0.267 & 0.6 & 0.1 \\ 0.333 & 0 & 0.133 & 0.533 \\ 0.2 & 0.267 & 0.133 & 0.4 \\ 0.1 & 0.6 & 0.2 & 0.1 \end{bmatrix}$$

9. Time units: jumps. States: the four lily pads numbered as in the description. Distances are symmetric, so the distance from pad 2 to 1 is the same as the distance from 1 to 2. From the verbal description, you can construct a distance matrix. To construct the transition probabilities consider each row separately. The frog does not jump until it changes pads, so the diagonal terms are all 0. The other terms in the same row are inversely proportional to the distance and must add to 1. So, for example, p_{12}, p_{13}, and p_{14} are $\frac{k}{6/5}$, $\frac{k}{2}$, and $\frac{k}{3/2}$ where k is some constant of proportionality. To determine k, add these up and set the sum equal to 1; then solve for k. The other rows are similar.

$$\mathbf{P} = \begin{bmatrix} 0 & \frac{5}{12} & \frac{1}{4} & \frac{1}{3} \\ \frac{5}{24} & 0 & \frac{7}{24} & \frac{1}{2} \\ \frac{1}{6} & \frac{7}{18} & 0 & \frac{4}{9} \\ \frac{1}{6} & \frac{1}{2} & \frac{1}{3} & 0 \end{bmatrix} = \begin{bmatrix} 0 & 0.417 & 0.25 & 0.333 \\ 0.208 & 0 & 0.292 & 0.5 \\ 0.167 & 0.389 & 0 & 0.444 \\ 0.167 & 0.5 & 0.333 & 0 \end{bmatrix}$$

11. Figure A2.1 shows the accounting problem. Time units: months. States 0–5, age of account in months, state 6 = bad debt, and state 7 = paid.

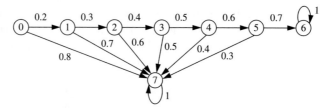

■ **FIGURE A2.1** The accounting problem

The transition matrix looks like this:

$$\mathbf{P} = \begin{bmatrix} 0 & 0.2 & 0 & 0 & 0 & 0 & 0 & 0.8 \\ 0 & 0 & 0.3 & 0 & 0 & 0 & 0 & 0.7 \\ 0 & 0 & 0 & 0.4 & 0 & 0 & 0 & 0.6 \\ 0 & 0 & 0 & 0 & 0.5 & 0 & 0 & 0.5 \\ 0 & 0 & 0 & 0 & 0 & 0.6 & 0 & 0.4 \\ 0 & 0 & 0 & 0 & 0 & 0 & 0.7 & 0.3 \\ 0 & 0 & 0 & 0 & 0 & 0 & 1 & 0 \\ 0 & 0 & 0 & 0 & 0 & 0 & 0 & 1 \end{bmatrix}$$

13. The states are 1 = intensive care, 2 = surgery, 3 = recovery, 4 = release, and 5 = die. (Yours may differ). We do not need a state for the emergency room because every patient enters there and proceeds with certainty to intensive care. Transitions follow the rules as shown in Figure A2.2.

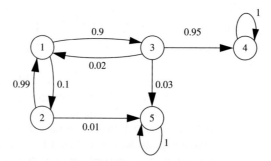

■ **FIGURE A2.2** The hospital problem

$$\mathbf{P} = \begin{bmatrix} 0 & 0.1 & 0.9 & 0 & 0 \\ 0.99 & 0 & 0 & 0 & 0.01 \\ 0.02 & 0 & 0 & 0.95 & 0.03 \\ 0 & 0 & 0 & 1 & 0 \\ 0 & 0 & 0 & 0 & 1 \end{bmatrix}$$

15. Figure A2.3 shows the solar heating problem. Time units: days. States 1 = sunny day, State 2 = cloudy day, State 3 = two cloudy days, and State 4 = three cloudy days.

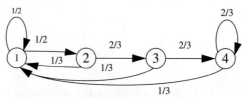

■ **FIGURE A2.3** The solar heating problem

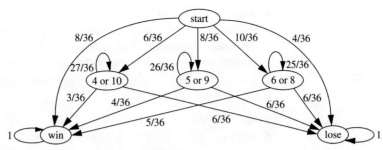

■ **FIGURE A2.4** The game of craps

17. Figure A2.4 shows the game of craps. Time units: rolls of the dice. You may have had more states, but the following is sufficient. The state (4, 10), for example, means that the first roll gives either 4 or 10 as the point. They can be lumped together because they have the same outward transition probabilities. (When you are experienced, you look for ways to reduce the number of states.) The transition probabilities come from the dice combination counts. For example, of the thirty-six possible outcomes from rolling a pair of dice, eight will give you a 7 or 11, for an immediate win. (Some of the fractions shown here can be reduced, but they are left unreduced to show the number of combinations out of thirty-six that lead to each outcome.)

Chapter 3: Markov Chain Calculations

1. **a.** Three-step transition probability, $p_{22}^{(3)} = 0.412$.
   b. Expected number of times in state 2 after seven transitions, $e_{31}^{(7)} = 2.29$.
   c. Steady-state probability, $\pi_1 = 0.3265$ (the others are: $\pi_2 = 0.3878$, $\pi_3 = 0.2857$).
   d. Mean first passage time, $m_{31} = 3.75$ days.
   e. Mean sojourn time, $h_3 = 2$ days.

3. **a.** A walk probability, $p_{12}p_{23}p_{34} = 0.004$.
   b. A steady-state probability, $\pi_1 = 0.333$.
   c. A mean first passage time, $m_{41} = 10$ days.
   d. A mean sojourn time, $h_1 = 5$ days.

5. This model is a good example of the use of probabilities to track a changing population. Probabilities can be interpreted as fractions and then converted to absolute quantities—in this case, acres.
   a. $12\pi_j$ million acres, for j—1, 2, 3, 4. To get the *number* of acres of each type, you need to multiple the total number by the fraction: $\pi^T = [0.387, 0.240, 0.230, 0.143]$, so amounts are [4.65, 2.88, 2.76, 1.72] in millions of acres.
   b. $e_{21}^{(5)} = 1.768$.
   c. Acreages: [3.81, 3.65, 3.66, 3.54] (These are obtained by multiplying $\mathbf{P}^{(2)}$ times the initial distribution for the last year that is known exactly to get the distribution two years later; then multiply by the total number of acres to convert from fractions to acres.)
   d. You have to use a probability distribution for the initial state. The table in the problem gives current acreages, from which $p_1^{(0)} = 6/12 = 0.5$, $p_2^{(0)} = 4/12 = 0.333$, $p_3^{(0)} = 1.5/12 = 0.125$, and $p_4^{(0)} = 0.5/12 = 0.042$. Use these to weight the three-step transition probabilities from the various initial states to state 4. Answer = 0.144.

e. Average number of years *between* fallow years: $m_{44} = 6.99$ years. (You might want to subtract 1 from this, depending upon your interpretation of the word "between." The mean recurrence time is almost seven years, but there would be six years of nonfallow crops in the years between, not counting the endpoints.)

7. a. $\max\{p_{1j}^{(2)}\} = p_{14}^{(2)} = 0.396$, so a grip is most likely.
   b. $m_{14} = 836/269 = 3.11$.
   c. $e_{41}^{(6)} = 1.045$.
   d. $\pi_2/\pi_3 = 0.288/0.237 = 1.22$ times as great.

9. a. $p_{23}^{(2)} = p_{21}p_{13} + p_{22}p_{23} + p_{23}p_{33} + p_{24}p_{43} = 7/32$.
   b. If the frog spends equal time between jumps, π_j would be the fraction of time spent on pad j, over the long run. There are several other interpretations.
   c. The frog either does not remember or does not care where he has been. The likelihood of jumping to any particular pad next depends only on the distance.

11. a. $a_{07} = 0.995$.
    b. $a_{16} = 0.025$.
    c. $p_{03}^{(3)} = 0.024$.
    d. Conditional $m_{07}^{(c)} = 1.279$ months. (The equations for conditional mean first passage times are usually quite nasty, but in this case are easy to solve if you start with $m_{57}^{(c)}$ and work backward through $m_{47}^{(c)}$, $m_{37}^{(c)}$, etc., to $m_{07}^{(c)}$.)

13. a. The first three states are transient, the last two are absorbing.
    b. We have two different initial states. The fraction who die is given by an absorption probability. So the answer is $1000(2/3 \times a_{15} + 1/3 \times a_{25}) = 0.0317$.
    c. The number of operations equal the number of times state 2 is visited. Again, you have to mix on the initial state, so $2/3 \times e_{12} + 1/3 \times e_{22} = 0.4088$.
    d. This is a hit probability for state 2, $f_{12} = 0.1018$.
    e. $d_1 = 2.265$.

15. a. This requires a mean first passage time, $m_{14} = 8.25$ days.
    b. This is just $m_{24} = 6.25$ days.
    c. This is a mean sojourn time $= 1.5$ days.
    d. This is a mean recurrence time $m_{44} = 3.75$ days.
    e. We need back up heat when in state 4. The fraction of time spent there is $\pi_4 = 4/15$. Hence the number of days in state 4 is $(120)(\pi_4)$. So expected cost is $(\$10)(120)(\pi_4) = \$320$.

17. a. The probability of winning is an absorption probability from the initial to the win state $= 0.4929$.
    b. The expected number of rolls to determine the outcome is $d_0 = 3.3758$.

19. a. Find the absorption probability $a_{16} = 0.916$.
    b. We need to count the expected number of times each transient state is visited, so we require the matrix **E**, which is:

$$\mathbf{E} = \begin{bmatrix} 1 & 0.69 & 0.2 & 0.23 & 0.21 \\ 0 & 1.01 & 0 & 0.04 & 0.17 \\ 0 & 0.1 & 1 & 0 & 0.42 \\ 0 & 0.3 & 0 & 1.01 & 0.15 \\ 0 & 0 & 0 & 0 & 1 \end{bmatrix}$$

Then we need to add across the first row (since all calls originate in state 1), but exclude the e_{11} value which simply counts the incoming call. The result is 1.38 transfers.

Chapter 4: Rewards on Markov Chains

1. **a.** Expected profit over next three trips starting from Baltimore = $v_2^{(3)}$ = \$424.60.
   b. Gain = \$115.83 per trip.
3. **a.** Expected profit over seven days, starting from good = $v_1^{(7)}$ = \$44,200.
   b. Gain = \$4,142.86 per day.
5. **a.** The reward matrix is

$$\mathbf{R} = \begin{bmatrix} 140 & 155 & 150 & 175 \\ 130 & 120 & 130 & 145 \\ 105 & 100 & 95 & 115 \\ -20 & -10 & -15 & 0 \end{bmatrix}$$

b. The immediate reward vector,

$$\mathbf{v} = \begin{bmatrix} 150 \\ 130 \\ 102 \\ -15 \end{bmatrix}$$

c. Gain per acre × 12 million acres = \$110.56 × 12,000,000 = \$1,326,751,000.

7. The states are deliberately numbered so that the matrices will be in the desired format. From the description, the reward matrix is

$$\mathbf{R} = \begin{bmatrix} 0 & -20 & 0 & 0 & 0 \\ 0 & 0 & -10 & 0 & -5 \\ -3 & -2 & 0 & 100 & -5 \\ 0 & 0 & 0 & 0 & 0 \\ 0 & 0 & 0 & 0 & 0 \end{bmatrix}$$

We calculate **v**:

$$\mathbf{v} = \begin{bmatrix} -20 \\ -9.5 \\ 79.4 \end{bmatrix}$$

The limiting result is obtained from the formula **Ev**. We first find

$$\mathbf{E} = (\mathbf{I} - \mathbf{Q})^{-1} = \begin{bmatrix} 1.052 & 1.156 & 1.040 \\ 0.052 & 1.156 & 1.040 \\ 0.058 & 0.173 & 1.156 \end{bmatrix}$$

then, by simple matrix multiplication,

$$\lim_{n \to \infty} \mathbf{v}^{(n)} = \mathbf{E}\mathbf{v} = \begin{bmatrix} 50.59 \\ 70.59 \\ 88.99 \end{bmatrix}$$

Since all new pieces start in state 1, we can say that the expected net profit when the part eventually comes out is \$50.59. This figure accounts for the possibilities of reworking, perhaps many times, and of loss due to scrap. The other two figures represent the expected profit for pieces "in process." For example, a piece that has reached state 2 is already worth \$20 more than a piece that is just starting out.

9. a. The reward/cost matrix is:

$$R = \begin{bmatrix} 0 & 20 & 0 & 0 & 0 \\ 0 & 0 & 10 & 0 & 10 \\ 4 & 3 & 0 & 1 & 1 \\ 0 & 0 & 0 & 0 & 0 \\ 0 & 0 & 0 & 0 & 0 \end{bmatrix}$$

b. The immediate reward/cost vector is (for transient states only):

$$v = \begin{bmatrix} 20 \\ 10 \\ 1.35 \end{bmatrix}$$

c. From **Ev**, the total expected cost of an entering part is $34.01.

11. a. Note that you are only rewarded when you receive payment; the amount depends upon when that happens. The reward matrix is

$$R = \begin{bmatrix} 0 & 0 & 0 & 0 & 0 & 0 & 0 & 1 \\ 0 & 0 & 0 & 0 & 0 & 0 & 0 & 1.02 \\ 0 & 0 & 0 & 0 & 0 & 0 & 0 & 1.04 \\ 0 & 0 & 0 & 0 & 0 & 0 & 0 & 1.06 \\ 0 & 0 & 0 & 0 & 0 & 0 & 0 & 1.08 \\ 0 & 0 & 0 & 0 & 0 & 0 & 0 & 1.1 \\ 0 & 0 & 0 & 0 & 0 & 0 & 0 & 0 \\ 0 & 0 & 0 & 0 & 0 & 0 & 0 & 0 \end{bmatrix}$$

b. The expected total reward is, from **Ev**, $1.00057. You can multiply this value per dollar times any original debt to get the value of an account.

c. We get the expected value of a three-month-old $1 account from **Ev** by reading the appropriate row. The answer is $0.846.

13. a. The reward matrix is:

$$R = \begin{bmatrix} 0 & 6400 & 6300 & 0 & 0 \\ 4400 & 0 & 0 & 0 & 4300 \\ 5400 & 0 & 0 & 5250 & 5300 \\ 0 & 0 & 0 & 0 & 0 \\ 0 & 0 & 0 & 0 & 0 \end{bmatrix}$$

b. The expected total cost is $13,000.

c. The expected remaining cost for a patient in recovery is $5,514.

Chapter 5: Continuous Time Markov Processes

1. a. Observe the traffic at a certain location, taking samples for the transition times among the three states. Use this data to calculate the sample means for the transition times. Take the reciprocal of each mean time to estimate the corresponding λ_{ij}.

b. $\frac{1}{\lambda_{LM}} = \frac{1}{4}$ hour. (This is just an expected time to make a particular transition.)

c. $\frac{1}{(\lambda_{ML}+\lambda_{MH})} = \frac{1}{5}$ hour. (This is a mean holding time.)
d. Because of forgetfulness, the same mean holding time, or 1/5 hour.
e. $m_{ML} = \frac{(\lambda_{MH}+\lambda_{HM})}{\lambda_{ML}\lambda_{HM}} = 1$ hour. (This is a mean first passage time; you have to set up the equations and solve them algebraically. It will get messy, but eventually will simplify.)
f. $\pi_H = \frac{8}{15}$.

3. Figure A2.5 shows a birth-death process with five states, defined by the number of rats eating (or the number of rats not eating, if you prefer). The transition rates are complicated by the fact that there are four rats. When the process is in state 0, for example, four rats are not eating. The rate at which one will start eating—the transition rate from 0 to 1—is four times the rate at which each rat starts, or 4/3 per hour.

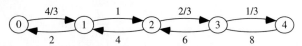

■ **FIGURE A2.5** The rat problem

The proportion of time no rats are eating is the steady-state probability for state 0 (if states are defined by the number eating). The full steady-state distribution (to check your calculations) is:

$$\pi_0 = 0.536 \quad \pi_1 = 0.357 \quad \pi_2 = 0.089 \quad \pi_3 = 0.015 \quad \pi_4 = 0.002$$

5. Figure A2.6 shows the a smoke detector problem. This problem is a bit subtle because there is extraneous information (the business about the rate of fires). It also requires the right kind of thinking about the states and transitions. The transition diagram is shown in Figure A2.6, with rates expressed in transitions per year (you could also use months as your time unit with appropriate scaling of the rates). You need to calculate the steady-state probability for the "bad" state, since that represents the fraction of time that the detector is out of service. (You do not want to multiply this probability by the rate at which fires occur. If the question were "what is the rate at which undetected fires occur?" you would have to do this multiplication.) $\pi_{bad} = \frac{2}{5}$

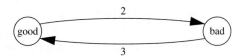

■ **FIGURE A2.6** The smoke detector problem

7. Figure A2.7 is used in problem 7.
   a. Mileage is analogous to time. Choose distance units to be 20,000 miles, so failure rate per tire = 1.0 (in units of failures per 20,000 miles). Alternatively, you could choose units of miles, 1,000 miles, 5,000 miles, or any reasonable standard value. The rate would have to be adjusted to match which distance units you use. Let the state equal the number of blowouts $\{0, 1\}$. Then $R(m) = p_{00}(m) = e^{-4m}$ (where m is measured in multiples of 20,000 miles).

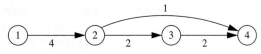

■ **FIGURE A2.7** The tire problem

b. Time to first blowout is negative exponentially distributed with parameter 4 (because there are four tires), so mean is 1/4, or 5,000 miles.

c. Let the states be {1 = all tires good, 2 = one tire failed, 3 = two tires failed on opposite sides, 4 = two tires on same side failed (zero or one failed on the other side)}.
An expression for Q(m) is $Q(m) = p_{11}(m) + p_{12}(m) + p_{13}(m)$, or $1 - p_{14}(m)$. The complete solution requires solving the differential equations:

$$\frac{d}{dm} p_{11}(m) = -4p_{11}(m)$$

$$\frac{d}{dm} p_{12}(m) = 4p_{11}(m) - 3p_{12}(m)$$

$$\frac{d}{dm} p_{13}(m) = 2p_{12}(m) - 2p_{13}(m)$$

$$\frac{d}{dm} p_{14}(m) = 5p_{12}(m) + 2p_{13}(m).$$

The answer (which requires solving the differential equations) is

$$Q(m) = e^{-4m} - 4e^{-3m} + 4e^{-2m}$$

d. $m_{03} = 11/12$ (of 20,000 miles), or 18,333 miles.

e. It is reasonable for chance blowouts, but after 20,000 miles, one would expect wear to become a factor. The presence of wear or aging effects would contradict the Markov assumption.

9. This question relies upon the "separation" property of a Poisson process. The arrival process for women (which is independent of that for men) is Poisson with a rate of two per hour (one-half of four, the rate for all customers). The fact that twenty male customers arrived during one eight-hour period is irrelevant. The expected number of women is λt, which translates to $2 \times 8 = 16$.

11. This is just a special case of the general continuous time Markov process with six states. You can solve the differential equations because they are essentially the same as they were for the Poisson process (except that there are only six of them). The first five equations look exactly like those for the Poisson process; the last one is different because the process is finite. If you cannot solve the differential equation for the last state directly, you can get it by realizing that for all t, the probability functions must add to 1. So you can get $p_{50}(t)$ as $1 - p_{55}(t) - p_{54}(t) - p_{53}(t) - p_{52}(t) - p_{51}(t)$. You need to find $p_{50}(8)$, which you get by finding the function $p_{50}(t)$, and then substituting the value 8 for the parameter t:

$$p_{50}(t) = 1 - e^{-t} - te^{-t} - \frac{t^2}{2!}e^{-1} - \frac{t^3}{3!}e^{-t} - \frac{t^4}{4!}e^{-t}$$

All of this evaluates (by calculator or spreadsheet) to:

$$p_{50}(8) = 0.900$$

13. Define the state as the number of units of blood (of this type) on hand. Changes occur one pint at a time, so it is a birth-death process with birth rate of four per day and death rate of three per day. The maximum state is N, which has to be determined. We can model the process for different values of N, assess the costs, and compare to determine the optimal N. (We do not need optimization theory here.)

The total cost of the system for any fixed value of N is the sum of holding costs and shortage costs. All of these are random variables, so we have to deal with expected values. The expected holding cost rate (in dollars per day) is $\$5 \times \boldsymbol{E}$(number of units on hand) $= 5 \times (0\pi_0 + 1\pi_1 + 2\pi_2 + \cdots + N p_N)$. The expected shortage cost rate is $30 \times$ rate of demand $\times P$(none on hand) $= 90\pi_0$. For the values given, it works out that the least cost rate is achieved for N = 1. A different solution might result if the costs or rates were different.

15. **a.** The state equals the number of molds available for use (including those in machines) and runs from 0 to 8. The transition diagram looks like Figure A2.8, where $\lambda =$ cleaning rate per worker and $\mu =$ forming mold usage rate per machine.

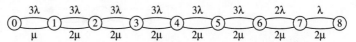

■ **FIGURE A2.8** The circulating molds problem

b. You will need the rates for the forming and cleaning processes (λ and μ), which you would get by estimating the average times (timing individual operations) and then taking the reciprocals.

c. You might come up with a different measure, but a good one would be the mean number of molds available, which corresponds to the mean state, or $0\pi_0 + 1\pi_1 + 2\pi_2 + \cdots + 8\pi_8$.

d. More molds in the system would increase the number of states. For example, nine molds would change the transition diagram to that shown in Figure A2.9.

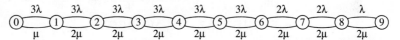

■ **FIGURE A2.9** One more mold

Another worker would change the rates on the original diagram, as shown in Figure A2.10.

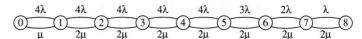

■ **FIGURE A2.10** One more worker

17. **a.** This problem can be modeled as either an ergodic or terminating process. If the state is the number of generators working, state 0 can be an absorbing state (if you consider it to be fatal) or it could be a recurrent state (if the repairman can still work without power). So you would have either as shown in Figure A2.11, or the same diagram without the arrow from 0 to 1. Here the rates are expressed with time measured in days. Notice also that the failure rate depends on the number working, but the repair rate is constant because there is only one repairman.

b. For part b we need to find m_{30} in the ergodic version or d_3 in the terminating one. (A conditional mean first passage time from 3 to 0 would also work, since there is only

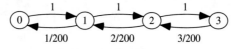

■ **FIGURE A2.11** The generator problem

one absorbing state.) The answer, if you care to work it out, is about 1,360,367 days—a comfortably long time for those who are living there.

Chapter 6: Queueing Models

1. a.

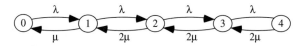

■ **FIGURE A2.12**   M/M/2/4

b.

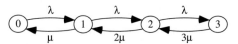

■ **FIGURE A2.13**   M/M/3/3

c.

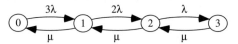

■ **FIGURE A2.14**   M/M/1/3/3

d.

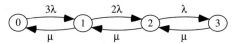

■ **FIGURE A2.15**   M/M/3/3/3

3. a. L = in-process inventory = 300 and W = time in factory = 6 weeks, so production rate = λ = 50 jobs per week or 2,600 jobs per year (if all 52 weeks are used).

b. Throughput = $\lambda = 3,000$ per day, or $3,000/24 = 125$ per hour, duration of stay = W = 1.5 hrs., so average occupancy = $L = (125)(1.5) = 187.5$ people.

c. $L = 6.4$, $L_q = 4.0$, so average number in service is $L - L_q = 2.4$. Hence, the average utilization per server is $2.4/3 = 0.8$.

5. M/M/2/10/10/FCFS.

7. a.

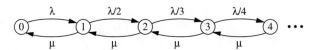

■ **FIGURE A2.16**   Discouraged arrivals

$$0 = -\lambda\pi_0 + \mu\pi_1$$
$$0 = \lambda\pi_0 - \left(\frac{\lambda}{2}+\mu\right)\pi_1 + \mu\pi_2$$
$$0 = \frac{\lambda}{2}\pi_1 - \left(\frac{\lambda}{3}+\mu\right)\pi_2 + \mu\pi_3$$
$$\cdots$$

These are birth-death type equations, so solve in the usual way. The normalizing equation produces an infinite series that is of the exponential form.
Solution: $\pi_j = \frac{\rho^j}{j!}e^{-\rho}$, where $\rho = \frac{\lambda}{\mu}$ (Note: same as M/M/∞, an interesting coincidence.)

b. λ is the arrival rate that would apply when there is no one in the system. It could be estimated by taking data on the inter-arrival times only when the system is empty. A better method would be to take data on all interarrival times, but note the number in the system at the time of arrival. Times taken when one customer is in the system estimate $\lambda/2$, and so forth. All of the observations can be weighted and combined to produce a single estimate of λ.

c. Solution is the Poisson distribution, so $L = \rho$. From this, it is easy to get $W_q = (\rho/R) - 1/\mu$. The main problem is to get R, the effective arrival (or throughput) rate, because it varies by state. To do this, weight the arrival rate for each state by the steady-state probability of being in that state, and sum (that is, compute a weighted average). After simplifying, $R = \mu(1 - e^{-\rho})$.

9 a. M/M/100/100 (finite capacity — spaces are the servers).

b.
$$0 = -\lambda\pi_0 + \mu\pi_1$$
$$0 = \lambda\pi_0 - (\lambda + \mu)\pi_1 + 2\mu\pi_2$$
$$0 = \lambda\pi_1 - (\lambda + 2\mu)\pi_2 + 3\mu\pi_3$$
$$\cdots$$
$$0 = \lambda\pi_{99} - 100\mu\pi_{100}$$
$$1 = \pi_0 + \pi_1 + \pi_2 + \pi_3 + \cdots + \pi_{100}.$$

The solution is the truncated Poisson distribution, shown on page 211.

c. L is the long-term average number of occupied spaces; W is the average length of time that a particular space is occupied by a single car (same as service time); L_q and W_q are 0, or not really pertinent, because there is no waiting room.

d. Adding spaces has some cost that may not be a linear function. That is, you might be able to add a few at reasonable cost, but the function would jump to a much higher value when you fill up all the space you have. Presumably, you can determine that function. On the benefit side, you need the queueing model to assess the value of additional spaces. If you can get the value to the shopping center of one more customer, you could multiply that by the expected addition to L when you go from M/M/100 to M/M/101. If benefit is greater than cost, figure the value of one more space by going to M/M/102, and so forth until benefit is less than cost.

e. If shopping trips are not negative exponential, nothing has to be changed because the M/G/c/c model has the same solution as the M/M/c/c.

11 Compare a M/M/2 to an M/M/3 with comparable rates. "Backlog of work" needs interpretation. It could be L_q, if it does not include the work in progress, or it could be L. From the executive's point of view, it is most likely L, since he or she would care about the total turnaround time, not just the time in queue. To establish comparable rates, assume λ is the same for both models. Let μ be the service rate of the class B secretary (the slower one), so that the rate of the class A secretary will be $3\mu/2$. Then, the traffic intensity for the class A M/M/2 will be $\rho = \lambda/3\mu$ (because in the denominator we use the number of servers times the service rate), and for the M/M/3 it will be $\rho = \lambda/3\mu$. In other words, we can use the same parameter value in both equations to compare values. You can get the formulas for L for the two models either by deriving them from the steady-state probabilities or by using the formulas in Table 6.3 on page 205 for L_q and adding B to get L. You will have to be careful with B, because it is (using Little's formula on the servers alone) equal to the rate of arrivals, λ, times the mean service time, which differs in the two cases. B is not just equal to ρ. It turns out that for all values of ρ, the L for the two-server system (the fast class A secretaries) is better than the L for the three-server system. You can prove that by charting the formulas in a spreadsheet or by algebraically comparing the two formulas. This is the reverse of the conclusion you would draw if you were comparing L_q, as the graph in Figure 6.17 on page 206 shows (for all values of ρ, more servers give a smaller L_q). The reason that it reverses is that the faster servers have a smaller B.

13. Arrival rate = 24 per hour (given). Current service rate is 30 per hour (from mean time of two minutes). Model is M/M/1. Concern is for waiting time, W. From the formula, appearing on page 196, W = 10 minutes. The first proposal is to change service rate to 3/5 per minute or 36 per hour (from changing the machine time from 1/2 minute to 1/6 minute; human time of 1.5 minutes is unchanged). W = 1/12 hours or 5 minutes. The second proposal is a M/M/2 with $\lambda = 24$, $\mu = 30$. Use value $\rho = 2/5$. To get W, you can solve steady-state equations, compute L, and then use Little's law, or you can use the result in Table 6.3 on page 205 for L_q, compute W_q using Little's law and add a mean service time to get from W_q to W. The final result is W = 2.38 minutes. For a total cost comparison, you can compare the cost per hour, since all costs are given as rates. For the current situation, the cost is $k_1 + k_3(10/60)$. For the faster machine, it will be $k_2 + k_3(6/60)$. For the double machine case, it is $2k_1 + k_3(2.38/60)$. Which one is smallest depends upon the values of k_1, k_2, and k_3.

15. System is $M/M/\infty$ because customers self-serve. (Note: Salespeople are *not* servers.) Solution is a Poisson distribution with mean $L = \lambda/\mu = 30/3 = 10$. So answer is ten salespeople.

17. Arrival rate of calls for ambulance = $(0.2)(2.4) = 0.48$ per day. Service rate = 2 per hour or 48 per day. Model is M/M/1 because there is only one ambulance and calls must wait. Probability server is busy = (1 − Probability system empty) = $\rho = 0.01$.

19. Use Little's law to determine the actual throughput rate, $R = L/W = 200/5 = 40$ per hour. The rate of attempted arrivals is 50 per hour, so $10/50 = 1/5$ or 20 percent of the customers are rejected.

Chapter 7: Networks of Queues

1. **a.** We have two M/M/1 subsystems in tandem, each with $\rho = 0.9$. For each, the W is $[\frac{1}{(1-\rho)}](\frac{1}{\mu}) = 10$ minutes.
   Each customer has to path through both subsystems, which can be treated as if they are independent, so the total time is 20 minutes.

   b. The arrival stream is divided in half, so the two stations at the top of the figure have an arrival rate of 0.9 and they will behave exactly as the system in problem 1 above. The bottom two stations are similar. An individual customer will see only one of those two identical subnetworks, so the total waiting time will be the same as in problem 1, that is, 20 minutes.

3. We have two M/M/2 queues in tandem. Each has $\rho = \frac{\lambda}{2\mu} = 0.9$, which means $W = 9.526$ for each, so the total time will be 19.053 minutes.

5. **a.** The traffic equations are:

$$\eta_1 = 0.1\eta_2 + 0.3\eta_3 + 0.6$$
$$\eta_2 = 0.15\eta_1 + 0.2$$
$$\eta_3 = 0.04\eta_1 + 0.2$$
$$\eta_4 = 0.01\eta_1 + 0.4\eta_2 + 0.1\eta_3.$$

After solving, the effective arrival rates are:

Sales	0.5824
Billing	0.2540
Support	0.1900
Supervisor	0.1264

b. The W_q's are (in minutes):

Sales	8.04
Billing	6.68
Support	88.77
Supervisor	8.59

7. **a.** Here is the computed table:

b. The utilization for both stations is 0.8.

Station	1	2
$\hat{\rho}_i$	2	2
c_i	1	1
0	1	1
1	2	4
2	4	12
3	8	32
4	16	80

9. The table showing the calculation of G (with the last row for part b) is:

Station	1	2
$\hat{\rho}_i$	6	6
c_i	2	3
0	1	1
1	6	12
2	18	72

3	54	306
4	162	1,098
5	486	3,654
6	1,458	11,682
7	4,374	36,486
8	13,122	112,338
9	39,366	342,774

a. Each of the two machines has a utilization of 0.974.

b. With nine molds in the system, the utilization of the machines becomes 0.983.

11. a. With N = 10, the utilization of the dummy server is 0.976.

b. The true utilization of the real server for the M/M/1 is 0.8. The value from the closed network approximation is 0.781.

c. The true value of L for the M/M/1 is 4. The value from the closed network approximation is 2.966.

Chapter 8: Using the Transition Diagram to Compute

1. a. The transition matrix is given in the answer to Chapter 2, Problem 1. The transition diagram (with loops omitted, since they are not used) is shown in Figure A2.17.

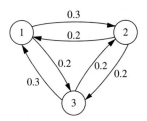

■ **FIGURE A2.17** The fishing problem

You can refer to Figures 8.13, 8.14, and 8.15 in Chapter 8 for the structure of the in-trees. All you have to do is to substitute the correct transition probabilities to get the steady-state probabilities: $\pi_1 = 0.3265$, $\pi_2 = 0.3878$, $\pi_3 = 0.2857$.

b. For the mean first passage times (any of them), we need all of the 2-forests. When the diagram has only three states, a 2-forest will have only one arc. Hence, there are a total of six 2-forests, corresponding to each of the six transitions from one state to another (ignoring loops). Their weights are 0.3, 0.2, 0.2, 0.2, 0.3, and 0.2. To calculate m_{31}, we must eliminate any of those 2-forests that do not have point 1 as a root (which would be the two that have arcs leaving point 1) and any that connect point 3 to point 1 (which would be just the one that contains the arc from point 3 to point 1). The sum of weights of all the remaining 2-forests is 0.6. That has to be divided by the sum of weights of all in-trees to point 1, which equals 0.16. The final result is $m_{31} = 3.75$ days, which matches the result found previously. If you want to try the same calculation using integer values, just multiply each probability by 10. The numerator will be 6 and the denominator will be 66, so the quotient is 0.375, which we have to multiply by 10 to correct for the scaling.

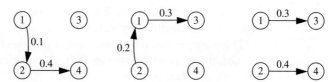

■ **FIGURE A2.18** All 2-forests

3. a. The absorption probability a_{14} requires the 2-forests (because there are two absorbing states), of which there are only the three shown in Figure A2.18.
Only the first of these three connects point 1 to point 4. The absorption probability is found as the ratio of the weight of all 2-forests connecting point 1 to point 4 to the total weight of all 2-forests. Hence, $a_{14} = 0.04/(0.04 + 0.06 + 0.12) = 0.182$.

b. e_{12} requires 3-forests, of which are there four, pictured in Figure A2.19.

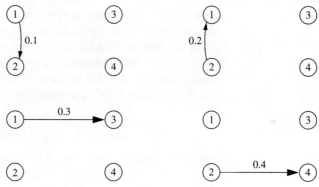

■ **FIGURE A2.19** All 3-forests

Of these four, only the first connects point 1 to point 2. The formula gives e_{12} as the ratio of the weight of all 3-forests in which point 1 is connected to point 2 to the total weight of all 2-forests, so $e_{12} = 0.1/(0.04 + 0.06 + 0.12) = 0.455$.

c. The hit probability f_{12} is obtained as the total weight of all 3-forests in which point 1 is connected to point 2 divided by the total weight of all 3-forests in which point 2 is a root. The former consists of a single 3-forest (the first), and the latter consists of two 3-forests (the first and third). Hence, $f_{12} = 0.1/(0.1 + 0.3) = 0.25$.

d. The conditional mean first passage time $m_{13}^{(c)}$ is found by forming the ratio of the total weights of all 3-forests to 2-forests (which in this case is 1/0.22) and then subtracting the ratio of the total weight of all 3-forests in which point 1 is connected to point 3 to the total weight of all 2-forests in which point 1 is connected to point 3 (which in this case is $0.3/(0.06 + 0.12)$). The result is $m_{13}^{(c)} = 2.88$.

5. Figure A2.20 shows is the transition diagram with loops removed and letters replacing the arc weights.

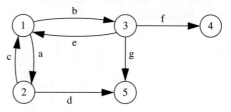

■ **FIGURE A2.20** The hospital problem

We will need all of the 2-forests having roots at point 4 and 5. There are seven, and they can be identified by their symbolic weights: ade, adf, adg, bcf, bcg, bdf, and bdg. (These are generated by considering every combination of one arc from each of the set {a,b}, the set {c,d}, and the set {e, f, g}, and eliminating those that have cycles; namely any that contain ac or be.)

a. For a_{15}, we form the ratio of the sum of weights of the 2-forests in which 1 is connected to 5 to the sum of the weights of all of them. In symbolic form, that ratio is:

$$a_{15} = \frac{ade + adf + adg + bcg + bdg}{ade + adf + adg + bcf + bcg + bdf + bdg}.$$

For a numerical value, you may substitute the original transition probabilities for the letters, or you may use any scaled version of those. Either way, $a_{15} = 0.0317$.

b. For e_{12} and f_{12}, we need 3-forests in which point 2 is a root (along with points 4 and 5, which must be roots because they are absorbing). To generate them, consider all combinations of one arc from each of the sets {a,b} and {e,f,g} and eliminate any that contain the cycle be. That leaves ae, af, ag, bf, and bg. For e_{12}, we form the ratio of the sum of weights of all of these 3-forests in which point 1 is connected to point 2 (which is just ae + af + ag) to the sum of the weights of all 2-forests (which is the same denominator as in a_{15} above). The result is:

$$e_{12} = \frac{ae + af + ag}{ade + adf + adg + bcf + bcg + bdf + bdg}$$

Now, if we substitute the original transition probabilities, we will get the correct numerical answer, which is $e_{12} = 0.1133$. If, however, we choose to use scaled values, we will have to adjust for the scaling by multiplying the result by whichever scaling value we used. For example, if we multiplied the original transition probabilities by 100 to get rid of the decimals, we will have to multiply the calculated value of e_{12} by 100 to get the correct result. You can tell that this scale adjustment will be necessary by observing that the size of the subgraphs in the numerator is one less than the size of the subgraphs in the denominator, so the scale factor will not automatically cancel out.

c. For f_{12}, we form the ratio of the sum of weights of 3-forests in which point 1 is connected to point 2 (the same numerator as above) to the sum of weights of all 3-forests in which point 2 is a root. That ratio is:

$$f_{12} = \frac{ae + af + ag}{ae + af + ag + bf + bg}$$

This result will come out the same ($f_{12} = 0.1018$) whether the transition probabilities are scaled or not. You can tell that at a glance by noting that the numerator terms have the same order (number of arcs) as the denominator terms.

6. This problem illustrates that solution in symbolic form is the same as numerical computation, using the graph theoretic formulas. The transition diagram is shown in Figure A2.21.

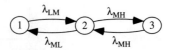

■ **FIGURE A2.21** The traffic problem

a. The diagram is simple enough to see the in-trees without drawing separate figures. There is only one in-tree to each point, consisting of the two arcs pointing to the point. Hence, the steady-state probabilities are:

$$\pi_1 = \frac{\lambda_{ML}\lambda_{HM}}{\lambda_{ML}\lambda_{HM} + \lambda_{LM}\lambda_{HM} + \lambda_{LM}\lambda_{MH}}$$

$$\pi_1 = \frac{\lambda_{LM}\lambda_{HM}}{\lambda_{ML}\lambda_{HM} + \lambda_{LM}\lambda_{HM} + \lambda_{LM}\lambda_{MH}}$$

$$\pi_1 = \frac{\lambda_{LM}\lambda_{MH}}{\lambda_{ML}\lambda_{HM} + \lambda_{LM}\lambda_{HM} + \lambda_{LM}\lambda_{MH}}.$$

b. For m_{21}, we need 2-forests in which point 1 is a root and point 2 is *not* connected to point 1. There are two, one containing just the arc from 2 to 3 and the other containing just the arc from 3 to 2. Hence, with no further ado,

$$m_{21} = \frac{\lambda_{MH} + \lambda_{HM}}{\lambda_{ML}\lambda_{HM}}$$

7. The transition diagram for the M/M/2/4 is shown in Figure A2.22.

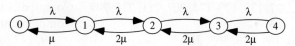

■ **FIGURE A2.22**   M/M/2/4

In any (finite) birth-death transition diagram, there is only one in-tree to each point; it consist of the arcs coming from the left to the point and from the right to the point. So, without even drawing any subgraphs, you can immediately write out the numerators of the steady-state probabilities, which gives an unnormalized solution

$$\hat{\pi}_0 = 8\mu^3 \quad \hat{\pi}_1 = 8\lambda\mu^3 \quad \hat{\pi}_2 = 4\lambda^2\mu^2 \quad \hat{\pi}_3 = 2\lambda^3\mu \quad \hat{\pi}_3 = \lambda^4$$

The final solution is obtained by dividing each of these by the sum of all of them.

INDEX

Numerics
1-graph, 260

A
Absolute state probability, 78, 160
Absorbing Markov chain, 97
Absorbing state, 42, 97, 109, 110, 131
Absorption probability, 100, 106, 160, 165, 265, 267
Acyclic graph, 256
Adjacency matrix, 255, 260
Aperiodic state, 114
Approximation, 127, 135, 173–174, 207, 211, 215, 247
Arrival process, 183
Arrival rate, 180, 184, 188, 189, 212, 215, 227
Axioms, 4

B
Bernoulli distribution, 21, 81
Bernoulli trials, 23, 28
Binomial distribution, 21, 22, 23, 26, 28, 63
Binomial process, 28, 64
Birth rate, 167
Birth-death process, 167, 193, 194, 230
Blocking, 204, 226
Branching, 226, 227–228
Brand switching, 54
Bundle, 42, 260

C
Case study
   Aging infrastructure, 64
   Appliance store inventory, 135
   Bakery shop queue, 180
   Call center, 48, 105, 132
   Day trading, 165
   Grain elevator queue, 209
   Manufacturing system design, 242
ccdf, 7
cdf, 7
Central Limit Theorem, 25, 27, 173

Channel capacity, 183
Chapman-Kolmogorov, 77, 88, 155, 161
Classification, 106, 160
Closed communicating class, 110
Closed queueing network, 229
Closed walk, 256
Coefficient of variation, 17, 174, 220
College progression example, 46, 59, 87, 96
Communicating class, 109, 110
Communicating states, 109, 110, 111, 114, 161, 256
Competing transitions, 152
Complement of an event, 3, 5
Complementary cumulative distribution function, 7, 9, 147
Conditional expectation, 15
Conditional mean first passage time, 104, 105, 160, 266, 267
Conditional probability, 5, 73
Conditional probability density function, 13
Conditional probability distribution function, 13
Connected graph, 256
Continuous random variable, 6, 8
Continuous sample space, 3
Continuous uniform distribution, 23
Correlation coefficient, 18
Cost, 32, 50, 119, 125, 132, 133, 135, 136, 200, 214, 242, 244–245
Covariance, 18
Cumulative distribution function, 7, 8
Customer, 19, 105, 132, 153, 169, 171, 173, 180, 183
Cycle in a graph, 256–257

D
Death rate, 167
Decomposition, 229
Defective distribution, 85
Density function, 9
Differential equation, 141, 145, 146, 155, 157–158, 169
Directed graph, 255
Discount factor, 134

Discounting, 133
Discrete probability distribution, 20
Discrete random variable, 6
Discrete sample space, 3
Discretizing time, 34
Dummy station, 241

E
Education, 57
Effective arrival rate, 189, 227–228
Elementary event, 3
Embedded Markov chain, 61, 154–155, 219
Empty set, 5
Ensemble average, 93
Epoch, 35, 142
Equal interval, 34
Ergodic, 86–89, 160
Ergodic Markov chain, 115, 126
Erlang, 174
Erlang distribution, 25, 174, 219
Erlang's lost call formula, 211
Event, 3
Event count, 34
Excel, 66, 77, 90, 211, 233, 271
Expectation, 14
Expectation of a function, 15
Expected duration, 99–100, 105, 160, 265, 267
Expected number of visits, 80–82, 86, 98–99, 104
Expected total reward, 121–124
Expected value, 14
Experiment, 3
External arrival rate, 227, 239

F
FCFS, 185–186, 217
FIFO, 185
Financial models, 61, 166
Finite capacity, 185
Finite source population, 185, 212–213
First moment, 14
First passage probability, 86, 159–160
First passage time, 83, 89
First return probability, 86
First return time, 83, 89
Flow equilibrium, 91, 192
Forgetfulness, 38, 149, 150, 151, 153, 197, 218

G
Gain rate, 127
Gambler's ruin, 60, 61

Gambling, 59–60
Geometric distribution, 22, 23, 83
Goodness-of-fit, 27
Graph, 254
Graph theory, 92
Growth, 58–59

H
Hazard function, 152
Higher order Markov chain, 39, 50–52, 62
Hit probability, 102–103, 160, 165, 265

I
Independence of events, 5
Independence of random variables, 12, 13
Infinite Markov chains, 111
Infinite series, 112, 131, 135, 193, 195, 196, 205, 210–211, 214
Infrastructure, 64
Inheritance, 64
Insensitivity, 229
Interest rate, 133
Internal arrival rate, 228, 230–232
Intersection of events, 3
In-tree, 257
Irreducible, 110, 161, 256
iThink, 158

J
Jackson networks, 226
Joint probability, 73, 230
Joint probability distribution function, 11, 38, 226, 238

K
Kendall's notation, 185
Khintchine, 173, 219
Kingman's approximation, 220, 248

L
Labeled graph, 254
Law of total probability, 6, 16, 18, 63
LCFS, 185, 217
Limited queue capacity, 202
Little's formula, 190, 196, 208, 217
Lumping states, 52

M
M/G/1, 219–220
M/M/1, 193–198

M/M/1/N, 202
M/M/1/N/N, 213
M/M/c, 226, 227
M/M/c/c, 204, 210
M/M/c/N, 204
Machine repairmen model, 214
MacLaurin's series, 156
Marginal probability density function, 12
Marginal probability distribution function, 12, 238–239
Marketing, 55
Markov, 32, 38
Markov assumption, 47, 55, 58, 62, 86, 149, 154–158
Markov chain, 38
Markov Chain Monte Carlo, 66
Markov Decision Processes, 119
Markov process, 140, 142
Markov property, 38
Mathematica, 158
MATLAB, 77, 158
Mean first passage time, 93, 96, 160, 163, 264, 267
Mean length of system, 188
Mean number of busy servers, 188, 237, 238
Mean of a distribution, 14
Mean queue length, 188, 200, 238
Mean recurrence time, 94, 96, 160, 163–164
Mean sojourn time, 83, 153–155, 160
Mean throughput rate, 188
Mean utilization, 188
Mean Value Analysis, 248
Mean waiting time in queue, 188
Mean waiting time in system, 188
Merging, 171, 227
Microsoft, 66, 77, 90, 211, 233, 271
Moments, 16
Multiple servers, 198, 204, 215
Music, 62–63
Mutually exclusive, 19, 75

N
Negative binomial distribution, 23, 83
Negative exponential density function, 149–150
Negative exponential distribution, 25, 26, 147, 150, 152, 154, 184, 192, 197
Network of queues, 225, 250
Non-null state (for def. of null), 112
Non-stationary Markov chain, 53

Normal distribution, 24–27
Normalization constant, 201, 202, 230, 235
Normalizing equation, 89, 90, 162
nth moment, 16
Null state, 112, 130
Number of servers, 183, 205

O
Odds, 2
Open product form queueing network, 226
Out-degree, 260
Overflow, 226

P
Pace, 237, 238
Partition, 5
Pascal distribution, 23
Path, 256
pdf, 7
Period, 114
Periodic Markov chain, 113
Periodic state, 114, 161
Phase-type distribution, 173, 174
Point spreads, 2
Poisson distribution, 22, 170, 211
Poisson process, 169, 170, 171, 173, 184, 226, 227
Pollaczek-Khintchine formula, 219
Pooling, 207
Population dynamics, 59
Positive recurrent state, 112
Power, 27
Powersim, 158
Principle of pooling, 207
Probability, 4
Probability density function, 9
Probability distribution, 6
Probability distribution function, 7
Product flow, 58
Product form, 226, 229
Production process, 139
Production process example, 96
Properties of means, 14
Properties of variances, 17

Q
QNA, 248
Queue, 184
Queue discipline, 185, 217
Queueing system, 181, 184

R

Random interval, 34
Random rewards, 124
Random variable, 6
Range, 16
Reachable, 109, 110, 164
Realization, 37, 182
Recurrence probability, 85
Recurrent state, 108, 111, 161
Reducible, 110, 112
Relative arrival rate, 230, 231, 234, 239, 243
Relative workload, 234, 244
Remaining time, 150
Renewal process, 173, 183
Reproductive, 25
Reward, 119
r-forest, 263
r-graph, 263
Robustness, 241
Root, 257
Routing matrix, 231
RSS, 185, 217

S

Sample space, 3, 4
Semi-Markov process, 83, 120, 126
Server, 183
Service process, 184
Service rate, 184, 194, 198, 214
Service time, 181, 182, 183, 227, 243
Set theory, 3
Simulation, 247
Size of a graph, 256
Social mobility, 57
Sojourn time, 82, 153
Source population, 185
Spanning subgraph, 256
Speech recognition, 62–63
Spreadsheet, 77, 79, 85, 123, 168, 211, 212, 214, 233, 236, 243
Standard deviation, 17
Standard normal, 24
State, 36
State probability, 78, 86, 160
State space, 36
Station, 225
Stationarity, 39, 47, 53, 55, 64, 86, 149, 154
Stationary, 39, 142, 183
Stationary distribution, 92
Steady-state, 92
Steady-state distribution, 88, 191
Steady-state equations, 162, 194
Steady-state probability, 90, 92, 96, 160, 187, 231, 264, 267
Steady-state vector, 88
Stella, 158
Stochastic matrix, 44
Stochastic process, 38, 143
Strongly connected, 256
Subgraph, 256
Superposition, 171

T

Taylor's series, 156
Terminating, 87
Terminating Markov chain, 96–98
Terminating Markov process, 130, 174
Terminating process, 46, 163–164, 264
Text analysis, 61
Throughput, 187–188, 237–238, 240
Time average, 98
Traffic equations, 227–228, 231
Traffic intensity, 198
Transient state, 97, 108, 110, 160
Transition, 40
Transition diagram, 41, 92, 142, 257, 256
Transition matrix, 43
Transition probability, 41, 74, 86
Transition probability function, 144, 157, 160
Transition probability matrix, 76
Transition rate, 141, 143, 146
Transition rate matrix, 144
Tree, 257
Trucking example
  expected number of visits, 81
  first passage probabilities, 85
  formulation, 33
  higher order Markov chain, 50–52
  lumpable states, 53
  mean first passage times, 94
  rewards, 121
  steady-state probabilities, 87
  transition probability, 74
  walk probability, 73
Truncated Poisson distribution, 211

U

Uniform distribution, 20, 204
Union, 3

Unnormalized solution, 229–232
Utilization, 188, 199, 238–240

V
Variance, 16
Vensim, 158
Virtual transition, 40, 142
Visual Basic, 216, 237, 239, 271–277

W
Waiting room, 181
Walk, 42, 256
Walk probability, 73, 87
Weight, 41, 254
Weight of a subgraph, 256
Whitt's approximation, 220, 248